Springer Geophysics

The Springer Geophysics series seeks to publish a broad portfolio of scientific books, aiming at researchers, students, and everyone interested in geophysics. The series includes peer-reviewed monographs, edited volumes, textbooks, and conference proceedings. It covers the entire research area including, but not limited to, applied geophysics, computational geophysics, electrical and electromagnetic geophysics, geodesy, geodynamics, geomagnetism, gravity, lithosphere research, paleomagnetism, planetology, tectonophysics, thermal geophysics, and seismology.

More information about this series at http://www.springer.com/series/10173

Tatsuhiko Saito

Tsunami Generation and Propagation

Springer

Tatsuhiko Saito
National Research Institute for Earth
Science and Disaster Resilience
Tsukuba, Ibaraki, Japan

ISSN 2364-9119 ISSN 2364-9127 (electronic)
Springer Geophysics
ISBN 978-4-431-56848-3 ISBN 978-4-431-56850-6 (eBook)
https://doi.org/10.1007/978-4-431-56850-6

Library of Congress Control Number: 2018954657

This Springer imprint is published by the registered company Springer Japan KK part of Springer Nature.
The registered company address is: Shiroyama Trust Tower, 4-3-1 Toranomon, Minato-ku, Tokyo 105-6005, Japan

Preface

In the study of tsunamis, researchers have traditionally analyzed tide-gauge records and surveyed inundation areas. In addition to these coastal observations, sea-bottom pressure gauges are now widely deployed in deep oceans. Offshore observations can detect tsunamis promptly near the source, and short-wavelength tsunamis can be clearly observed using these records. After the 2011 Tohoku-Oki earthquake, observations were carried out at array stations inside the earthquake focal area in order to detect tsunamis more rapidly. At the stations inside the focal area, tsunami signals are recorded simultaneously with seismic waves.

The tsunami data have changed dramatically as the methods of observation have changed. A theory should also be developed for the effective use of the new records. We can use dispersive theory to analyze offshore records. Seismic-wave theory would be helpful for analyzing the tsunami records inside the focal area. At the forefront of tsunami research, new application studies and numerical methods are advancing greatly, as reported in journals. However, description of fundamental theories and the derivation of basic equations are often skipped in the reports, but deriving the equations is not always straightforward. Textbooks may be a more suitable venue for explaining the fundamental theories.

When I started doing tsunami research, it was difficult for me to find books that explain in detail how to derive fundamental equations and how to apply numerical methods from tsunami generation to propagation, although there are many excellent textbooks about fluid dynamics and tsunamis at present. This book focuses on the quantitative modeling of earthquake tsunamis using real data and mathematical representations. Considering that seismic waves cannot be neglected in tsunami observations, we treat both fluid dynamics and elastic dynamics. I tried to reasonably organize the seismology and tsunami research to construct a theoretical framework for tsunami generation due to earthquakes.

I could not have completed this book without the help and encouraging words of my supervisors and colleagues. Kenji Satake carefully reviewed the draft and gave important comments. I thank Takashi Furumura, Yuichiro Tanioka, Takuto Maeda, Hiroyuki Kumagai, and Eiichi Fukuyama. Their comments and suggestions are very important in the study of earthquakes and tsunamis. Kentaro Emoto and Tatsuya Kubota helped me very much with their comments about seismic waves and tsunamis. Daisuke Inazu, Toshitaka Baba, Hiroaki Tsushima, Ryota Hino, Takayuki Miyoshi, Satbyul Kim, Shunsuke Takemura, and Nelson Pulido worked with me on tsunami, for which I am very grateful. I express my deepest gratitude to Haruo Sato. His enthusiasm for his research and teaching helped make geophysics enjoyable for me. I thank all those who held discussions with me and provided valuable comments on the research.

Tsukuba, Japan Tatsuhiko Saito

Contents

Chapter 1
Introduction

Abstract Tsunami research has been advanced by developments in theory and in the methods of observation. Coastal records such as tide gauges and inundation surveys have played an important role for a long time. Offshore observations have been undertaken widely across deep oceans. After the 2011 Tohoku-Oki earthquake, dense and wide offshore tsunami observation was begun. We now expect tsunami generation to be observed inside the focal areas of huge earthquakes. New observations require new theoretical frameworks. In Sect. 1.1, we briefly introduce the development of offshore observation using ocean-bottom pressure gauges. Section 1.2 illustrates the 2011 Tohoku-Oki earthquake tsunami. This is to show what we know and what we do not know about that huge earthquake and tsunami. Section 1.3 introduces a new observation network that was deployed after the Tohoku-Oki earthquake. This observation network is designed for the detection of tsunamis inside the focal area. In Sect. 1.4, we present the focus of this book. The wide use of ocean-bottom pressure gauges and the construction of new observation network can greatly advance our understanding of tsunami. This book is aimed to illustrate theoretical frameworks in analyzing these records.

Keywords Fluid dynamics · Elastic dynamics · Offshore observations · The 2011 Tohoku-Oki earthquake

1.1 Offshore Tsunami Observations

Tide gauge records have played a significant role in tsunami studies. However, tide gauges, which are located inside harbors and bays, can record only tsunamis that are near the coast. Those records are considerably affected and distorted by the complex geometries of bays and harbors. Practically, it is difficult to extract detailed information regarding the tsunami source and propagation only from coastal observations because the site corrections for harbors and bays are not straightforward.

If a tsunami can be observed in an offshore region before it enters harbors and coastal areas, this greatly contributes to our understanding of tsunamis. For example,

T. Saito, *Tsunami Generation and Propagation*, Springer Geophysics,
https://doi.org/10.1007/978-4-431-56850-6_1

to reveal the generation mechanism, we should observe tsunamis near or inside the earthquake focal region. If we try to model the propagation precisely in the deep ocean, we should use records that are free from strong site effects. Even if we intend to study the mechanisms of tsunami inundation into coasts, we should first describe the tsunami in the deep ocean and compare the offshore tsunami with the tsunami near the coast. Although the amplitudes hence the signals in the offshore region are much smaller than those near coasts, there are many merits to offshore observations.

There are various kinds of sensors that can observe tsunamis in offshore regions, such as GNSS buoy and satellite altimeter techniques. These are already in use and are expected to be deployed more widely in the near future (e.g., Kato et al. 2000; Inazu et al. 2016). Among the sensors used in offshore observations, the ocean-bottom pressure gauge is one of the most popular technologies in tsunami studies at present.

The development of ocean-bottom pressure gauges started in the 1960s, and some field trials were conducted in the 1980s. The history of this development is well summarized in Bernard and Meinig (2011), Mungov et al. (2013), and Rabinovich and Eble (2015). The experiments proved that ocean-bottom pressure gauges could clearly record tsunamis even if the signal was very small. Figure 1.1 shows an ocean-bottom pressure record during the Petatlan earthquake (M_W 7.6) in 1979 observed off the California Peninsula (Filloux 1982). The record showed the pressure change caused by the high-frequency seismic waves (Rayleigh wave) and low-frequency tsunami. The tsunami amplitude was very small (less than 1 cm in height), but it was clearly recorded due to the significant difference between the dominant periods of the seismic waves and tsunami.

The Deep-ocean Assessment and Reporting of Tsunamis (DART), a real-time tsunami monitoring system, is one of the most representative tsunami observation networks that uses ocean-bottom pressure gauges (e.g., Bernard and Meinig 2011). It was developed by the Pacific Marine Environmental Laboratory (PMEL) in the 1980s. The DART system has progressed from the first generation to the fourth generation (Fig. 1.2). Stations are now deployed not only in the Pacific Ocean but also other oceans (Fig. 1.3). The data obtained all over the world are available in real time and are publicly accessible on the Internet. Thanks to this data set, many researchers can analyze the ocean-bottom pressure change in the deep sea, and this has greatly contributed to the advancement of tsunami research. DART was designed as a tsunami early warning system (Titov et al. 2005). Before the introduction of DART, tsunami warnings were usually issued mainly based on the earthquake magnitude as estimated by seismogram analysis (e.g., Tatehata 1997). On the other hand, by using DART, tsunami prediction is being done without estimating the earthquake magnitude but rather by estimating the initial tsunami height distribution by analyzing ocean-bottom pressure records. This predicts the tsunami height and inundation near the coast more reliably (e.g., Tang et al. 2012).

Observations made using seafloor cables can record sea-bottom pressure changes with mm-order resolution in height with a high-frequency sampling rate (~1–100 Hz) in real time. In Japan, the Meteorological Research Institute (MRI) and National Research Institute for Earth Science and Disaster Resilience (NIED)

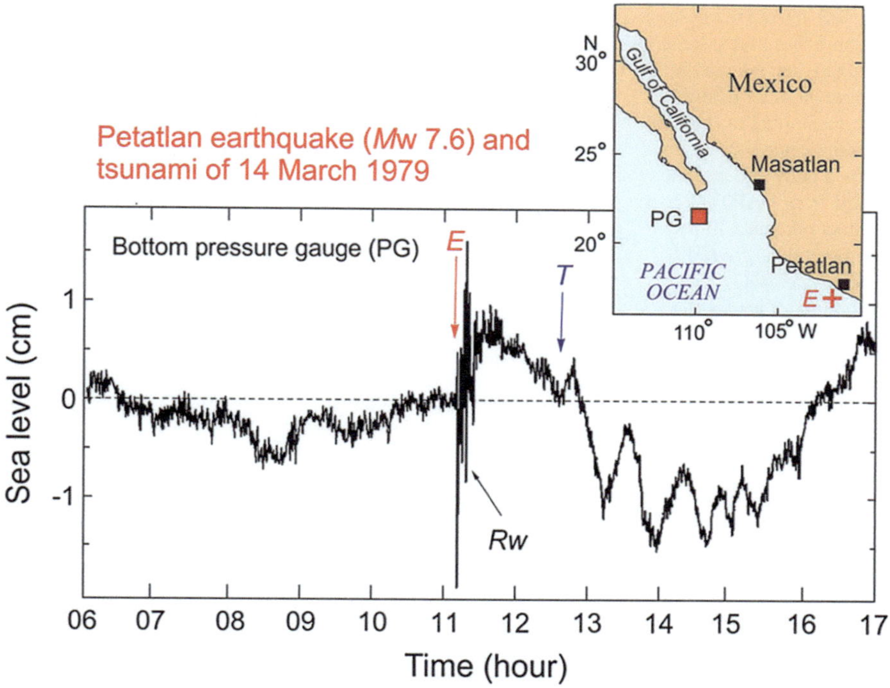

Fig. 1.1 An ocean-bottom pressure record during the 1979 Petatlan earthquake (Mw 7.6) off the southern coast of Mexico. Arrows marked E and T show the earthquake origin time and the arrival time of the tsunami, respectively. Rw indicates the Rayleigh wave. The map indicates the locations of the earthquake epicenter (cross) and the ocean-bottom pressure gauge (square). (Modified from Filloux 1982) (Rabinovich and Eble 2015, copyright by Springer)

deployed tsunami and seismic observation stations using seafloor cables (e.g., Eguchi et al. 1998). The Japan Agency for Marine-Earth Science and Technology (JAMSTEC) is one of the leading institutes operating the stations connected with seafloor cables (e.g., Hirata et al. 2002). Tsushima and Ohta (2014) well summarized the offshore tsunami observation systems in Japan. Figure 1.4 shows cable-based ocean-bottom pressure gauges deployed off Shikoku and off Hokkaido by JAMSTEC. Those ocean-bottom pressure gauges were used for tsunami detection (e.g., Baba et al. 2004) and also for coseismic slip detection (Mikada et al. 2006). Baba et al. (2006) estimated the afterslip distribution of the 2003 Tokachi-Oki earthquake using the ocean-bottom pressure gauges as geodetic sensors. The JAMSTEC also developed the Dense Oceanfloor Network system for Earthquakes and Tsunamis (DONET), which includes thermometers, hydrophones, and seismographs in addition to pressure gauges (Fig. 1.5). A high-frequency sampling rate of tsunami and seismic observations is useful for rapid source estimation for earthquakes and tsunamis that occur near coasts. The DONET data is also used to correctly determine earthquake hypocenter locations and the mechanisms of small earthquakes (e.g., Nakano et al. 2015). Determining the hypocenters of small earthquakes is useful for estimating the plate geometry.

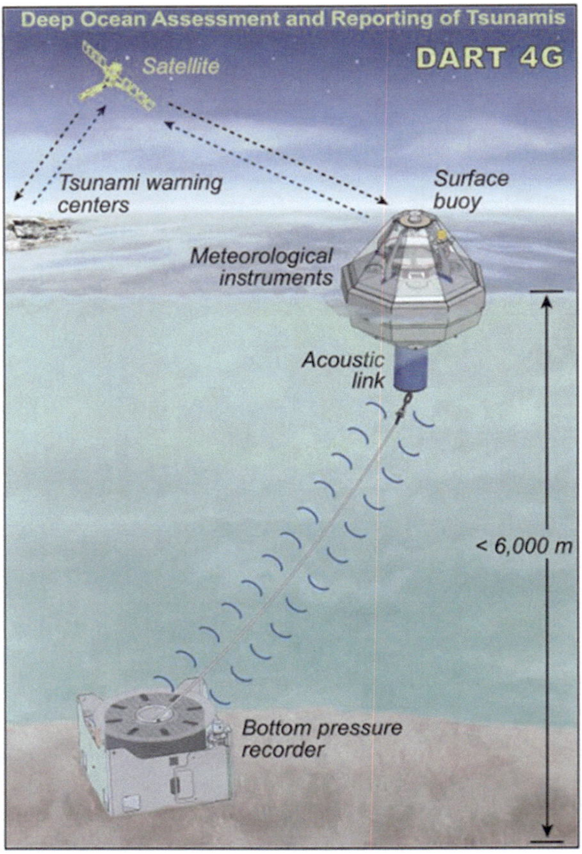

Fig. 1.2 DART4G system. Ocean-bottom pressure change is recorded by a bottom pressure recorder and transmitted to tsunami warning centers through a surface buoy and satellite. (Cited from DART4G product brochure, http://nctr.pmel.noaa.gov/Pdf/brochures/dart4G_Brochure.pdf accessed 2017-08-15)

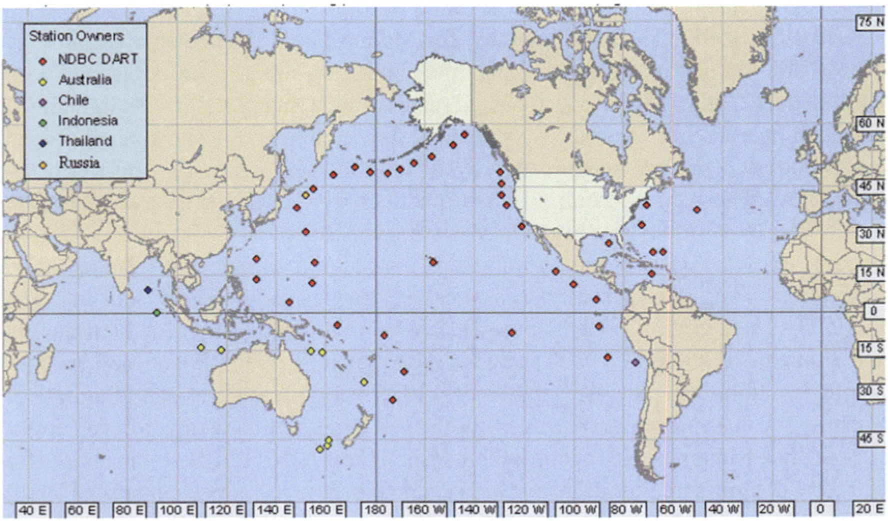

Fig. 1.3 DART station distribution. (Cited from http://nctr.pmel.noaa.gov/Dart/ accessed 2017-08-15)

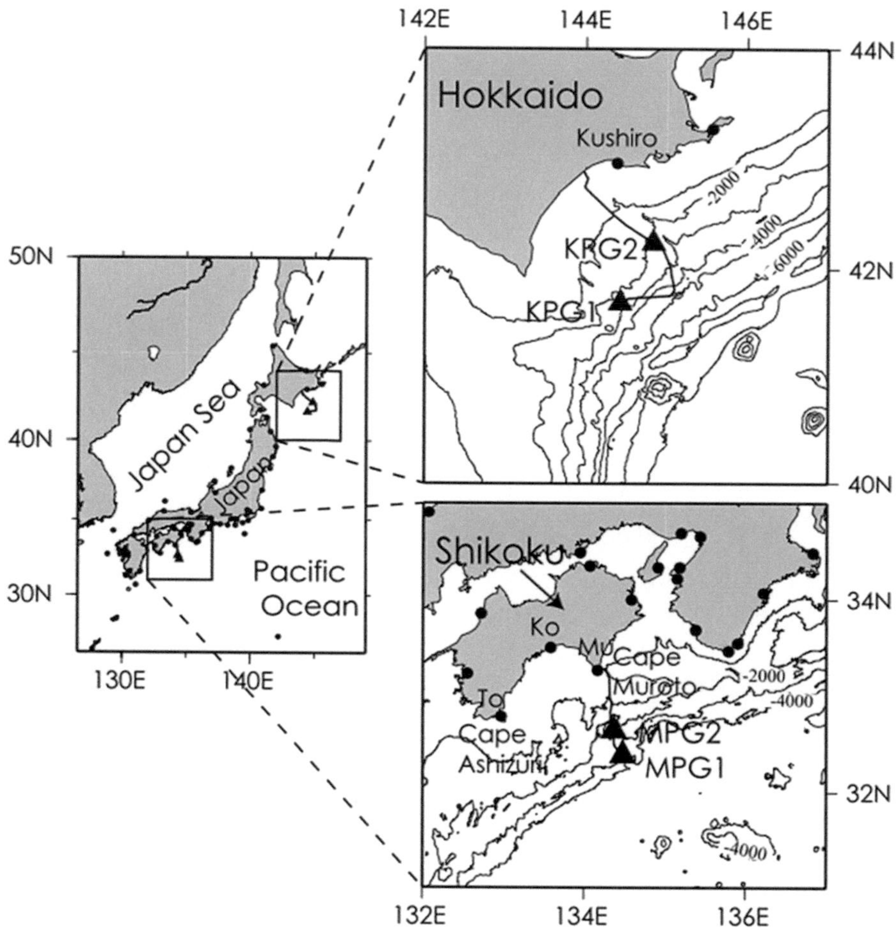

Fig. 1.4 Location of the ocean-bottom pressure gauges (triangles) of JAMSTEC. The lines connecting the pressure gauges offshore represent the main submarine cables. (Baba et al. 2004, copyright by the American Geophysical Union)

1.2 The 2011 Tohoku-Oki Earthquake

Offshore observation technology has greatly advanced tsunami research. However, we could not effectively mitigate the disaster caused by the 2011 Tohoku-Oki earthquake tsunami. One of the reasons for this devastating disaster is that we do not fully understand what a tsunami is. Some people might think that tsunamis are well understood as a topic in natural science. This is partly true because the tsunami propagation process is very precisely reproduced by fluid dynamics with reliable bathymetry data. It is easier to reproduce tsunami propagation than seismic waves because simulated seismic waves inevitably have model errors due to the larger

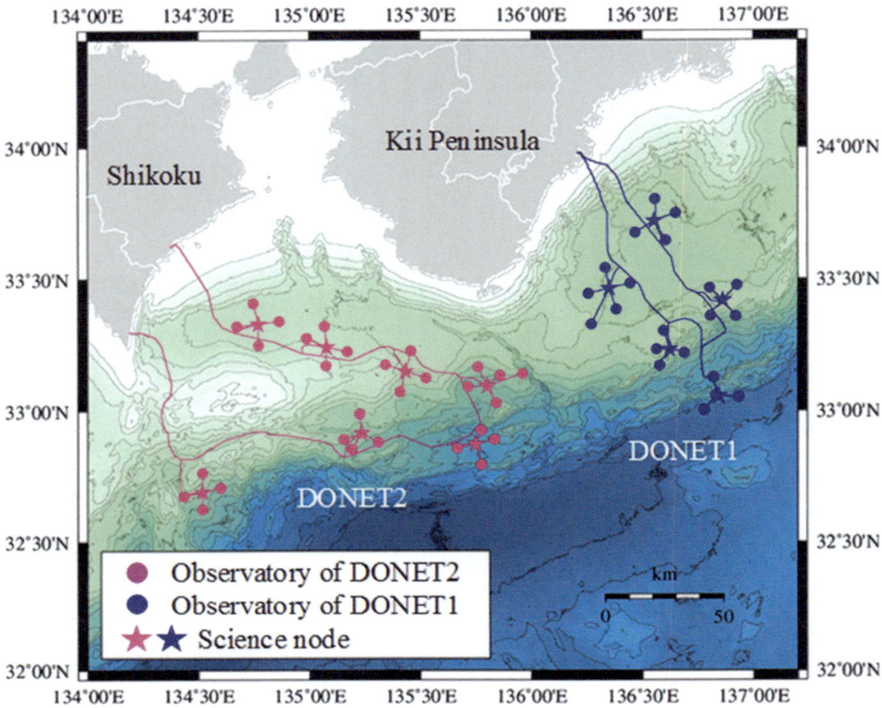

Fig. 1.5 Offshore tsunami and seismic stations in Japan, DONET1 and DONET2. (Cited from http://www.jamstec.go.jp/donet/e/index.html access 2017-08-15)

uncertainties of the subsurface structure. However, we should note that the tsunami generation process is much less understood than the propagation process. This is mainly because we could not make enough observations inside the generation region.

A large amount of high-quality seismic and tsunami data was obtained for the 2011 Tohoku-Oki earthquake. By analyzing these data, researchers are gradually coming to understand the Tohoku-Oki earthquake. Summarizing our present view of the Tohoku-Oki earthquake may clarify what we know and what we do not know about this huge earthquake and tsunami.

The earthquake was so huge (M_W 9.0) that estimating its precise characteristics using routine analyses just after the earthquake was not so straightforward. After examining the data carefully and conducting detailed field investigations, researchers have gradually constructed a description of the huge Tohoku-Oki earthquake. Although there are still some debates about the Tohoku-Oki earthquake, one plausible scenario is as follows.

Off the Pacific coast of northeastern Honshu, Japan, the Pacific plate subducts beneath the North American plate at a rate of ~8 cm/year toward the west. There has been active seismicity including small and large earthquakes along the plate

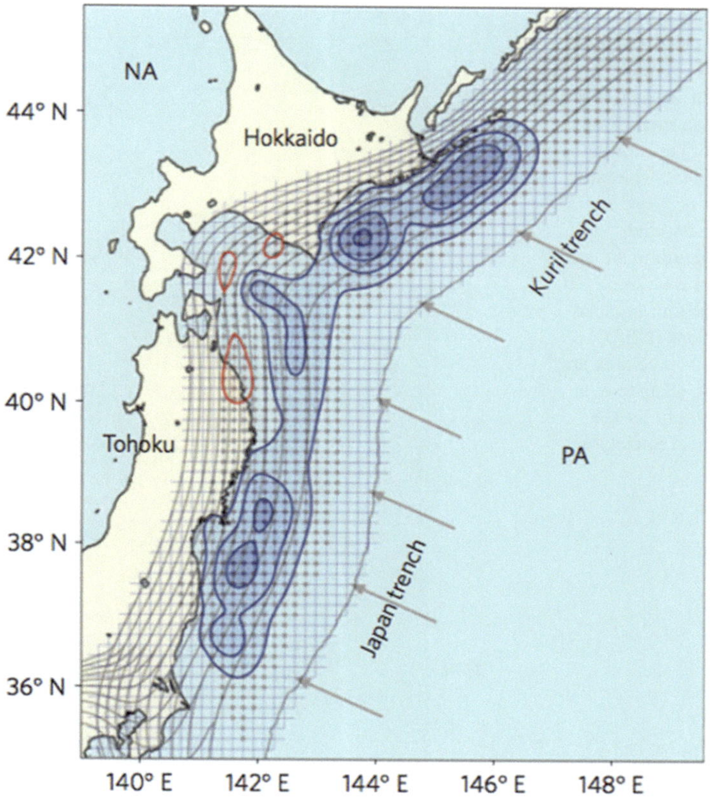

Fig. 1.6 Slip-deficit rate distribution. The blue and red contours show the slip-deficit and slip-excess rates at intervals of 3 cm/year. The arrows indicate the relative plate motion (Hashimoto et al. 2009, copyright by Springer)

boundary. A historical document (titled *Sandai Jitsuroku*) reported that there was a huge earthquake in 869 AD, which is referred to as the Jogan earthquake (e.g., Imamura 1934). Tsunami deposit surveys revealed that the magnitude was larger than M_W 8.6 (Namegaya and Satake 2014). After the Jogan earthquake, huge earthquakes and tsunamis caused devastating damage repeatedly in this area including the 1611 Keicho Sanriku earthquake. Although the 1611 Keicho Sanriku earthquake was one of the largest such events in Japan, the source location and size are not well known. A more recent huge earthquake was the 1896 Meiji Sanriku earthquake (M_W 8.4, e.g., Tanioka and Satake 2001; Satake et al. 2017). While some earthquakes often released the stress accumulated along the plate interface, the stress gradually but steadily was increasing due to the subducting Pacific plate.

We cannot measure the total stress (or slip deficit) accumulated along the plate interface. However, the rate of slip deficit can be detected from the crustal deformation observed by Global Navigation Satellite System (GNSS) observations (Fig. 1.6). The rate of the slip deficit around the source area of the Tohoku-Oki

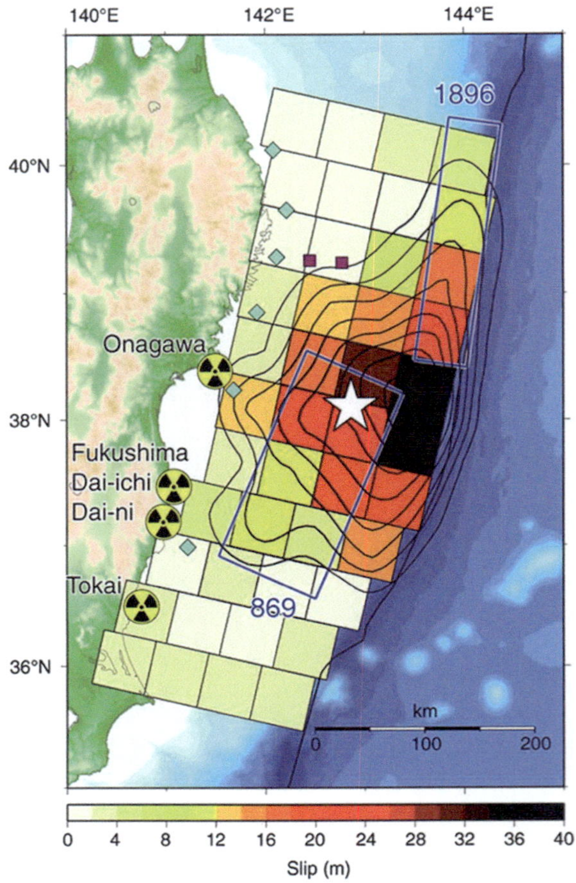

Fig. 1.7 The slip distribution of the 2011 Tohoku-Oki earthquake. The contour interval is 4 m. The white star indicates the epicenter of the rupture imitation point. The fault models of the 1896 Sanriku earthquake and the 869 Jogan earthquake are shown with blue lines. The locations of four nuclear power stations (NPS) around the source area are also shown. (Satake et al. 2013, copyright by the Seismological Society of America)

earthquake is ~6 cm/year (e.g., Hashimoto et al. 2009). If we assume that the rate was constant from 896 to 2011, the total slip deficit is estimated to be ~67 m.

On March 11, 2011, the accumulated slip deficit was abruptly released. This was the Tohoku-Oki earthquake. Figure 1.7 shows the slip distribution. The rupture initiated from the deep part of plate interface and reached the trench, resulting in a whole rupture of the seismogenic zone from the deep to shallow parts (e.g., Fujii et al. 2011; Koketsu et al. 2011). The deep rupture area corresponded to the slip area of the 869 Jogan earthquake and the shallow rupture area to that of the 1896 Meiji Sanriku earthquake (Fujii et al. 2011; Satake et al. 2013; Satake et al. 2017). The magnitude of the Tohoku-Oki earthquake was M_W 9.0. This was the third largest event in the world and the largest event in Japan among the earthquakes for which seismograms are available. The seismic waves of the Tohoku-Oki earthquake propagated globally. Strong ground motion was observed all over Japan (Fig. 1.8)

Fig. 1.8 Snapshots of seismic wave propagation during the 2011 Tohoku-Oki earthquake (M_W 9.0), Japan. The amplitudes of ground-velocity motion are shown for times of 60, 110, 160, 210, 260, and 310 s. The star shows the hypocenter of this earthquake. (Furumura et al. 2011, copyright by Springer)

(e.g., Furumura et al. 2011; Suzuki et al. 2011). Submarine landslide might also have been triggered during the earthquake (e.g., Tappin et al. 2014).

This huge earthquake caused a large vertical displacement of ~10 m at the sea bottom. A massive volume of seawater was uplifted by the sea-bottom displacement inside the focal area. This is the source of the tsunami. A large tsunami arrived at the coasts all over northeastern Honshu, Japan. A tsunami more than 5 m in height was observed even at some offshore stations (Fig. 1.9), and a large tsunami inundated the Pacific coasts of Japan (Fig. 1.10). The tsunami was so large that it was observed not only in Japan but also on the coasts of other Pacific countries. Across the Pacific Ocean, a tsunami was detected by DART ocean-bottom pressure gauges (Fig. 1.11). The DART successfully estimated the tsunami source energy and predicted the tsunami height arriving on the coasts of Hawaii (Tang et al. 2012). The tsunami propagated with the velocity of ~0.2 km/s. It arrived on the coasts of Chile about 1 day after the earthquake occurred (e.g., Inazu and Saito 2013; Watada et al. 2014). The tsunami energy was so huge that the energy was trapped within the Pacific Ocean for a long time and a higher signal level than usual remained throughout the Pacific Ocean for more than 1 week (Saito et al. 2013).

When the earthquake occurred, the Japan Meteorological Agency (JMA) rapidly determined the magnitude of the earthquake as M_{JMA} 7.9. This was estimated from

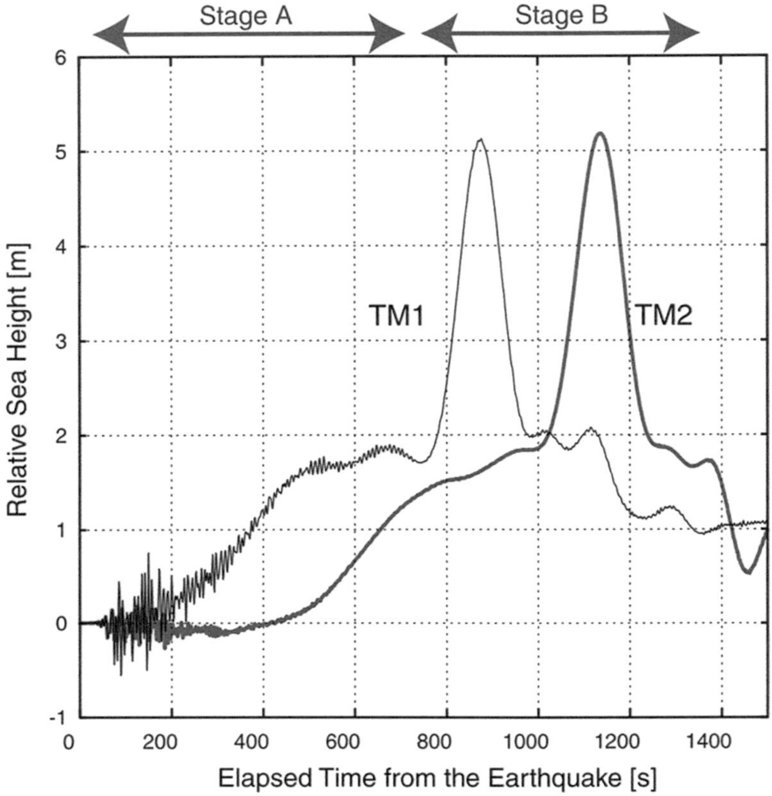

Fig. 1.9 Relative sea height trace recorded at TM1 (thin line) and TM2 (thick line) stations for the 2011 Tohoku-Oki earthquake. (Maeda et al. 2011, copyright by Springer)

the amplitude of the observed seismograms, but not from a waveform analysis. According to this rapid estimation of the magnitude, the JMA issued "Major Tsunami" warnings for the coasts of Japan (Iwate, Miyagi, and Fukushima prefectures) within 3 min of the earthquake occurrence (Ozaki 2011). However, the estimated M_{JMA} of 7.9 was much smaller than the moment magnitude of M_W 9.0 defined by the final slip of the earthquake. As a result, the tsunami height was estimated to be considerably smaller than the tsunami that actually arrived on the coasts.

This may show the limitation of the present forecasting system using only the seismic wave amplitude or the seismic intensity. Considering the success of tsunami forecasting using the DART system (e.g., Titov et al. 2005; Tang et al. 2012), ocean-bottom pressure gauges are one of the main candidates for improving the tsunami warning system. It is a challenge to build a more reliable tsunami forecasting system and algorithm for the coasts near the hypocenter where a tsunami is expected to arrive within 5–30 min. To accomplish this, observations made adjacent to the focal area and an understanding of the tsunami generation process are both important.

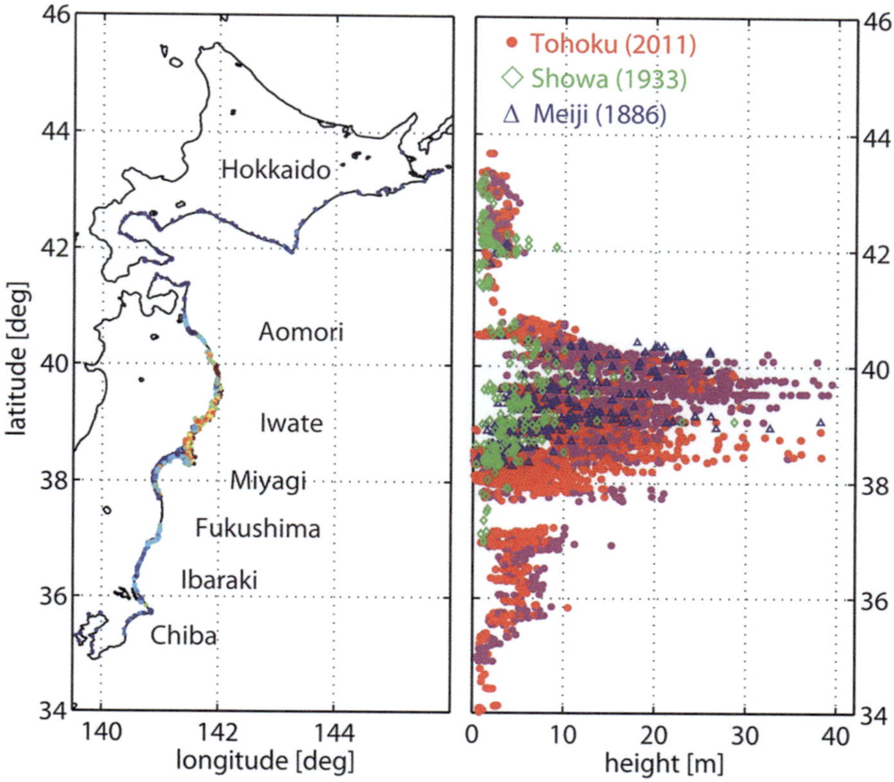

Fig. 1.10 Maximum measured local tsunami heights plotted versus latitude with previous tsunami records (circle, 2011 Tohoku tsunami; diamond, 1933 Sanriku tsunami; triangle, 1896 Meiji Sanriku tsunami). (Mori et al. 2011, copyright by the American Geophysical Union)

1.3 Observation Inside Earthquake Focal Area

After the Tohoku-Oki earthquake, the National Research Institute for Earth Science and Disaster Resilience (NIED) started to construct a dense and wide observation network named Seafloor observation network for earthquake and tsunamis along the Japan Trench (S-net), in and around the source region of the Tohoku earthquake (Fig. 1.12). This network is intended to detect seismic and tsunami signals more rapidly and to issue more reliable seismic and tsunami warnings in the event of huge future earthquakes. The S-net is a new observation that can observe tsunamis inside the focal area with densely deployed sensors. Algorithms suitable for this observation have been newly proposed and extensively developed since the Tohoku-Oki earthquake (e.g., Tsushima et al. 2012; Ohta et al. 2012; Maeda et al. 2015; Yamamoto et al. 2016; Tanioka 2017).

Dense ocean-bottom observation networks such as S-net and DONET can observe the tsunami generation process inside the focal area. This situation is totally

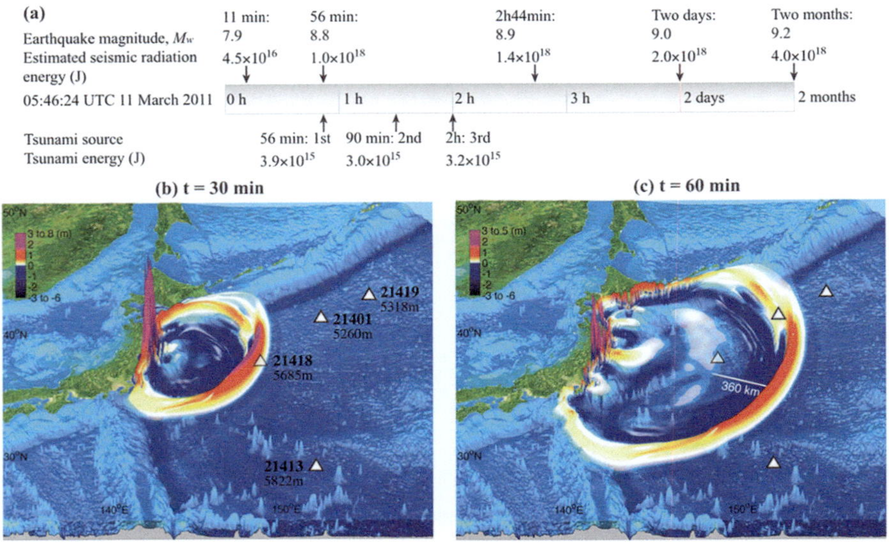

Fig. 1.11 (**a**) Timeline for assessment of the 2011 Tohoku-Oki earthquake and tsunami. (**b** and **c**) Snapshots of tsunami propagation at the elapsed times of 30 and 60 min after the earthquake. (Tang et al. 2012, copyright by the American Geophysical Union)

different from that before the Tohoku-Oki earthquake. The data inside the focal area would contain important information about the tsunami source that cannot be obtained by observations made far from the source. Therefore, it is important to revisit the mechanisms of tsunami generation and propagation in order to fully interpret those records. In particular, in order to theoretically understand the mechanism of tsunami generation, the theories of elastic dynamics are necessary in addition to fluid dynamics. Actually, ocean-bottom pressure records contain considerable amounts of seismic and ocean acoustic waves (e.g., Nosov 1999, Nozov and Kolesov 2007, Matsumoto et al. 2012). It would be more reasonable to employ elastic dynamics to analyze these waves.

1.4 Focus of This Book

Tsunami science involves a broad range of fields such as geology, engineering, and social science (e.g., Levin and Nosov 2009; Satake 2015). This book treats the geophysical aspect of tsunamis. Our understanding of the geophysics of tsunamis has been developed mainly based on tsunami observations made at coastal sites and deep ocean sites. This trend is likely to continue in the future. There are already many excellent reviews and books about tsunami generation and propagation (e.g., Okal 1988; Geist 1998; Levin and Nosov 2009; Satake 2015). However,

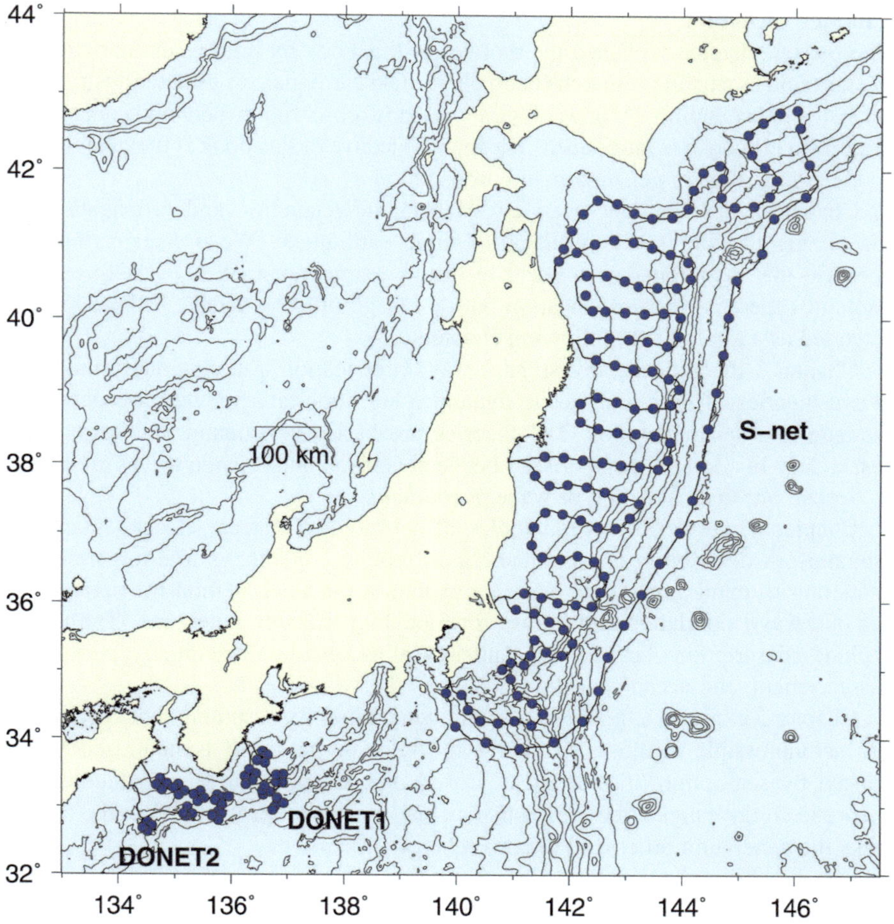

Fig. 1.12 Offshore tsunami and seismic stations in Japan (DONET1, DONET2, and S-net). Observed data are transmitted to data centers by seafloor cables

observations made inside the focal area would substantially drive the development of the tsunami theory. Therefore, it is useful at present to revisit the mechanisms of tsunami generation and propagation to prepare for the use of dense and deep ocean observation networks in the next generation. This book focuses on this point based on fluid and elastic dynamics.

This book devotes more pages to the topics of seismology and elastic dynamics than other tsunami books. We explain tsunami generation and propagation using mathematical equations. At the same time, we show real data and observed waveforms as much as possible because the mathematical models are based on these data. We try to reasonably organize seismology and tsunami research (or elastic dynamics and fluid dynamics) to give a theoretical framework in order to obtain additional viewpoints on the phenomenon of tsunami generation.

Chapter Structure

The present chapter explained the motives of this book by briefly summarizing the background of tsunami research. The Tohoku-Oki earthquake is used to illustrate our present understanding. A new offshore tsunami observation network constructed after the earthquake is introduced. The main objective of this book is the quantitative modeling of tsunami generation and propagation.

Chapter 2 describes an overview of tsunami generation and propagation by visualizing the simulation results of a huge earthquake. We use an earthquake scenario that is anticipated to occur in southwestern Japan. We concisely explain tsunami generation and propagation but do not go into the details. This chapter is intended as a guide to the following chapters.

Chapter 3 theoretically investigates wave propagation in a continuum medium. Wave theories of fluid and elastic dynamics are important for the propagation of tsunami and seismic waves. The theories are basics of tsunami and earthquake researches. In addition, comparisons between tsunami and seismic waves are useful to deepen our understanding of wave phenomena.

Chapter 4 introduces several topics of seismology that are closely related to tsunamis. A description of earthquake fault motion is given, as it is important for modeling tsunami generation. We also explain a practical method for performing seismic wave simulation and its use for modeling tsunami generation. The topics include a description of earthquake faults, a scaling law of fault geometry, permanent displacement, and ocean acoustic waves.

Chapter 5 explains a theory of tsunami generation. By assuming that seawater is an incompressible medium, we derive the intrinsic nature of tsunami generation caused by sea-bottom displacement. Based on mathematical representations, we theoretically investigate the mechanism of the generation process. Also, the theory links the generation process to the propagation process.

Chapter 6 describes methods for calculating tsunami propagation with realistic bathymetry. Depending on the conditions and the order of the approximations, various types of 2-D tsunami equations are derived from the 3-D equation of motion. We explain how to numerically solve these tsunami equations in practice. Some simulation results are also shown as examples.

Chapter 7 summarizes tsunami generation and propagation. Also, we briefly mention additional important topics that we do not fully treat in this book.

References

Baba T, Hirata K, Kaneda Y (2004) Tsunami magnitudes determined from ocean-bottom pressure gauge data around Japan. Geophys Res Lett 31:L08303. https://doi.org/10.1029/2003GL019397

Baba T, Hirata K, Hori T, Sakaguchi H (2006) Offshore geodetic data conducive to the estimation of the afterslip distribution following the 2003 Tokachi-oki earthquake. Earth Planet Sci Lett 241(1):281–292. https://doi.org/10.1016/j.epsl.2005.10.019

Bernard EN, Meinig C (2011) History and future of deep-ocean tsunami measurements. In: Proceedings of Oceans'11 MTS/IEEE, Kona. IEEE, Piscataway, NJ, 19–22 September, 2011

Eguchi T, Fujinawa Y, Fujita E, Iwasaki S, Watanabe I, Fujiwara H (1998) A real-time observation network of ocean-bottom-seismometers deployed at the Sagami trough subduction zone, Central Japan. Mar Geophys Res 20:73–94

Filloux JH (1982) Tsunami recorded on the open ocean floor. Geophys Res Lett 9(1):25–28. https://doi.org/10.1029/GL009i001p00025

Fujii Y, Satake K, Sakai S, Shinohara M, Kanazawa T (2011) Tsunami source of the 2011 off the Pacific coast of Tohoku earthquake. Earth Planet Sp 63(7):55. https://doi.org/10.5047/eps.2011.06.010

Furumura T, Takemura S, Noguchi S, Takemoto T, Maeda T, Iwai K, Padhy S (2011) Strong ground motions from the 2011 off-the Pacific-Coast-of-Tohoku, Japan (Mw = 9.0) earthquake obtained from a dense nationwide seismic network. Landslides 8(3):333. https://doi.org/10.1007/s10346-011-0279-3

Geist EL (1998) Local tsunamis and earthquake source parameters. Adv Geophys 39:117–209. https://doi.org/10.1016/S0065-2687(08)60276-9

Hashimoto C, Noda A, Sagiya T, Matsu'ura M (2009) Interplate seismogenic zones along the Kuril-Japan trench inferred from GPS data inversion. Nat Geosci 2:141–144. https://doi.org/10.1038/ngeo421

Hirata K et al (2002) Real-time geophysical measurements on the deep seafloor using submarine cable in the southern Kurile subduction zone. IEEE J Ocean Eng 27(2):170–181. https://doi.org/10.1109/JOE.2002.1002471

Imamura A (1934) Past tsunamis of the Sanriku coast. Jpn J Astron Geophys 11:79–93

Inazu D, Saito T (2013) Simulation of distant tsunami propagation with a radial loading deformation effect. Earth Planet Sp 65(8):835–842. https://doi.org/10.5047/eps.2013.03.010

Inazu D, Waseda T, Hibiya T, Ohta Y (2016) Assessment of GNSS-based height data of multiple ships for measuring and forecasting great tsunamis. Geosci Lett 3:25. https://doi.org/10.1186/s40562-016-0059-y

Kato T et al (2000) Real-time observation of tsunami by RTK-GPS. Earth Planet Sp 52(10):841–845. https://doi.org/10.1186/BF03352292

Koketsu K et al (2011) A unified source model for the 2011 Tohoku earthquake. Earth Planet Sci Lett 310(3):480–487. https://doi.org/10.1016/j.epsl.2011.09.009

Levin BW, Nosov M (2009) Physics of tsunamis, vol 327. Springer, Dordrecht

Maeda T, Furumura T, Sakai S, Shinohara M (2011) Significant tsunami observed at ocean-bottom pressure gauges during the 2011 off the Pacific coast of Tohoku earthquake. Earth Planet Sp 63(7):53. https://doi.org/10.5047/eps.2011.06.005

Maeda T, Obara K, Shinohara M, Kanazawa T, Uehira K (2015) Successive estimation of a tsunami wavefield without earthquake source data: a data assimilation approach toward real-time tsunami forecasting. Geophys Res Lett 42(19):7923–7932. https://doi.org/10.1002/2015GL065588

Matsumoto H, Inoue S, Ohmachi T (2012) Dynamic response of bottom water pressure due to the 2011 Tohoku earthquake. J Disaster Res 7(7):468–475. https://doi.org/10.20965/jdr.2012.p0468

Mikada H et al (2006) New discoveries in dynamics of an M8 earthquake-phenomena and their implications from the 2003 Tokachi-oki earthquake using a long term monitoring cabled observatory. Tectonophysics 426(1):95–105. https://doi.org/10.1016/j.tecto.2006.02.021

Mori N, Takahashi T, Yasuda T, Yanagisawa H (2011) Survey of 2011 Tohoku earthquake tsunami inundation and run-up. Geophys Res Lett 38:L00G14. https://doi.org/10.1029/2011GL049210

Mungov G, Eblé M, Bouchard R (2013) DART® tsunameter retrospective and real-time data: a reflection on 10 years of processing in support of tsunami research and operations. Pure Appl Geophys 170:1369–1384. https://doi.org/10.1007/s00024-012-0477-5

Nakano M, Nakamura T, Kaneda Y (2015) Hypocenters in the Nankai trough determined by using data from both ocean-bottom and land seismic networks and a 3D velocity structure model: implications for seismotectonic activity. Bull Seismol Soc Am 105(3):1594–1605. https://doi.org/10.1785/0120140309

Namegaya Y, Satake K (2014) Reexamination of the AD 869 Jogan earthquake size from tsunami deposit distribution, simulated flow depth, and velocity. Geophys Res Lett 41:2297–2303. https://doi.org/10.1002/2013GL058678

Nosov MA (1999) Tsunami generation in compressible ocean. Phys Chem Earth, Part B: Hydrol Oceans Atmos 24(5):437–441. https://doi.org/10.1016/S1464-1909(99)00025-8

Nosov MA, Kolesov SV (2007) Elastic oscillations of water column in the 2003 Tokachi-oki tsunami source: in-situ measurements and 3-D numerical modeling. Nat Hazards Earth Syst Sci 7(2):243–249. https://doi.org/10.5194/nhess-7-243-2007

Ohta Y et al (2012) Quasi real-time fault model estimation for near-field tsunami forecasting based on RTK-GPS analysis: application to the 2011 Tohoku-Oki earthquake (Mw 9.0). J Geophys Res: Solid Earth 117:B02311. https://doi.org/10.1029/2011JB008750

Okal EA (1988) Seismic parameters controlling far-field tsunami amplitudes: a review. Nat Hazards 1:67. https://doi.org/10.1007/BF0016822

Ozaki T (2011) Outline of the 2011 off the Pacific coast of Tohoku earthquake (Mw 9.0). Earth Planet Sp 63:57. https://doi.org/10.5047/eps.2011.06.029

Rabinovich AB, Eblé MC (2015) Deep-ocean measurements of tsunami waves. Pure Appl Geophys 172:3281–3312. https://doi.org/10.1007/s00024-015-1058-1

Saito T, Inazu D, Tanaka S, Miyoshi T (2013) Tsunami coda across the Pacific Ocean following the 2011 Tohoku-Oki earthquake. Bull Seismol Soc Am 103(2B):1429–1443. https://doi.org/10.1785/0120120183

Satake K (2015) Tsunamis, treatise on geophysics, 2nd edn. Elsevier, Oxford, pp 477–504

Satake K, Fujii Y, Harada T, Namegaya Y (2013) Time and space distribution of coseismic slip of the 2011 Tohoku earthquake as inferred from tsunami waveform data. Bull Seismol Soc Am 103(2B):1473–1492. https://doi.org/10.1785/0120120122

Satake K, Fujii Y, Yamaki S (2017) Different depths of near-trench slips of the 1896 Sanriku and 2011 Tohoku earthquakes. Geosci Lett 4:33. https://doi.org/10.1186/s40562-017-0099-y

Suzuki W, Aoi S, Sekiguchi H, Kunugi T (2011) Rupture process of the 2011 Tohoku-Oki mega-thrust earthquake (M9. 0) inverted from strong-motion data. Geophys Res Lett 38:L00G16. https://doi.org/10.1029/2011GL049136

Tang L et al (2012) Direct energy estimation of the 2011 Japan tsunami using deep-ocean pressure measurements. J Geophys Res: Oceans 117:C08008. https://doi.org/10.1029/2011JC007635

Tanioka Y, Seno T (2001) Sediment effect on tsunami generation of the 1896 Sanriku tsunami earthquake. Geophys Res Lett 28(17):3389–3392. https://doi.org/10.1029/2001GL013149

Tanioka Y (2017) Tsunami simulation method assimilating ocean bottom pressure data near a tsunami source region. Pure Appl Geophys 175:721. https://doi.org/10.1007/s00024-017-1697-5

Tappin DR et al (2014) Did a submarine landslide contribute to the 2011 Tohoku tsunami? Mar Geol 357:344–361. https://doi.org/10.1016/j.margeo.2014.09.043

Tatehata H (1997) The new tsunami warning system of the Japan meteorological agency. In: Perspectives on tsunami hazard reduction. Springer, Dordrecht, pp 175–188

Titov VV et al (2005) Real-time tsunami forecasting: challenges and solutions. In: Developing tsunami-resilient communities. Springer, Dordrecht, pp 41–58

Tsushima H, Hino R, Tanioka Y, Imamura F, Fujimoto H (2012) Tsunami waveform inversion incorporating permanent seafloor deformation and its application to tsunami forecasting. J Geophys Res Solid Earth 117:B03311. https://doi.org/10.1029/2011JB008877

Tsushima H, Ohta Y (2014) Review on near-field tsunami forecasting from offshore tsunami data and onshore GNSS data for tsunami early warning. J. Disaster Res 9(3):339–357. https://doi.org/10.20965/jdr.2014.p0339

Watada S, Kusumoto S, Satake K (2014) Traveltime delay and initial phase reversal of distant tsunamis coupled with the self-gravitating elastic earth. J Geophys Res Solid Earth 119:4287–4310. https://doi.org/10.1002/2013JB010841

Yamamoto N, Aoi S, Hirata K, Suzuki W, Kunugi T, Nakamura H (2016) Multi-index method using offshore ocean-bottom pressure data for real-time tsunami forecast. Earth Planet Sp 68(1):128. https://doi.org/10.1186/s40623-016-0500-7

Chapter 2
Overview of Tsunami

Abstract When a large earthquake occurs in an offshore region, a tsunami is generated. The generation and propagation can be mathematically described based on equation of motion. This chapter visualizes and overviews tsunami generation and propagation to obtain an overall grasp of tsunamis and elucidate their fundamental nature. Section 2.1 illustrates the tsunami generation and propagation caused by a huge earthquake. We use a dynamic rupture scenario constructed for an anticipated huge earthquake in the Nankai Trough, Japan. Section 2.2 shows a simple model for the generation and propagation. While the generation is basically independent of gravity, gravity is the main force to move tsunami. Section 2.3 describes the fundamental properties of the propagation. Tsunami propagation depends on the wavelength and sea depth. Also, when the sea depth is very shallow, tsunami becomes to show nonlinear characteristics. Section 2.4 summarizes the main points of tsunami generation and propagation.

Keywords Earthquake rupture · Seismic wave · Permanent displacement · Nonlinear tsunami

2.1 Earthquakes and Tsunami: A Possible Scenario in Nankai Trough, Japan

In order to obtain an overview of tsunamis, we propose a huge earthquake scenario that can possibly occur in southwestern Japan as an example, and we simulate and visualize the results.

2.1.1 Nankai Trough, Japan

At the Nankai Trough (Fig. 2.1), the Philippine Sea Plate subducts underneath the Eurasian Plate. Great (M $\sim$ 8) earthquakes have occurred repeatedly along the plate

T. Saito, *Tsunami Generation and Propagation*, Springer Geophysics,
https://doi.org/10.1007/978-4-431-56850-6_2

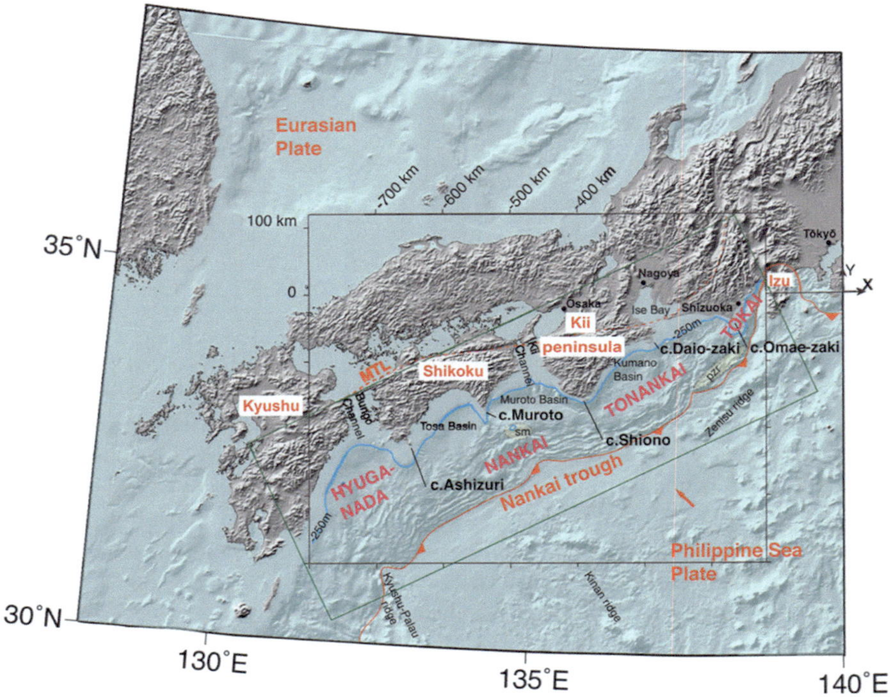

Fig. 2.1 Nankai Trough in southwest Japan. The abbreviations "sm," "pzr," and "c" stand for subducted seamount, subducted paleo-Zenisu ridge, and cape, respectively (Hok et al. 2011, copyright by the American Geophysical Union)

interface with a recurrence interval on the order of 100 years (e.g., Ando 1975; Kumagai 1996). The rupture process and area of each historical earthquake are not clearly known. The 1707 Hoei earthquake was believed to involve a rupture of the whole area from Hyuganda to the Tokai area (Furumura et al. 2011). The 1944 Tonankai (M 7.9) and the 1946 Nankai (M 8.0) earthquakes caused serious tsunami damage along the Pacific coast of Japan around this area.

2.1.2 *Earthquake Rupture Model*

We used an earthquake rupture scenario constructed by a numerical simulation (Hok et al. 2011). Since an earthquake occurs when the stress reaches or excesses a critical level (strength), the earthquake size is strongly controlled by the stress accumulated along the plate interface between the Philippine Sea Plate and the Eurasian Plate. By using the rate of the stress accumulation (or slip deficit) estimated by the geodetic data and assuming that the stress has been constantly accumulated for 100 years, Hok et al. (2011) estimated the accumulated stress along the plate boundary. They

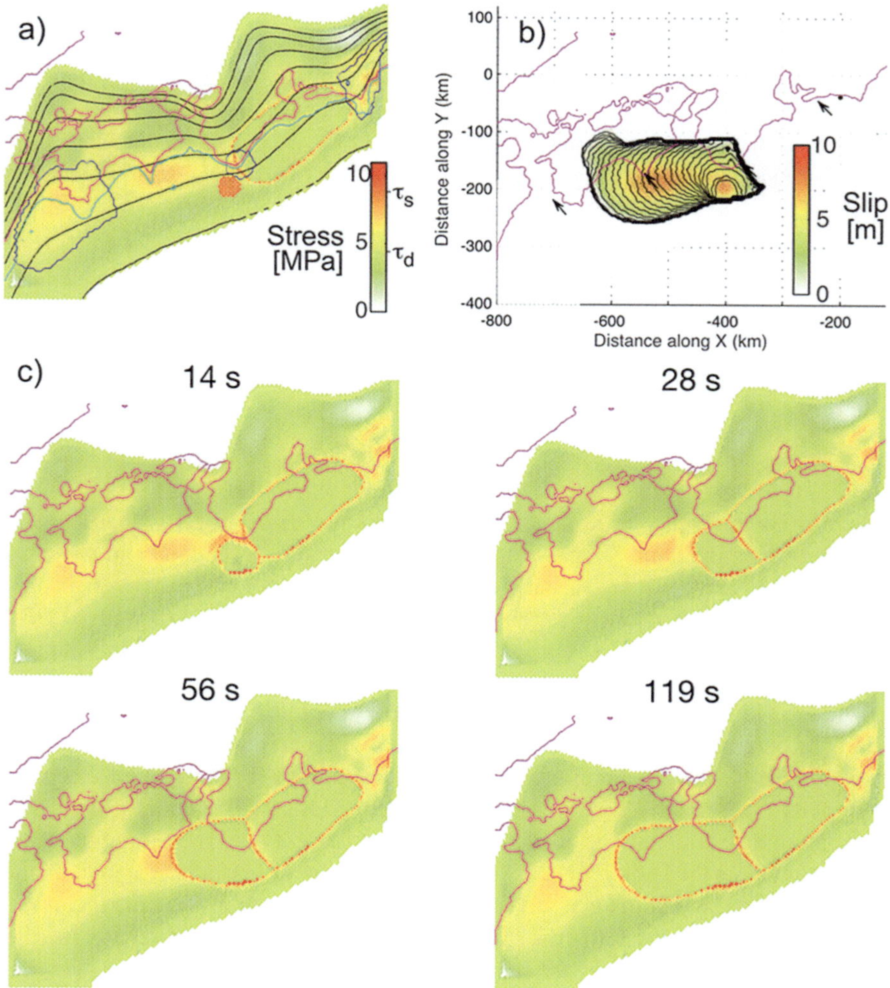

Fig. 2.2 Earthquake fault model used in the tsunami generation and propagation simulation. The fault model was derived by a dynamic rupture simulation (Hok et al. 2011). (**a**) Initial shear stress distribution. The overshoot area of the stress for initiation appears in red. Lines indicate 10-km-depth contours of plate boundary (black), coast line (pink), and sea depth of 250 m (light blue). (**b**) Final slip distribution and rupture time contours every 5 seconds. Arrows denote the slip direction of the slab. (**c**) Snapshots of the stress during the rupture propagation (Hok et al. 2011, copyright by the American Geophysical Union)

used this stress distribution as an initial condition of their dynamic rupture simulation. The spatial and temporal variation of the dynamic rupture was simulated based on the equation of motion and the frictional constitutive law along the plate interface.

Figure 2.2 shows the temporal change of the stress distribution calculated by the rupture simulation. This represents a rupture scenario that is similar to the 1946 Nankai earthquake that occurred after the 1944 Tonankai earthquake. The rupture started at the high stress portion off the southern tip of the Kii Peninsula (Fig. 2.2a). The rupture propagated westward releasing the stress accumulated in the offshore region (see the snapshots at the elapsed times of 14, 28, and 56 s in Fig. 2.2c). At the elapsed time of 119 s, the rupture ended at the central part of Shikoku.

One of the most important quantities that characterizes an earthquake is the moment magnitude M_w. The moment magnitude is defined by the seismic moment M_0 as

$$M_w = \frac{1}{1.5}(\log M_0[\mathrm{N\ m}] - 9.1) \tag{2.1}$$

and the seismic moment M_0 is given by

$$M_0 = \mu \iint_S [u(\xi)]dS(\xi) = \mu DS \tag{2.2}$$

where μ is the rigidity or shear modulus of the solid earth, $[u(\xi)]$ is the final slip distribution on the fault, S represents the fault area, and D is the final slip averaged over the fault area. For the scenario shown in Fig. 2.2, the fault area, average slip, rigidity, and seismic moment are about $S = 260 \times 130$ km^2, $D = 4$ m, $\mu = 40$ GPa, and $M_0 = 5.2 \times 10^{21}$ N m, respectively. The moment magnitude of this rupture model is M_w 8.4.

Another important aspect of the earthquake rupture is its time scale. There are two kinds of parameters that characterize the time scale. One is the rupture time. The rupture time is the total duration of the rupture. Figure 2.2 indicates that the rupture ends at the elapsed time (time measured from rupture initiation) of about 120 s. Thus, the rupture time is 120 s. The other is the rise time. The rise time is the duration of the slip at a point on the fault. Figure 2.3 shows a time history of the slip at a point on the fault. The slip starts at the elapsed time of 60 s and increases as the time increases. After an elapsed time of 100 s, the slip becomes almost constant. In this case, the rise time is about 40 s (=100 s − 60 s).

The moment magnitude and the time scales of the rupture are important parameters in tsunami study. The moment magnitude strongly controls the permanent sea-bottom displacement and the magnitude of the tsunami. Roughly speaking, the

Fig. 2.3 An example of the time history of the slip at a point on the fault model

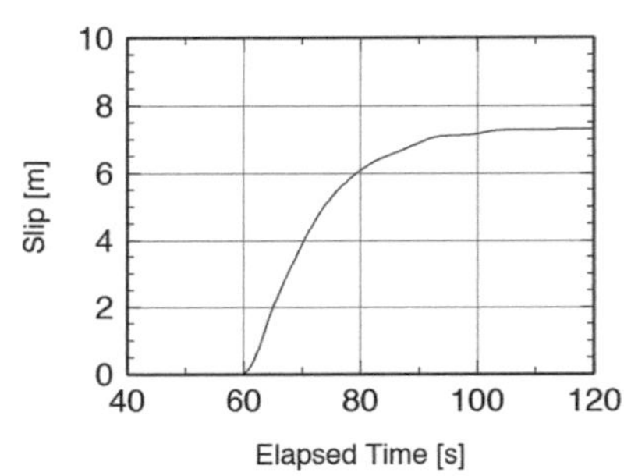

magnitude of the tsunami is almost determined by the moment magnitude. The tsunami is not strongly affected by the time scales of the rupture. Nevertheless, the time scale is definitely important in tsunami studies because it controls the excitation of seismic waves. Seismic waves play various roles in tsunami studies. By analyzing seismic waves, we can estimate the earthquake magnitude and location. We often use the earthquake focal information to predict the height of the tsunami on the coast for the purpose of issuing warnings. Besides, in the next-generation tsunami observation, which detects tsunamis inside the focal area, seismic waves can be noise for tsunami signals.

- Earthquakes are characterized by static and kinematic parameters.
- The seismic moment ($M_0 = \mu DS$) describes the size of the earthquake.
- The rupture time and rise time describe the kinematics of the earthquake fault.

2.1.3 A Two-Step Method for Simulating Tsunami Generation and Propagation

Tsunami generation and propagation from the earthquake rupture is simulated by a two-step method (Fig. 2.4). In the first step, a seismic wave simulation is conducted with the earthquake rupture model used as input data. The simulation numerically calculates the sea-bottom motion. The second step is a tsunami simulation using the sea-bottom motion calculated in the first step as the tsunami source. The details are explained in Appendix A.

2.1.4 Seismic Waves and Permanent Sea-Bottom Displacement

The seismic wave simulation is conducted using the earthquake rupture model shown in Fig. 2.2 as the source. The slip on the fault (dislocation) is equivalent to the body force distribution and is represented as a stress glut in the seismic wave simulations. The details of the earthquake fault and seismic wave propagation are

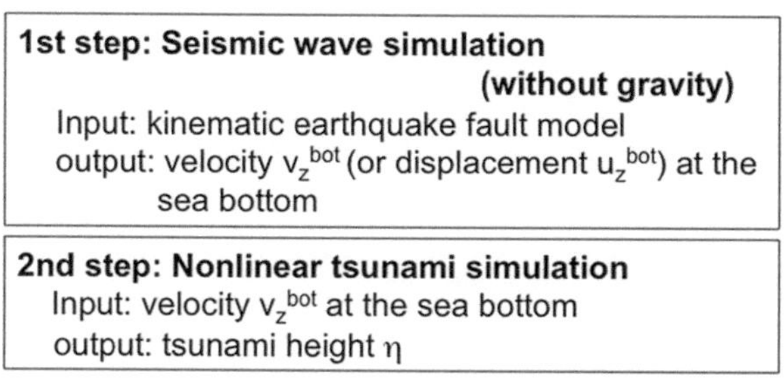

Fig. 2.4 A flowchart of the simulation of tsunami generation due to an earthquake and its propagation

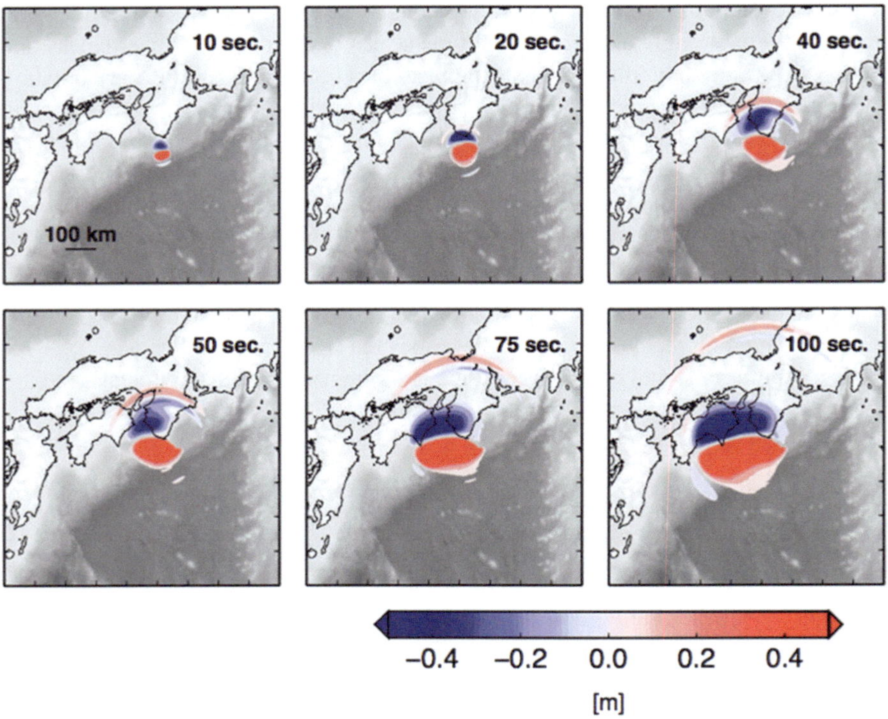

Fig. 2.5 Results of seismic wave simulations using the 1946 Nankai-like earthquake rupture model of Hok et al. (2011). Vertical displacements at the ground surface and at the sea bottom at various elapsed times

described in Chap. 4: *Earthquakes*. Figure 2.5 shows the simulated vertical displacement at the ground surface and the sea bottom for various elapsed times. At the elapsed time of 10 s, the vertical displacement begins to appear on the sea bottom. The displacement extends westward as the time increases. During the rupture from 40 to 100 s, the seismic wave propagating northward is clearly recognized. The propagation velocity is ~3 km/s (the wave propagates about 200 km during the time of 40–100 s). This wave is a seismic surface wave called a Rayleigh wave. The displacement caused by seismic waves is transient, i.e., displacement returns to zero after the passage of seismic waves. In addition to this transient Rayleigh wave displacement, there is a large vertical displacement that does not propagate but persists above the earthquake focal area (see a snapshot at the elapsed time of 100 s).

Figure 2.6 shows the distribution of the vertical displacement when enough time elapsed (300 s). This distribution persists permanently, and so it is referred to as permanent displacement. A significant uplift higher than 0.5 m extends ~200 km in length and ~100 km in width in the offshore region, and subsidence larger than 0.5 m is recognized in the land area.

Both the Rayleigh wave and the permanent displacement cause significant sea-surface height and sea-bottom displacement changes. Note that the tsunami is

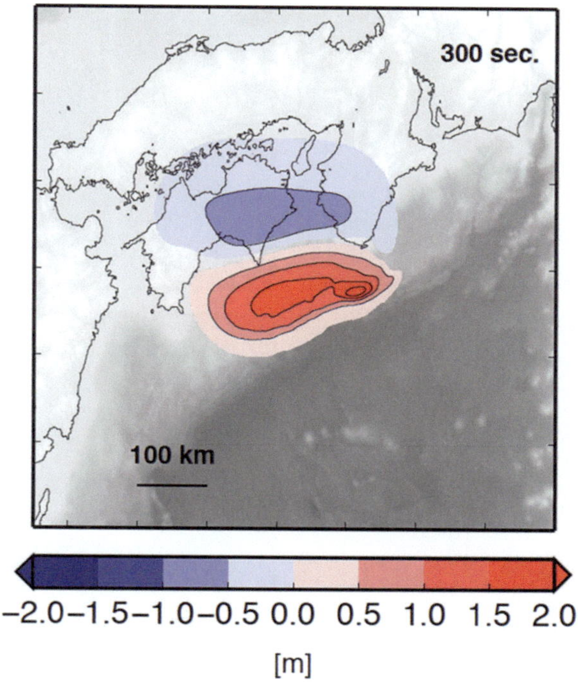

Fig. 2.6 Vertical displacement distribution at the elapsed time of 300 s. This is considered a permanent displacement due to the earthquake

excited mainly by the permanent displacement, but not by the Rayleigh wave. The permanent displacement sustains the sea surface uplifted until the sea-surface displacement collapses as a tsunami. On the other hand, Rayleigh waves do not sustain the sea-surface displacement but just oscillate the sea surface.

- Seismic waves and permanent displacement are excited by an earthquake.
- Seismic waves represent transient displacements propagating a long distance.
- Permanent displacement appears only near the earthquake fault. This functions as the tsunami source.

2.1.5 Tsunami Generation and Propagation

By using the sea-bottom motion calculated by the seismic wave simulation (Fig. 2.5), we conducted tsunami simulation. In this simulation, we calculated the sea-surface height change or vertical displacement at the sea surface.

Figure 2.7 shows the sea-surface height distribution calculated by the tsunami simulation. At the elapsed time of 1.5 min, the sea surface is uplifted more than ~0.5 m for an area extending ~200 km in length and ~100 km off Shikoku. This is caused by the permanent sea-bottom displacement by the earthquake. A wave

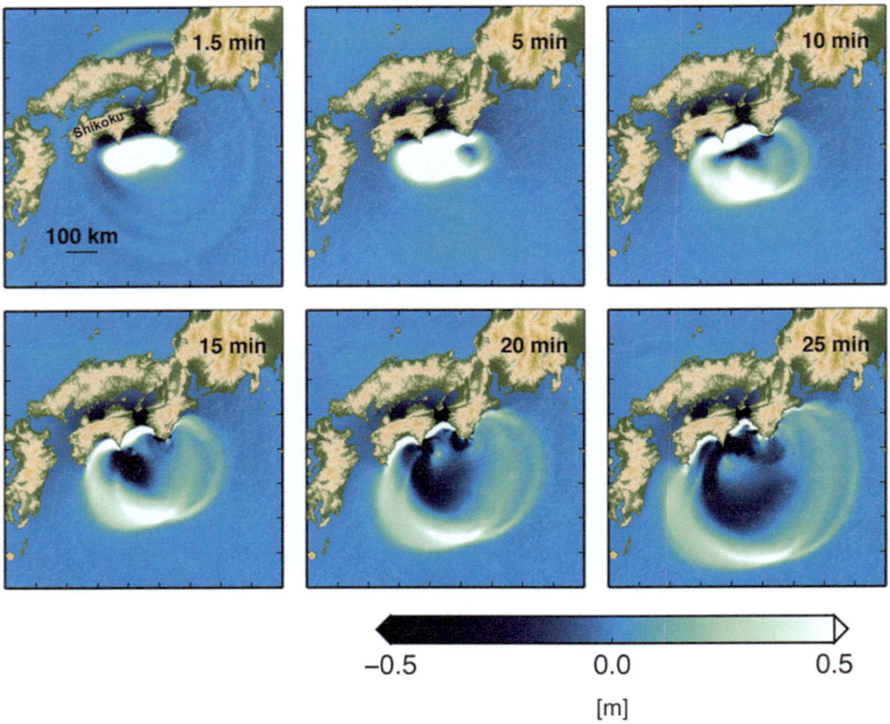

Fig. 2.7 Results of tsunami simulations using the 1946 Nankai-like earthquake rupture model of Hok et al. (2011). Sea-surface height distribution at various elapsed times

propagating cylindrically outward from the hypocenter is also recognized at this time $t = 1.5$ min. This is the Rayleigh wave.

At $t = 5$ min, the sea-surface displacement by the Rayleigh wave already moved away from the simulation region. The Rayleigh wave does not contribute to tsunami excitation. Only the sea-surface displacement caused by the permanent sea-bottom displacement exists. The uplifted sea surface cannot keep its shape due to gravity, and the sea surface collapses with increasing the time. As a result, the sea-surface displacement propagates as a tsunami. This clearly illustrates that gravity plays an important role in tsunami propagation. If there was no gravity, the tsunami would not propagate, and the sea surface uplifted by the sea-bottom deformation would maintain its shape permanently. Hence, it is natural that the tsunami propagation speed is controlled by the gravitational acceleration $g_0(=9.8\ \mathrm{m/s^2})$. Actually, the tsunami velocity c is given by using the gravitational acceleration as

$$c = \sqrt{g_0 h}, \tag{2.3}$$

where h is the sea depth. We will show this derivation in Chap. 3: *Propagation of Tsunami and Seismic Waves*. Equation (2.3) predicts that a tsunami would

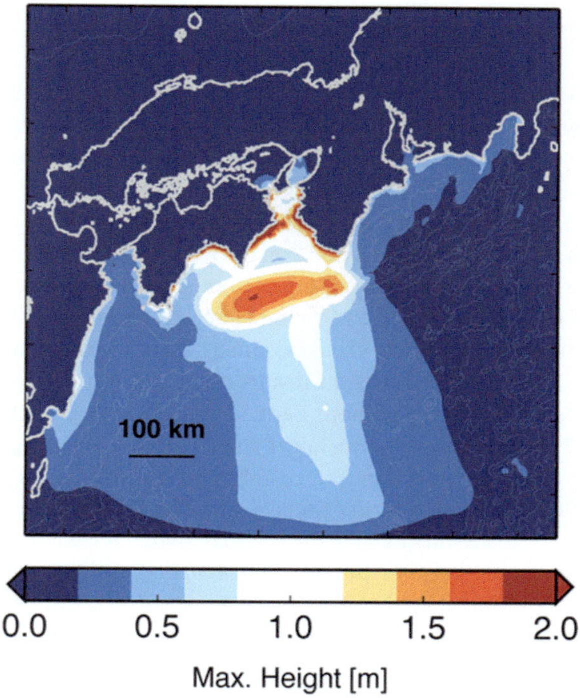

Fig. 2.8 Maximum sea-surface height distribution

propagate faster in the deep ocean and slower in a shallow sea. Figure 2.7 also shows that the tsunami propagates in the offshore region with higher velocity. In the Pacific Ocean, the average sea depth is ~4000 m. At that depth, the tsunami propagates with a velocity of ~200 m/s. This is much slower than the seismic waves (seismic waves propagate with a velocity of a few km/s). When a tsunami propagates in a shallow sea with a depth of ~50 m, it propagates more slowly, with a velocity of ~20 m/s.

In general, when the wave propagates more slowly, the wave amplitude increases so as to conserve the energy flux along the ray of the wave (e.g., Sect. 3.2.3 *Energy Density and Energy Flux Density for Incompressible Fluid*). Tsunamis then show height amplification when they propagate from a deep (fast) to shallow sea (slow). Figure 2.8 shows the maximum tsunami height distribution of the simulation. A tsunami height higher than 1.5 m appeared near coasts as well as at the source area. Even if the tsunami height is lower in the offshore region, it becomes amplified when the tsunami approaches the coast.

- Permanent sea-bottom displacement and gravity cause tsunamis.
- Tsunamis propagate with a velocity of about < ~100 m/s at a depth of 1000 m.
- Tsunamis propagate faster in deeper oceans.
- Tsunamis propagate slower and the height becomes larger when tsunamis approach coasts.

2.1.6 Slow Rupture

Many earthquakes show a relatively rapid slip on the fault characterized by a rupture with a short duration. Some earthquakes show a relatively slow slip on the fault characterized by a long rise time and a long rupture time (e.g., Kanamori and Kikuchi 1993). We simulated a slow rupture by extending the time scale of the fault model two times longer than in the original model. Figure 2.9 shows the time history of the slip at a point on the fault model for a slow rupture (solid line) together with that for the original standard rupture (dashed line). The start of the slip changed from 60 to 120 s, the rise time increased from 40 to 80 s, and the slip rate (the slope during the rise time) became small, while the final slip did not change.

Figure 2.10 shows the results of the seismic wave simulation with the earthquake fault of a long duration. Permanent uplifted sea-bottom displacement higher than 0.5 m and subsidence lower than −0.5 m occur near the earthquake fault. This permanent displacement distribution is identical to the original one shown in Fig. 2.6. However, the excitation of the Rayleigh wave (seismic wave) is considerably smaller than that in the original model. It is hard to recognize the Rayleigh wave in Fig. 2.10. This is because the excitation of the seismic wave is strongly controlled by the slip rate rather than by the slip itself (we will treat the excitation of the seismic waves in Chap. 4: *Earthquakes*). Hence, even though the two rupture models have the same seismic moment, the Rayleigh wave becomes small if the moment rate is smaller due to a long source duration.

Figure 2.11 shows the tsunami simulation results with an earthquake fault of a long duration. Basically, tsunami propagation was almost the same as in Fig. 2.7. The difference in the rise times of 40 s and 80 s did not significantly affect the results of tsunami simulation. Actually, we obtained similar maximum amplitude distributions in the cases of short- and long-duration earthquake ruptures (Fig. 2.12).

Tsunamis caused by short-duration and long-duration earthquake ruptures are almost similar, whereas seismic waves are quite different. We call the earthquake that excites tsunami much larger than the one expected from the seismic wave amplitude as a tsunami earthquake (Kanamori 1972). Therefore, the earthquake with long duration can be a tsunami earthquake. Also, we should note that the

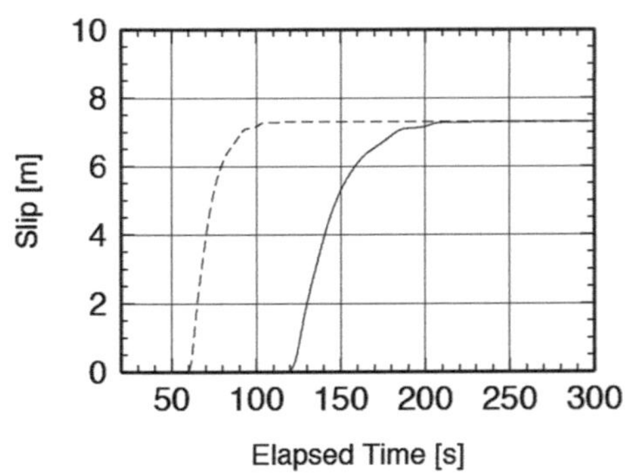

Fig. 2.9 Time history of the slip at a point on the fault model. The cases of short duration (dashed line) and long duration (solid line). The short duration corresponds to the 1946 Nankai-like earthquake rupture model of Hok et al. (2011). The long-duration fault model is obtained by multiplying the time axis by two

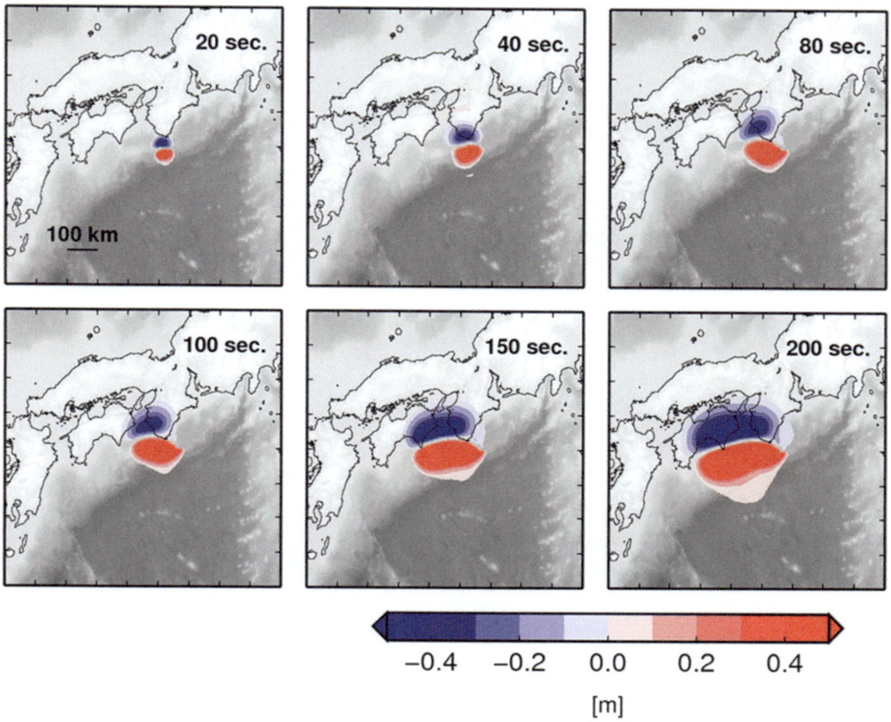

Fig. 2.10 Results of seismic wave simulations using the long-duration earthquake rupture model (corresponding to the black line in Fig. 2.9). Vertical displacements at the ground surface and at the sea bottom at various elapsed times

records can be quite different between short- and long-duration earthquakes. Figure 2.13 shows synthesized ocean-bottom pressure records at a station (indicated by a white triangle) for short-duration and long-duration rupture models. We use mH_2O as the unit of the pressure change instead of Pa. A pressure of 1 mH_2O refers to the hydrostatic pressure at the sea bottom due to water that is 1 m deep, i.e., 1 mH_2O = 9.8 MPa. For a short-duration rupture (blue line), a large-amplitude Rayleigh wave is recognized during the elapsed time from 0 to ~300 s. The tsunami then arrives at an elapsed time of ~300 s. The tsunami is ~0.5 m height. For a long-duration rupture (red line), the pressure change caused by the Rayleigh wave is considerably smaller than that for the short-duration rupture. These findings indicate that the tsunamis are similar but the ocean-bottom pressure records are substantially different. This is crucially important for tsunami early warnings, as it would be difficult to predict tsunami height just by looking at the records. We should discriminate between the tsunami and seismic wave components. This discrimination is easy when the observation station is located far from the source, because seismic waves arrive at observation points much earlier than the tsunami. However, when the observation station is inside the source region, the seismic waves are overlapped with tsunamis. If we misunderstand the seismic waves as the tsunami, the tsunami

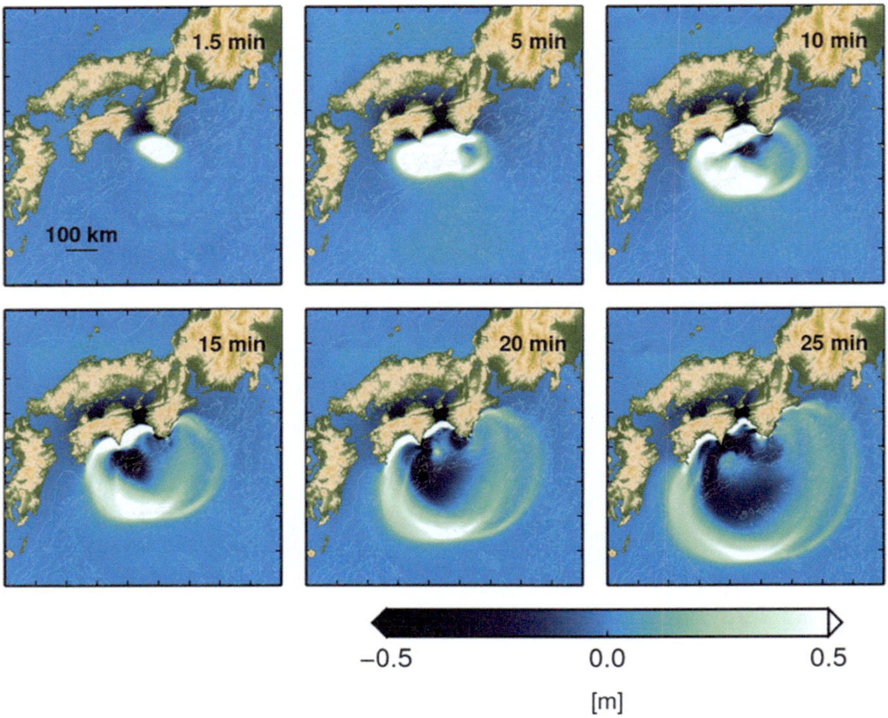

Fig. 2.11 Results of tsunami simulations using the long-duration earthquake rupture model (corresponding to black line in Fig. 3.8). Sea-surface height distribution at various elapsed times

prediction would be incorrect. We investigate the details of the ocean-bottom pressure records in Sect. 5.4.2: *Synthesis of Ocean-Bottom Pressure Records*.

- Excitation of seismic waves strongly depends on the time scale of the fault rupture.
- Excitation of the tsunami does not strongly depend on the time scale of the fault rupture.
- Observed records look quite different due to seismic waves even if the same tsunami signals are included.

2.2 Tsunami Generation

In order to understand the mechanisms underlying tsunamis, we investigate the generation and propagation processes separately by assuming simple model settings.

We consider tsunami generation due to sea-bottom displacement where incompressible fluid and a constant sea depth of h_0 are supposed. In that case, the tsunami generation and propagation caused by the sea-bottom displacement is given by

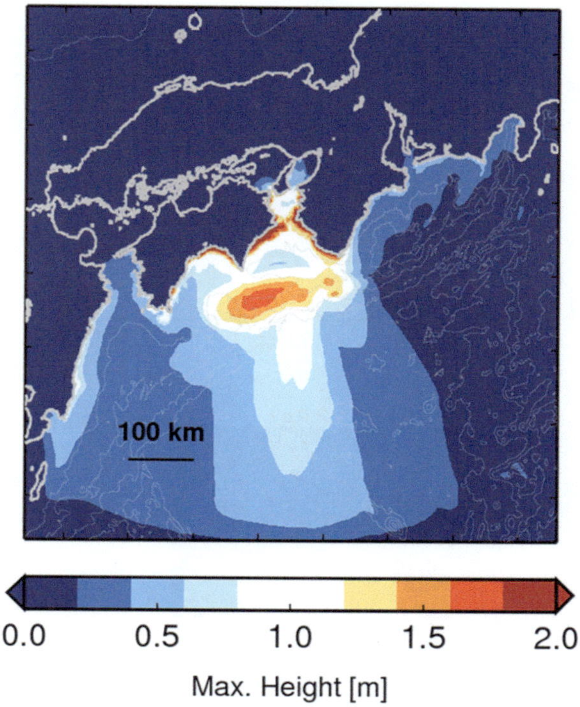

Fig. 2.12 Maximum sea-surface height distribution for the long-duration earthquake rupture model

analytical solutions (e.g., Takahashi 1942; Kajiura 1963; Saito 2013, see also Chap. 5: *Tsunami Generation*). By using the analytical solutions, we visualize the tsunami generation and propagation caused by an earthquake in Fig. 2.14.

When an earthquake occurs, the sea bottom deforms largely above the earthquake fault (elapsed times of 20 s and 60 s). At the same time, a seismic wave radiated from the fault propagates horizontally. The seismic wave is clearly recognized at elapsed times of 60, 80, and 100 s. The seismic wave propagates fast (about 4 km/s) and moves away from the earthquake fault region swiftly, while the permanent sea-bottom displacement remains near the earthquake fault region.

The sea-bottom deformation displaces the seawater, and the sea surface is uplifted (elapsed times of 20 and 60 s). The permanent sea-bottom deformation keeps the sea surface uplifted in the focal area for a short time (elapsed times of 60–80 s). However, as the time elapses (100–200 s), gravity causes the uplifted sea surface to collapse. The collapsed water volume displaces a massive water volume in the horizontal direction, and the sea-surface displacement propagates as a long-wavelength wave. This water wave is the tsunami. Note that the sea-surface height change caused by the propagating seismic waves should not be referred to as a tsunami because the driving force of the wave propagation is different from that of the tsunami. Seismic wave propagation is mainly due to the elasticity of the medium. We consider the tsunami to be a surface water wave caused by gravity.

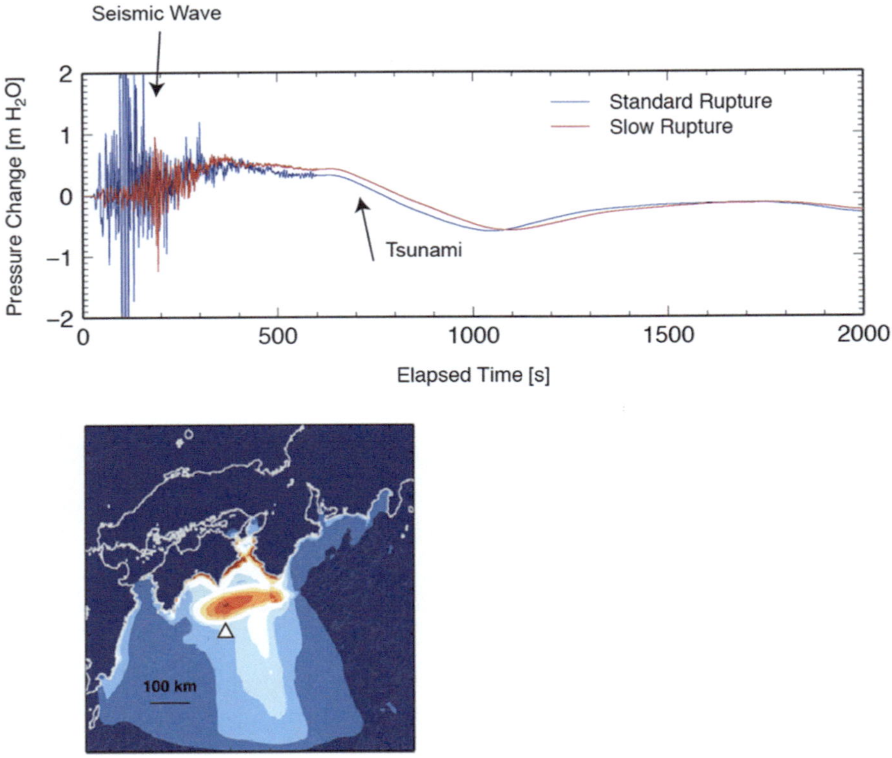

Fig. 2.13 The waveforms of the short-duration (blue) and long-duration (red) earthquake fault models. The temporal change of the sea-surface height off Muroto is plotted (the location is indicated by a triangle on the map)

- Excitation of the tsunami is strongly controlled by the permanent sea-bottom displacement.
- The tsunami is caused by the collapse of the sea-surface height distribution due to gravity.
- Gravity causes tsunami propagation, while elasticity causes seismic wave propagation.

2.3 Tsunami Propagation

Tsunami propagation is described as long-wavelength surface-wave propagation through an incompressible fluid with a free surface under gravity. The details of the theoretical background and the practical numerical methods are treated in Chap. 3: *Propagation of Tsunami and Seismic Waves* and Chap. 6: *Propagation Simulation*. By using the simulation results, this section illustrates the fundamental nature of tsunami propagation.

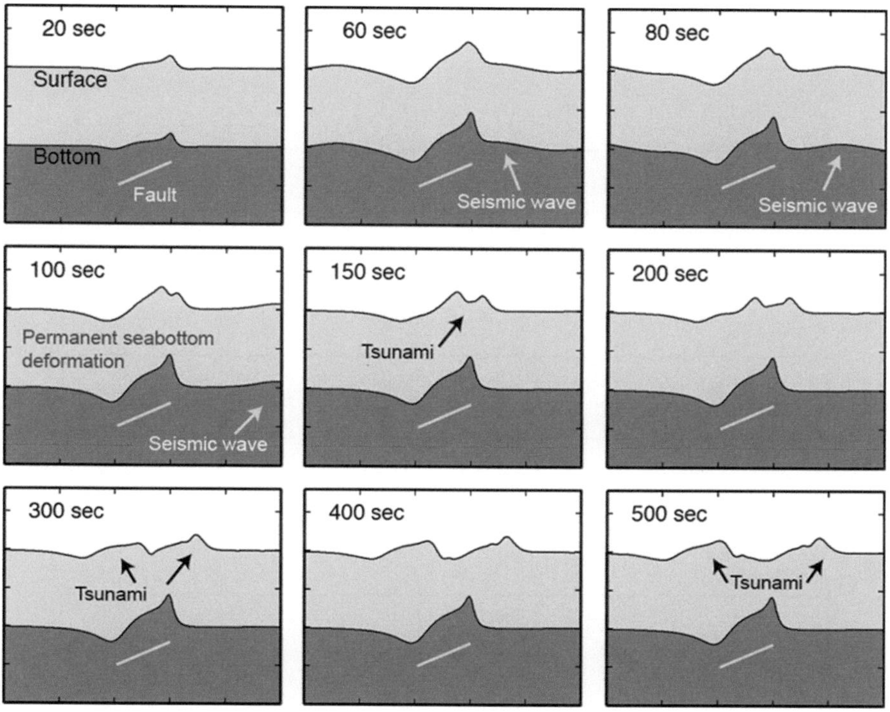

Fig. 2.14 Tsunami generation and propagation caused by an earthquake fault. Seismic waves radiated from the fault propagate horizontally (elapsed times from 20 to 100 s). The sea surface is uplifted by the sea-bottom deformation. As time elapses (after 80 s), the uplifted sea surface collapses due to gravity. The sea-surface height propagates as a tsunami (elapsed times from 150 to 500 s)

2.3.1 Propagation Speed

Figure 2.15a displays the tsunami propagation in a shallow sea (depth of 1 km). At the elapsed time of 0 s, the tsunami height distribution has a peak at a distance of 0 km. As time elapses, the tsunami propagates both rightward and leftward. At the elapsed times of 500, 1000, and, 2000 s, the peak arrives at distances of 50, 100, and 200 km, respectively. The propagation velocity is then estimated to be 0.1 km/s. On the other hand, Figure 2.15b shows the tsunami propagation in a deeper sea (depth of 4 km). At elapsed times of 500 and 1000 s, the peak arrives at distances of 100 and 200 km, respectively. The propagation velocity is 0.2 km/s. The tsunami propagates faster in deeper oceans. The tsunami phase velocity c_0 is approximately given by $c_0 = \sqrt{g_0 h_0}$ where $g_0 = 9.8$ m/s^2 is the gravitational acceleration and h_0 is the sea depth when the tsunami wavelength is much longer than the sea depth (the derivation is shown in Chap. 3).

In principle, for waves (not only tsunamis but also other waves such as elastic waves), the restoring force is fundamentally important. Since the tsunami phase

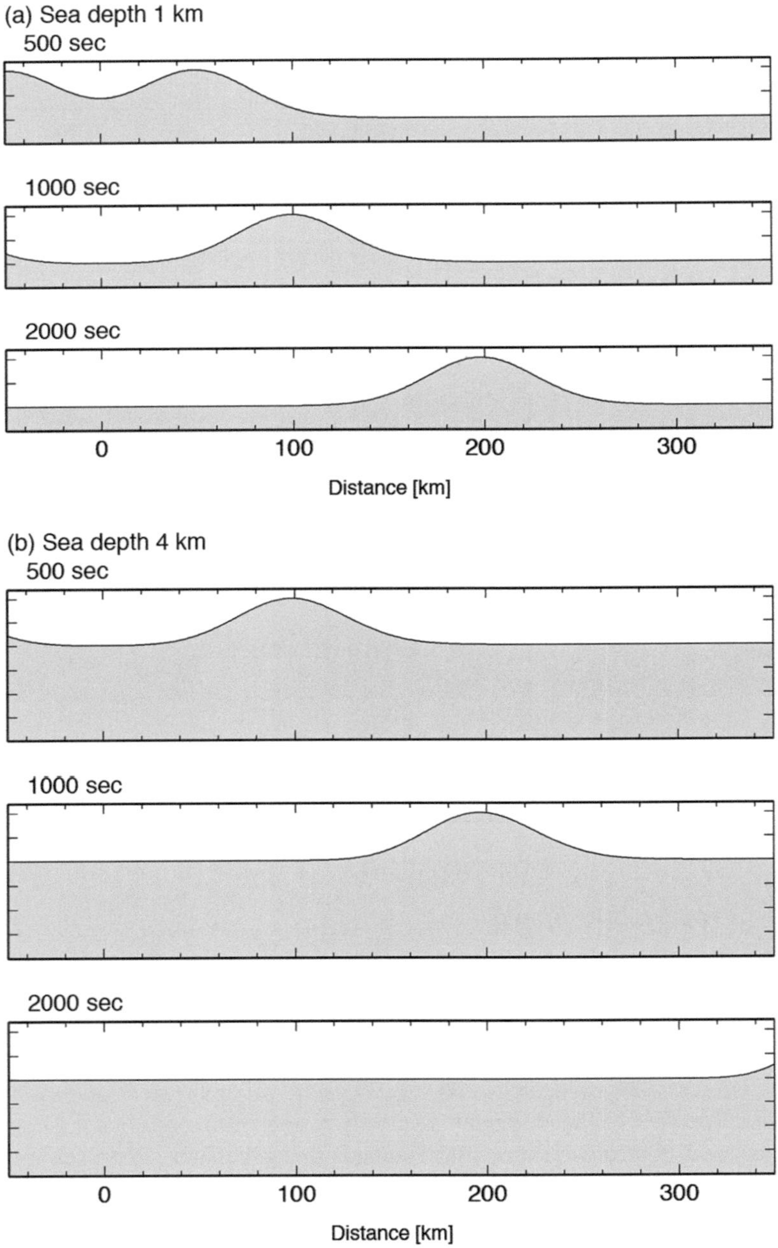

Fig. 2.15 Tsunami propagation in (**a**) shallow sea (1 km depth) and (**b**) deep sea (4 km depth)

velocity is represented using gravitational acceleration g_0, it can easily be realized that the gravitational force is the restoring force of tsunami propagation. The gravitational force is determined not only by g_0 but also by the water volume. Hence, it is reasonable that the phase velocity be represented using the sea depth h_0 in addition to g_0. We should note that the tsunami propagation velocity is given by $c_0 = \sqrt{g_0 h_0}$ only when the tsunami wavelength is much longer than the sea depth. However, in general, the phase velocity depends on the tsunami wavelength and tsunami height as well.

The tsunami propagations of short- and long-wavelength tsunamis are compared in Fig. 2.16. We set the initial height distribution (elapsed time of 0 s) by using a Gaussian function as

$$\eta_0(x,t) = \frac{4}{\sqrt{\pi}L_c}\exp\left(-\frac{16x^2}{L_c^2}\right), \tag{2.4}$$

where a parameter L_c controls the tsunami wavelength. In the case of the long-wavelength tsunami ($L_c = 40\,\mathrm{km}$) (Figure 2.16a), the tsunami height distribution keeps the same shape during the propagation (compare the height distribution at the elapsed times of 500 and 1500 sec). On the other hand, in the case of the short-wavelength tsunami ($L_c = 4\,\mathrm{km}$) (Figure 2.16b), the tsunami height distribution changes as the travel distance increases. We also find that the longer-wavelength component leads to faster propagation by noticing that the long-wavelength tsunami propagates a longer distance in the same time (see 1500 s in Figure 2.16b). This shows that the velocity changes according to the wavelength. This is called dispersion.

The water wave theory (we will treat this in Chap. 3*: Propagation of Tsunami and Seismic Waves*) predicts that the wavenumber k and the angular frequency ω satisfy

$$\omega = \sqrt{g_0 k \tanh(kh_0)}. \tag{2.5}$$

When the relation between k and ω is given by Eq. (2.5), the phase velocity c is calculated as

$$\begin{aligned} c = \frac{\omega}{k} &= \sqrt{g_0 h_0}\sqrt{\frac{\tanh(kh_0)}{kh_0}} \\ &= \begin{cases} \sqrt{\dfrac{g_0}{k}} & kh_0 \gg 1 \\ \sqrt{g_0 h_0} & kh_0 \ll 1 \end{cases} = \begin{cases} \sqrt{\dfrac{g_0 \lambda}{2\pi}} & \lambda \ll h_0 \\ \sqrt{g_0 h_0} & \lambda \gg h_0, \end{cases} \end{aligned} \tag{2.6}$$

where we used an asymptotic relation

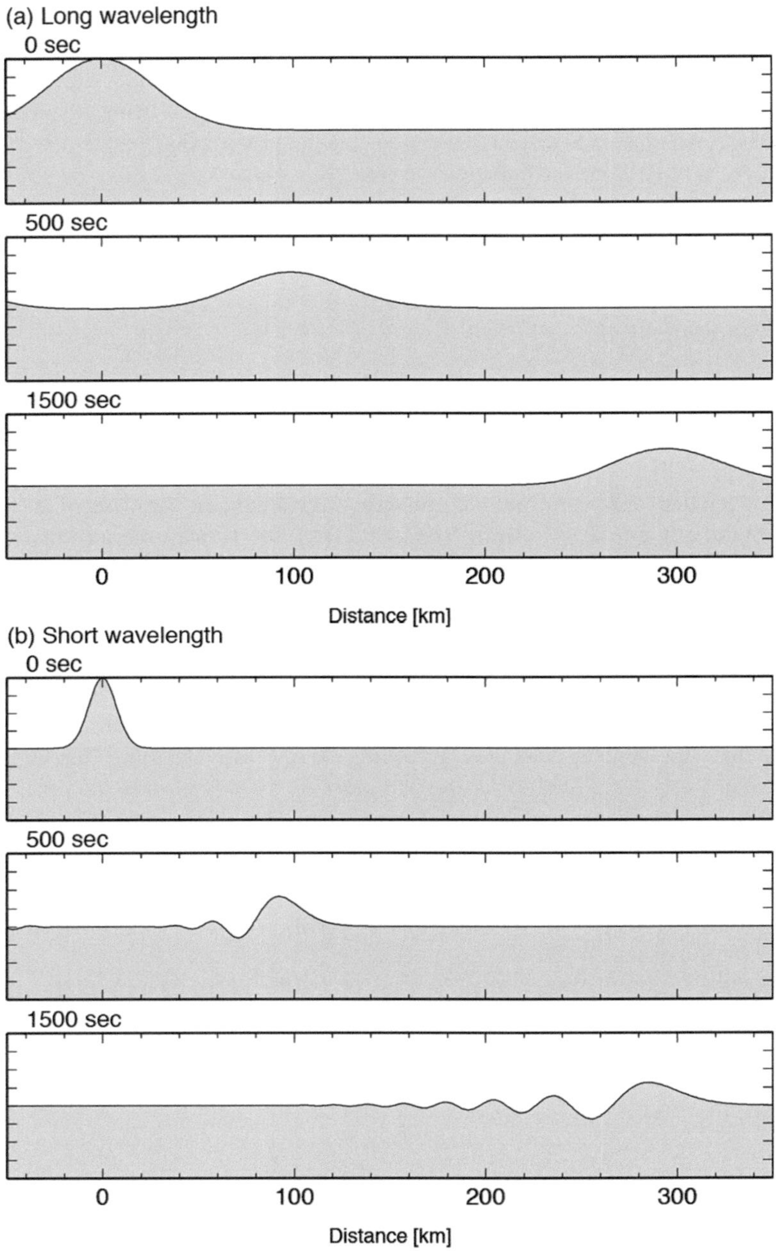

Fig. 2.16 Tsunami propagation. (**a**) Long-wavelength tsunami (L_c = 40 km) and (**b**) short-wavelength tsunami (L_c = 4 km) for a sea depth of 4 km

$$\tanh(kh_0) = \begin{cases} 1 & kh_0 \gg 1 \\ kh_0 & kh_0 \ll 1. \end{cases}$$

Equation (2.6) indicates that the phase velocity depends on the wavelength $\lambda = 2\pi/k$. The phase velocity represented by $c = \sqrt{g_0 h_0}$ is an approximation when the tsunami wavelength is much greater than the sea depth h_0 ($\lambda \gg h_0$). Figure 2.17 shows the phase velocity as a function of the tsunami wavelength.

Equation (2.6) also indicates that a "long" wave means that the wavelength is long compared to the sea depth h_0. When the tsunami wavelength is short enough, the velocity is given by $\sqrt{g_0\lambda/2\pi}$, which is independent of the sea depth h_0. The velocity depends only on the tsunami wavelength.

Why does the tsunami speed depend on the sea depth h_0 and the tsunami wavelength λ? The dependence of the phase velocity on the wavelength (Fig. 2.17) can be interpreted as follows. First, we consider a short wavelength ($\lambda \ll h_0$). In this case, because the wavelength is very small compared to the sea depth, the wave motion does not occur in a deep part of the sea but only in a shallow part of the sea layer. Therefore, the volume of the water in motion is limited to the shallow part of the sea. Since the gravitational force is proportional to the volume in motion, the gravitational force is relatively small when the wavelength is short. When the gravitational force is small, the propagation velocity becomes slow. The phase velocity is hence slower for shorter-wavelength tsunami. As the tsunami wavelength becomes longer, the volume of the moving water gets larger. As a result, the gravitational force gets larger, and the phase velocity increases with increasing wavelength. However, when the wavelength greatly exceeds the sea depth, the volume of moving water does not increase but stops growing due to the finite sea depth h_0. When the tsunami wavelength is much longer than the sea depth h_0

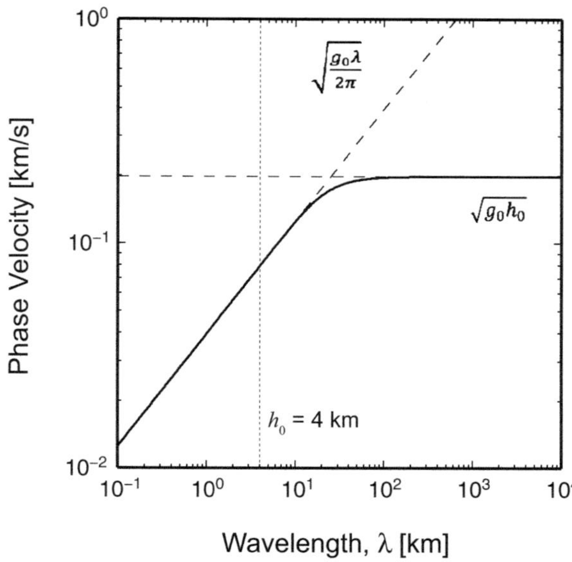

Fig. 2.17 Tsunami phase velocity as a function of the tsunami wavelength at a sea depth of 4 km

($\lambda \gg h_0$), the phase velocity becomes independent of the wavelength and becomes dependent on the sea depth.

As shown above, the restoring force is a key to wave propagation. For seismic waves, the restoring force originates from the elasticity of the medium. Hence, a softer medium results in a smaller restoring force and a slower wave speed. Note that this is an interpretation and that these mechanisms are not treated quantitatively in this chapter. In general, we should rely on theory and mathematical operations in order to derive rigorous solutions. We perform a quantitative treatment in Chap. 3: *Propagation of Tsunami and Seismic Waves*.

2.3.2 Amplification

The tsunami propagation speed changes depending on the sea depth when the sea depth is shallow compared to the tsunami wavelength. At the same time, the tsunami height also changes depending on the sea depth. As the tsunami approaches the coast propagating across a shallower sea, the tsunami height becomes greater. A simple case is shown in Fig. 2.18. The peak tsunami height at the time of 0 s is located at a sea depth of 4 km. The sea depth linearly decreases as it approaches the coast, and the sea depth is zero at the coast. The coast is located at a distance of 300 km from the point where the tsunami peak is located at the time of 0 s. As time elapses, the tsunami approaches the coast. Since the sea depth is shallower near the coast, the tsunami propagation speed becomes slower nearer the coast. The tsunami propagates about 100 km in the elapsed time from 0 to 600 s, whereas it propagates about 50 km in the elapsed time from 1500 to 2000 s. The tsunami height becomes greater as the

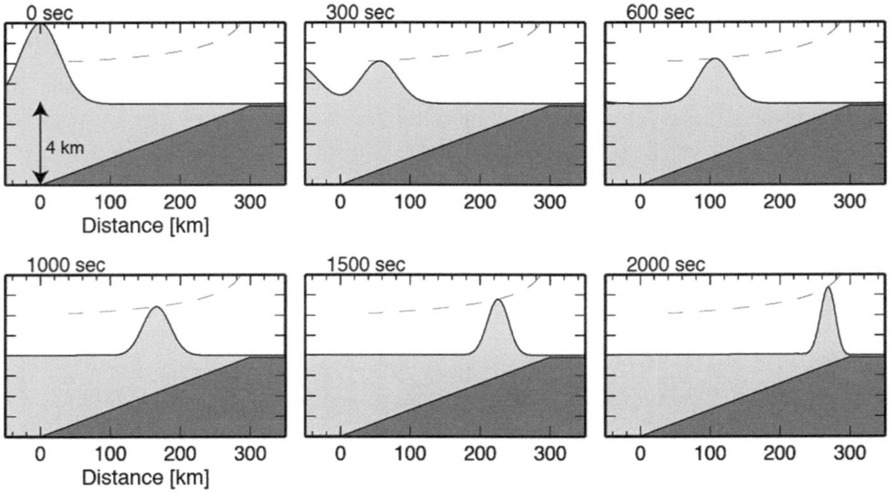

Fig. 2.18 Tsunami amplification near coast. The sea depth at a distance of 0 km is 4 km. The sea depth decreases linearly with the distance. The coast is located at a distance of 300 km. The dashed line is the track of the peak tsunami height

tsunami approaches the coast. The tsunami height at the coast is about two times larger than that located offshore. It should also be noted that the dominant tsunami wavelength becomes shorter when the tsunami approaches the coast.

The tsunami height η and the velocity c are closely related with the conservation of the tsunami energy flux. The tsunami energy flux density is given by

$$J = \rho_0 g_0 |\eta|^2 c = \rho_0 g_0 |\eta|^2 \sqrt{g_0 h_0}, \tag{2.7}$$

when the tsunami wavelength is much greater than the sea depth. The derivation is explained in Sect. 3.2.3 *Energy Density and Energy Flux Density for Incompressible Fluid*. The energy flux density is an important quantity from the viewpoint of energy conservation. When we assume that the reflected tsunami does not appear during tsunami propagation, we may consider that the energy flux density J is constant irrespective of the location and the sea depth. This can be expressed as

$$|\eta(x_1)|^2 \sqrt{g_0 h(x_1)} = |\eta(x_2)|^2 \sqrt{g_0 h(x_2)}, \tag{2.8}$$

and

$$|\eta(x_2)| = \left|\frac{h(x_1)}{h(x_2)}\right|^{\frac{1}{4}} |\eta(x_1)| \tag{2.9}$$

where $\eta(x)$ and $h(x)$ are the tsunami height and the sea depth at the location x, respectively. When the point x_2 is located near the coast and the sea depth is shallower than the point x_1, i.e., $h(x_2) < h(x_1)$, the tsunami height $\eta(x_2)$ near the coast is then larger than $\eta(x_1)$ in the offshore region. This is often referred to as Green's law (Green's law is different from Green's function which is often used in mathematical physics). Green's law is derived in an idealistic situation where a reflected wave does not exist. However, in practice, a reflected tsunami usually appears (e.g., Hayashi 2010). Hence, Eq. (2.9) is not always quantitatively correct, but it provides a good estimation of tsunami amplification mechanisms.

2.3.3 *Nonlinearity*

When a tsunami propagates across a shallow sea, it has a nonlinear nature. Figure 2.19a shows the tsunami propagation across a shallow sea with a depth of 10 m. The tsunami height distributions at various elapsed times are plotted together. The tsunami propagates rightward. The peaks are located at about 5, 10, and 15 km at the times of 500, 1000, and 1500 s, respectively. The velocity is ~10 m/s, which is almost the same as the theoretical phase velocity $c = \sqrt{g_0 h_0}$ derived based on the linear long-wave equations. Some features of the nonlinear long-wave tsunami propagation can be seen in Figure 2.19a. One is that the shape of the tsunami

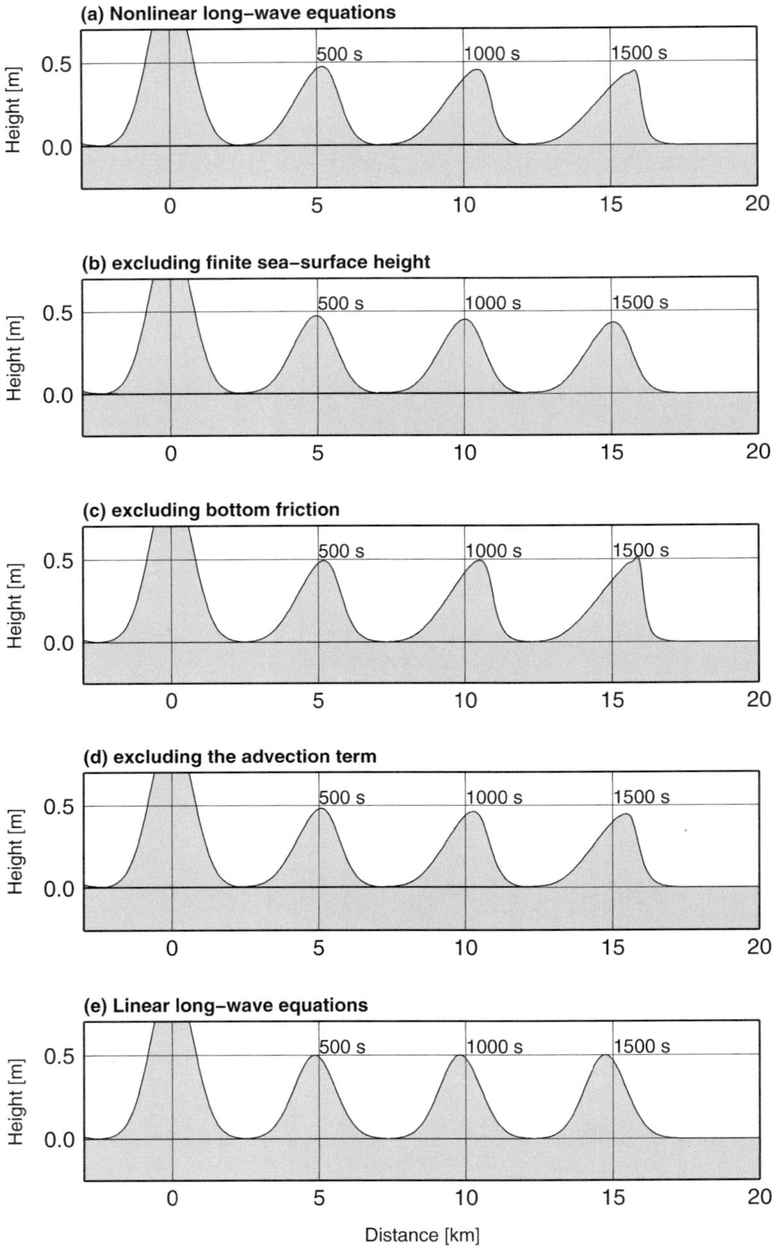

Fig. 2.19 Simulation results of tsunami propagation across a shallow sea (depth of 10 m). The tsunami height distributions at the elapsed times of 0, 500 1000, and 1500 s are plotted. (**a**) A simulation using nonlinear long-wave equations given by Eqs. (2.10) and (2.11), (**b**) using nonlinear long-wave equations but omitting the tsunami height η in the second term in Eq. (2.10), (**c**) using nonlinear long-wave equations with the bottom friction ($C_f = 0$) in Eq. (2.11), (**d**) using nonlinear long-wave equations but omitting the advection term in Eq. (2.11), and (**e**) using linear long-wave equations

deforms more as the travel distance increases. The slope of the front side becomes steeper than the back side (or the head of the tsunami falls forwardly) during the propagation. Another is that the peak height slightly decreases with increasing travel distance.

The nonlinear tsunami propagation shown in Figure 2.19a was simulated based on the 1-D nonlinear long-wave equations

$$\frac{\partial\eta(x,t)}{\partial t}+\frac{\partial}{\partial x}\{[h(x)+\eta(x,t\ )]u(x,t)\}=0, \tag{2.10}$$

and

$$\frac{\partial u(x,t)}{\partial t}+u\frac{\partial u(x,t)}{\partial t}=-g_0\frac{\partial\eta(x,t)}{\partial x}-C_f\frac{u^2}{h+\eta} \tag{2.11}$$

where $\eta(x,t)$ is the spatial and temporal variation of the sea-surface height or tsunami, $u(x,t)$ is the horizontal velocity of the water particle, h is the sea depth, g_0 is the gravitational acceleration, and C_f is the friction coefficient. The derivation of these equations is shown in Chap. 6*: Propagation Simulation*. These equations are nonlinear for the tsunami height $\eta(x,t)$. The superposition principle does not hold for $\eta(x,t)$: for example, when $\eta_1(x,t)$ and $\eta_2(x,t)$ satisfy the equations of system, the function $\eta_1(x,t)+\eta_2(x,t)$ does not always satisfy the equations of the system. The nonlinearity is caused by three terms: the second term in Eq. (2.10), the second term on the left-hand side of Eq. (2.11), and the second term on the right-hand side of Eq. (2.11). If those nonlinear terms are excluded, the equations become linear as follows:

$$\frac{\partial\eta(x,t)}{\partial t}+\frac{\partial}{\partial x}\{h(x)u(x,t)\}=0 \tag{2.12}$$

$$\frac{\partial u(x,t)}{\partial t}=-g_0\frac{\partial\eta(x,t)}{\partial x} \tag{2.13}$$

The second term in Eq. (2.10), $\partial_x\{[h(x)+\eta(x,t\ )]u(x,t)\}$, indicates the finiteness of the sea-surface height $\eta(x,t\ )$ with respect to the sea depth $h(x)$. When the sea-surface height $\eta(x,t)$ is much smaller than the sea depth, Eq. (2.10) can be approximated as Eq. (2.12). Figure 2.19b shows the simulation result when Eq. (2.12) is used instead of Eq. (2.10). The height distribution is different from those in Figure 2.19a and b. The tsunami front becomes steep due to the finiteness of the sea-surface height $\partial_x\{[h(x)+\eta(x,t\ )]u(x,t)\}$. When the finite tsunami height is included, the tsunami propagation velocity is theoretically given by

$$c\sim\sqrt{g_0h_0}(1+(3\eta)/(2h_0)), \tag{2.14}$$

when the sea depth is constant $h(x)=h_0$ (see Appendix B). Higher tsunamis (large η) propagate faster than lower tsunamis (small η). Therefore, if considering the finite tsunami height, the tsunami peak (high tsunami) propagates faster than the other parts (low tsunami). This mechanism causes the peak to fall forwardly during the propagation.

The second term on the right-hand side of Eq. (2.11), $-C_f\ u^2/(h+\eta)$, represents the bottom friction. When this term is neglected in the nonlinear long-wave equations (Eq. (2.11)), we obtain the result shown in Figure 2.19c. The peak amplitude does not decrease in Figure 2.19c, whereas the peak amplitude decreases with the travel distance in Figure 2.19a. This term represents the energy dissipation which causes the attenuation of the tsunami. Equation (2.11) indicates that the term contributes to the temporal change of the horizontal velocity as

$$\frac{\partial u(x,t)}{\partial t} \sim -C_f \frac{u^2}{h+\eta} \tag{2.15}$$

if the other terms are neglected. Equation (2.15) indicates that the horizontal velocity decreases more with increasing time when the horizontal velocity is larger or the water column height of $h+\eta$ is smaller. Since the horizontal velocity change per unit time is proportional to u^2, the system is nonlinear. Therefore, a larger-amplitude tsunami attenuates at a higher rate.

The second term on the left side of Eq. (2.11), $u\ \partial u(x,t)/\partial t$, is referred to as the advection term. This term comes from the total time derivative

$$\frac{Du}{Dt} = \frac{\partial u}{\partial t} + u\frac{\partial u}{\partial x}. \tag{2.16}$$

This term is excluded from Eq. (2.11). The simulation result is plotted in Figure 2.19d. The difference between Figure 2.19a and d is small. The tsunami peak also bends forward in Figure 2.19d, but the peak bends more in Figure 2.19a.

Finally, the simulation result of the linear long-wave equations is plotted in Figure 2.19e. Contrary to the result of the nonlinear long-wave equations, the peak does not bend forward nor does the peak become attenuated when the linear long-wave equations are used. The propagation speed is slightly slower.

2.3.4 Inundation

Tsunami inundation and tsunami run-up into land areas are also nonlinear. Since the process should be closely related to the three-dimensional water flow and the excitation of turbulence, the process is considerably complicated compared to offshore propagation. It is not straightforward to describe the process rigorously. While it is not rigorously correct in the three-dimensional water flow, we can approximate the inundation process as a moving boundary condition in the 2-D horizontal propagation simulation. Figure 2.20 shows the simulation results of the

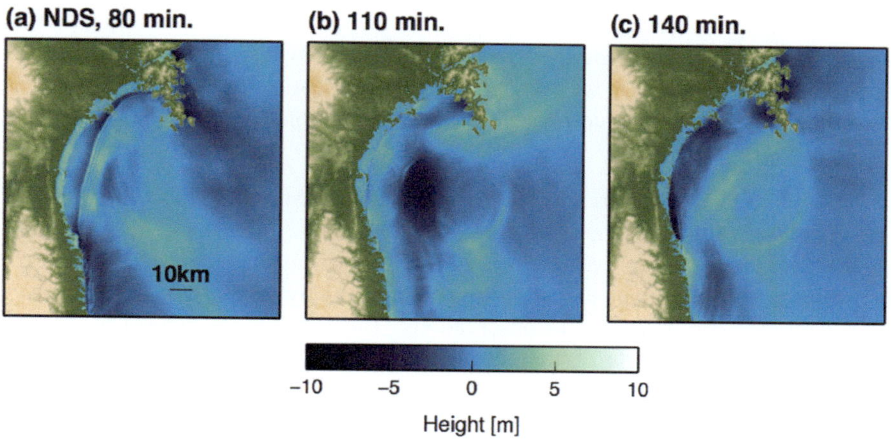

Fig. 2.20 Simulation results for inundation around Sendai Plain during the 2011 Tohoku-Oki earthquake. Nonlinear dispersive tsunami equations are used for the simulation. (Saito et al. 2014, copyright by the American Geophysical Union)

inundation around Sendai Plain for the 2011 Tohoku-Oki earthquake, where the inundation is reproduced as a moving boundary condition. The tsunami simulation method is described in Chap. 6: *Propagation Simulation*. In the 2011 Tohoku-Oki earthquake, the tsunami reached ~5 km from the coast on Sendai Plain. This 2-D simulation roughly reproduces the inundation of the Tohoku-Oki earthquake.

- The sea depth controls tsunami propagation.
- The propagation process also depends on the tsunami wavelength.
- Near coasts, nonlinear nature appears: a higher tsunami propagates faster and the energy dissipates.
- Tsunami inundation is a nonlinear process. It can be modeled as a moving boundary condition between land and sea.

2.4 Points of Tsunami Generation and Propagation

In this chapter, we have illustrated the general and fundamental features of tsunamis by using simulation results and some simple equations. The following points summarize the noteworthy features of tsunamis:

- Permanent displacement at the sea bottom controls tsunami size.
- Gravity collapses the initial sea-surface height distribution and works as the main force to move tsunami.
- Tsunami velocity is primarily determined by sea depth, but it also depends on the wavelength and tsunami height.

- Tsunamis propagate considerably slower (~0.1 km/s for sea depth of 1 km) than seismic wave propagation (~3 km/s for Rayleigh wave).
- Tsunami propagation is well described by linear theories in deep oceans, whereas nonlinear theories are necessary for the propagation in shallow seas near the coast.

Offshore tsunami observation is being developed, and we expect that tsunami observations inside the generation area will be in practical use in the near future. In such cases, the seismic waves and tsunami coexist inside the generation area. Therefore, it is important to know the mechanisms by which the seismic waves affect the tsunami and tsunami observations. The following attributes of seismic waves are closely related to tsunami studies.

- Not only tsunamis but also seismic waves excite sea-surface displacement.
- The excitation of seismic waves strongly depends on the time scale of the earthquake rupture.

In the following chapters, we illustrate the mechanisms of these tsunami features based on fundamental theories with mathematical derivations.

Appendices

Appendix A: Simulation Method of Tsunami Generated by an Earthquake

There are various methods for simulating tsunamis. This appendix explains one of the methods for simulating tsunami induced by an earthquake. The method involves two steps (Fig. 2.21). In the first step, we conduct a simulation of seismic waves caused by an earthquake. Then, we conduct tsunami simulation across the sea using the seismic simulation result as input data.

First, we perform seismic wave simulation in 3-D space. A finite earthquake fault is divided into numerous small subfaults. Each small subfault is approximately represented by a point dislocation source. The point dislocation source is given by

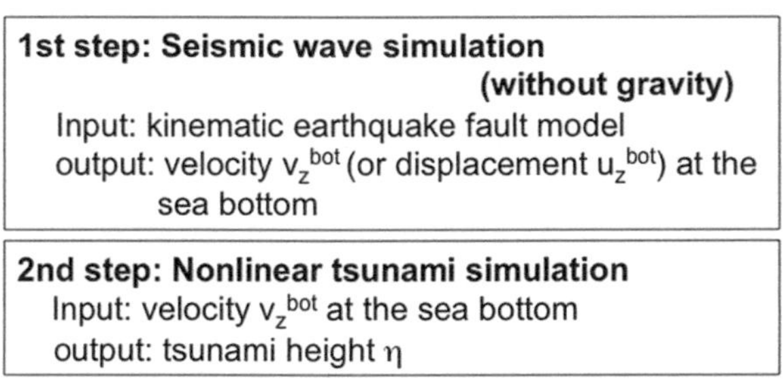

Fig. 2.21 A flowchart of tsunami generation and propagation excited by an earthquake (same as Fig. 2.4)

a stress change or stress glut τ_{ij}^{s} (see Chap. 4: *Earthquakes*). The equation of motion in the earth medium is

$$\rho \frac{\partial^2 u_i}{\partial t^2} = \frac{\partial}{\partial x_j}\left(\tau_{ij} + \tau_{ij}^{s}\right), \tag{A.1}$$

where ρ is the density, u_i is the displacement, and τ_{ij} is the stress tensor. A constitutive law or the generalized Hooke's law gives the relation between the displacement and the stress as

$$\tau_{ij} = \lambda \delta_{ij} u_{k,k} + 2\mu\left(u_{i,j} + u_{j,i}\right) \tag{A.2}$$

for an isotropic medium, where λ and μ are referred to as Lamé parameters. We numerically simulate the spatial and temporal distribution of the motion $u_i(\mathbf{x}, t)$ based on Eqs. (A.1) and (A.2) by using, for example, the finite difference method (the method is described in Sect. 4.3, *Seismic Wave Simulation*). We obtain the vertical velocity at the sea bottom $v_z^{\mathrm{bot}} = \dot{u}_i\left(x, y, z^{\mathrm{bot}}, t\right)$ where z^{bot} indicates the location of the sea bottom.

In the second step, we use the velocity at the sea bottom $v_z^{\mathrm{bot}}(x, y, t)$ as source of tsunami. According to the incompressible fluid theory (see Chap. 5: *Tsunami Generation*), when the vertical displacement $u_z^{\mathrm{bot}}(x, y)$ occurs at the sea bottom, the sea-surface height is given by

$$\eta_0(x, y) = \frac{1}{(2\pi)^2} \iint_{-\infty}^{\infty} dk_x dk_y \exp\left[i\left(k_x x + k_y y\right)\right] \frac{\tilde{u}_z^{\mathrm{bot}}\left(k_x, k_y\right)}{\cosh(kh_0)}, \tag{A.3}$$

where $\tilde{u}_z^{\mathrm{bot}}\left(k_x, k_y\right)$ is the 2-D Fourier transform of the vertical displacement at the bottom $u_z^{\mathrm{bot}}(x, y)$. Therefore, the sea-surface elevation $\Delta\eta(x, y, t)$ caused by the sea-bottom displacement during a fractional duration Δt at the time t is given by

$$\Delta\eta(x, y, t) = \frac{1}{(2\pi)^2} \iint_{-\infty}^{\infty} dk_x dk_y \exp\left[i\left(k_x x + k_y y\right)\right] \frac{\tilde{v}_z^{\mathrm{bot}}\left(k_x, k_y, t\right)}{\cosh(kh_0)} \Delta t, \tag{A.4}$$

where $\tilde{v}_z^{\mathrm{bot}}\left(k_x, k_y, t\right)$ is the 2-D Fourier transform of the vertical velocity at the bottom $v_z^{\mathrm{bot}}(x, y, t)$. The sea-surface elevation $\Delta\eta(x, y, t)$ is added as $\eta(x, y, t) = \eta^*(x, y, t) + \Delta\eta(x, y, t)$ where $\eta^*(x, y, t)$ is the tsunami height distribution numerically calculated at each time step in the simulation. We use the 2-D nonlinear long-wave equations

$$\frac{\partial \eta(x, y, t)}{\partial t} + \frac{\partial}{\partial x}\left[(h + \eta)v_x^{\mathrm{av}}\right] + \frac{\partial}{\partial \bar{y}}\left[(h + \eta)v_y^{\mathrm{av}}\right] = 0, \tag{A.5}$$

$$\frac{\partial v_x^{\mathrm{av}}(x,y,t)}{\partial t} + v_x^{\mathrm{av}} \frac{\partial v_x^{\mathrm{av}}}{\partial x} + v_y^{\mathrm{av}} \frac{\partial v_x^{\mathrm{av}}}{\partial y} + g_0 \frac{\partial \eta}{\partial x} = 0, \tag{A.6}$$

$$\frac{\partial v_y^{\mathrm{av}}(x,y,t)}{\partial t} + v_x^{\mathrm{av}} \frac{\partial v_y^{\mathrm{av}}}{\partial x} + v_y^{\mathrm{av}} \frac{\partial v_y^{\mathrm{av}}}{\partial y} + g_0 \frac{\partial \eta}{\partial y} = 0. \tag{A.7}$$

The parameter η is the tsunami or vertical displacement at the sea surface, v_x^{av} and v_y^{av} are the horizontal velocity averaged from the sea bottom to the sea surface, is the sea depth, and g_0 is the gravitational acceleration. We numerically calculate the spatial and temporal evolution of the tsunami $\eta(x, y, t)$ based on Eqs. (A.5), (A.6), and (A.7) by using the finite difference method (see Chap. 6: *Propagation Simulation*).

Appendix B: Phase Velocity in Nonlinear Long-Wave Equations: The First-Order Approximation Method

One-dimensional nonlinear tsunami propagation in a sea with a constant depth h_0 is described by the following equations:

$$\frac{\partial \eta(x,t)}{\partial t} + \frac{\partial}{\partial x}\{[h_0 + \eta(x,t)]u(x,t)\} = 0, \tag{B.1}$$

and

$$\frac{\partial u(x,t)}{\partial t} + u\frac{\partial u(x,t)}{\partial x} + g_0 \frac{\partial \eta(x,t)}{\partial x} = 0, \tag{B.2}$$

where $\eta(x, t)$ is tsunami height, $u(x, t)$ is horizontal velocity, and g_0 is the gravitational acceleration.

The tsunami height and horizontal velocity are written as

$$\eta(x,t) = \eta^0(x,t) + \eta^1(x,t), \text{and } u(x,t) = u^0(x,t) + u^1(x,t) \tag{B.3}$$

where $\eta^0(x, t)$ and $u^0(x, t)$ satisfy the following linear equations:

$$\frac{\partial \eta^0}{\partial t} + h_0 \frac{\partial u^0}{\partial x} = 0, \tag{B.4}$$

and

$$\frac{\partial u^0}{\partial t} + g_0 \frac{\partial \eta^0}{\partial x} = 0. \tag{B.5}$$

We assume that $|\eta^1| \ll |\eta^0|$ and $|u^1| \ll |u^0|$. In other words, we consider a situation in which the tsunami propagation is roughly described by the linear equations but also includes the nonlinear effects.

The linear equations of (B.4) and (B.5) give a wave equation for η^0 as

$$\frac{\partial^2 \eta^0}{\partial x^2} - \frac{1}{c_0^2} \frac{\partial^2 \eta^0}{\partial t^2} = 0, \tag{B.6}$$

where c_0 is given by $c_0 = \sqrt{g_0 h_0}$. A plane wave $\eta^0 = \exp[-i\omega(t - x/c_0)]$ satisfies Eq. (B.6) where c_0 works as the phase velocity. Substituting $\eta^0 = \exp[-i\omega(t - x/c_0)]$ into (B.5), we obtain

$$\frac{\partial u^0}{\partial t} = -g_0 \frac{i\omega}{c_0} e^{-i\omega\left(t - \frac{x}{c_0}\right)}. \tag{B.7}$$

Then, we suppose u^0 as

$$u^0 = \frac{g_0}{c_0} e^{-i\omega\left(t - \frac{x}{c_0}\right)} = \frac{g_0}{c_0}\eta^0 = \frac{c_0}{h_0}\eta^0. \tag{B.8}$$

The $\eta^0 = \exp[-i\omega(t - x/c_0)]$ and $u^0 = (c_0/h_0)\eta^0$ satisfy Eqs. (B.4) and (B.5). Substituting $u^0 = (c_0/h_0)\eta^0$ into Eq. (B.4), we obtain

$$\frac{\partial \eta^0}{\partial t} + c_0 \frac{\partial \eta^0}{\partial x} = 0. \tag{B.9}$$

Substituting (B.3) and (B.8) into (B.2) gives

$$\frac{\partial}{\partial t}\left(\frac{c_0}{h_0}\eta^0 + u^1\right) + \left(\frac{c_0}{h_0}\eta^0 + u^1\right)\frac{\partial}{\partial t}\left(\frac{c_0}{h_0}\eta^0 + u^1\right) + g_0 \frac{\partial}{\partial x}\left(\eta^0 + \eta^1\right) = 0, \tag{B.10}$$

and we calculate as

$$\begin{aligned} \frac{\partial u^1}{\partial t} = & -\frac{c_0}{h_0}\frac{\partial \eta^0}{\partial t} - \left(\frac{c_0}{h_0}\right)^2 \eta^0 \frac{\partial \eta^0}{\partial x} - u^1 \frac{c_0}{h_0}\frac{\partial \eta^0}{\partial t} \\ & - \eta^0 \frac{c_0}{h_0}\frac{\partial u^1}{\partial t} - u^1 \frac{\partial u^1}{\partial t} - g_0 \frac{\partial \eta^0}{\partial x} - g_0 \frac{\partial \eta^1}{\partial x}. \end{aligned} \tag{B.11}$$

Considering that $|\eta^1| \ll |\eta^0|$ and $|u^1| \ll |u^0|$ and neglecting the small terms containing $|\eta^1|$ or $|u^1|$ on the right-hand side, we approximate (B.11) as

$$\frac{\partial u^1}{\partial t} \approx -\frac{c_0}{h_0}\frac{\partial \eta^0}{\partial t} - \left(\frac{c_0}{h_0}\right)^2 \eta^0 \frac{\partial \eta^0}{\partial x} - g_0 \frac{\partial \eta^0}{\partial x}. \tag{B.12}$$

Using Eq. (B.9), we calculate (B.12) as

$$\begin{aligned} \frac{\partial u^1}{\partial t} &\approx \frac{c_0^2}{h_0}\frac{\partial \eta^0}{\partial x} - \left(\frac{c_0}{h_0}\right)^2 \eta^0 \frac{\partial \eta^0}{\partial x} - g_0 \frac{\partial \eta^0}{\partial x} \\ &= -\left(\tfrac{c_0}{h_0}\right)^2 \eta^0 \frac{\partial \eta^0}{\partial x} \\ &= \frac{c_0}{h_0^2}\eta^0 \frac{\partial \eta^0}{\partial t} \\ &= \frac{c_0}{2h_0^2}\frac{\partial}{\partial t}(\eta^0)^2. \end{aligned} \tag{B.13}$$

As a result, we obtain

$$u^1 \approx \frac{c_0}{2h_0^2}\left(\eta^0\right)^2. \tag{B.14}$$

Substituting (B.8) and (B.14) into (B.1), we calculate

$$\begin{aligned} \frac{\partial \eta(x,t)}{\partial t} &= -h_0 \frac{\partial}{\partial x}\left[\frac{c_0}{h_0}\eta^0 + \frac{c_0}{2h_0^2}(\eta^0)^2\right] - \frac{\partial}{\partial x}\left[(\eta^0+\eta^1)\left(\frac{c_0}{h_0}\eta^0 + u^1\right)\right] \\ &= -c_0\frac{\partial \eta^0}{\partial x} - \frac{c_0}{2h_0}\frac{\partial}{\partial x}(\eta^0)^2 - \frac{c_0}{h_0}\frac{\partial}{\partial x}(\eta^0)^2 \\ &\quad - \frac{\partial}{\partial x}\left(\eta^1 \frac{c_0}{h_0}\eta^0\right) - \frac{\partial}{\partial x}(u^1\eta^0) - \frac{\partial}{\partial x}(\eta^1 u^1). \end{aligned} \tag{B.15}$$

If we neglect the smaller terms on the right-hand side, we obtain

$$\frac{\partial \eta(x,t)}{\partial t} \approx -c_0 \frac{\partial \eta^0}{\partial x} - \frac{3c_0}{2h_0}\frac{\partial}{\partial x}\left(\eta^0\right)^2.$$

This is rewritten as

$$\frac{\partial \eta(x,t)}{\partial t} \approx -\frac{\partial}{\partial x}\left[\left(c_0 + \frac{3\eta^0}{2h_0}\right)\eta^0\right]. \tag{B.16}$$

Comparing (B.16) with (B.9), we find that $\eta(x, t)$ propagates with the phase velocity as

$$c \approx \left(c_0 + \frac{3\eta^0}{2h_0}\right)\eta^0. \tag{B.17}$$

References

Ando M (1975) Source mechanisms and tectonic significance of historical earthquakes along the Nankai Trough Japan. Tectonophysics 27(2):119–140. https://doi.org/10.1016/0040-1951(75)90102-X

Furumura T, Imai K, Maeda T (2011) A revised tsunami source model for the 1707 Hoei earthquake and simulation of tsunami inundation of Ryujin Lake, Kyushu, Japan. J Geophys Res 116: B02308. https://doi.org/10.1029/2010JB007918

Hayashi Y (2010) Empirical relationship of tsunami height between offshore and coastal stations. Earth Planets Space 62(3):269–275. https://doi.org/10.5047/eps.2009.11.006

Hok S, Fukuyama E, Hashimoto C (2011) Dynamic rupture scenarios of anticipated Nankai-Tonankai earthquakes, Southwest Japan. J Geophys Res 116(B12):B12319. https://doi.org/10.1029/2011JB008492

Kajiura K (1963) The leading wave of a tsunami. Bull Earthquake Res Inst 41:535–571

Kanamori H (1972) Mechanism of tsunami earthquakes. Phys Earth Planet Inter 6:346–359. https://doi.org/10.1016/0031-9201(72)90058-1

Kanamori H, Kikuchi M (1993) The 1992 Nicaragua earthquake: a slow tsunami earthquake. Nature 361:714–716. https://doi.org/10.1038/361714a0

Kumagai H (1996) Time sequence and the recurrence models for large earthquakes along the Nankai trough revisited. Geophys Res Lett 23(10):1139–1142. https://doi.org/10.1029/96GL01037

Saito T (2013) Dynamic tsunami generation due to sea-bottom deformation: analytical representation based on linear potential theory. Earth Planets and Space 65:1411–1423. https://doi.org/10.5047/eps.2013.07.004

Saito T, Inazu D, Miyoshi T, Hino R (2014) Dispersion and nonlinear effects in the 2011 Tohoku-Oki earthquake tsunami. J Geophys Res Oceans 119:5160–5180. https://doi.org/10.1002/2014JC009971

Takahashi R (1942) On seismic sea waves caused by deformations of the sea bottom. Bull Earthquake Res Inst 20:357–400 (in Japanese with English abstract)

Chapter 3
Propagation of Tsunami and Seismic Waves

Abstract Tsunamis and seismic waves are generated by earthquakes. We often reflexively discriminate the waves into tsunami and seismic waves. However, what is the difference between tsunami and seismic waves? What is a tsunami? This is a very simple but meaningful question. This chapter investigates the propagation process of tsunami and seismic waves. Section 3.1 introduces the general mathematics that is commonly used in wave theory. The basic theory involves continuum mechanics. Section 3.2 illustrates how we define tsunami. The fundamental nature of tsunami including the dispersion, particle motion, velocity distribution, and energy flux is theoretically investigated. Section 3.3 contains an investigation of seismic waves based on elastic dynamics. We define *P*, *S*, and Rayleigh waves. Also, ocean acoustic waves are classified as seismic wave. In Section 3.4, we summarize the similarities and differences between tsunami and seismic waves.

Keywords Tsunami · Seismic wave · Ocean acoustic wave · Equation of motion · Constitutive law

3.1 Governing Equations

3.1.1 Stress and Strain

In order to study waves, we need to devise a method for representing force and deformation mathematically. The stress and strain tensors are used to represent the force and deformation in continuum media (e.g., Aki and Richards 2002; Shearer 2009).

3.1.1.1 Stress

The force in a continuum medium is described by the stress tensor. In 3-D Cartesian coordinates (x, y, z), the stress tensor is given by

T. Saito, *Tsunami Generation and Propagation*, Springer Geophysics,
https://doi.org/10.1007/978-4-431-56850-6_3

$$\boldsymbol{\tau} = \begin{pmatrix} \tau_{xx} & \tau_{xy} & \tau_{xz} \\ \tau_{yx} & \tau_{yy} & \tau_{yz} \\ \tau_{zx} & \tau_{zy} & \tau_{zz} \end{pmatrix}. \tag{3.1}$$

Each component of the stress tensor represents the force per unit area. Consider a virtual cube in a continuum medium as shown in Fig. 3.1a. The component τ_{ij} represents the stress (force per unit area) in the ith direction exerted on the plane of which the normal vector is in the jth direction. For example, τ_{xy} represents force per unit area in the x direction exerted on the plane of which the normal vector is in the positive y direction. In particular, the component of the stress that is normal to the plane (e.g., τ_{xx}) is called normal stress, and the stress that is parallel to the plane (e.g., τ_{xy}) is called shear stress. The force per unit area exerted on a plane is given by a vector, which is referred to as traction. When the normal vector of a plane is given by $\bar{\mathbf{n}} = (n_x, n_y, n_z)$ ($\bar{\mathbf{n}}$ is a unit vector; $|\bar{\mathbf{n}}| = 1$), the traction $\mathbf{T}(\bar{\mathbf{n}})$ on the plane is given by

$$\mathbf{T}(\bar{\mathbf{n}}) = \boldsymbol{\tau}\bar{\mathbf{n}} = \begin{pmatrix} \tau_{xx} & \tau_{xy} & \tau_{xz} \\ \tau_{yx} & \tau_{yy} & \tau_{yz} \\ \tau_{zx} & \tau_{zy} & \tau_{zz} \end{pmatrix} \begin{pmatrix} n_x \\ n_y \\ n_z \end{pmatrix}. \tag{3.2}$$

In a continuum medium, the traction $\mathbf{T}(\bar{\mathbf{n}})$ depends on the plane direction $\bar{\mathbf{n}}$ in addition to the location $\mathbf{x}$. We need to specify not only the location $\mathbf{x}$ but also the plane direction $\bar{\mathbf{n}}$ on which plane the force acts when specifying the force given by the stress tensor in a continuum medium.

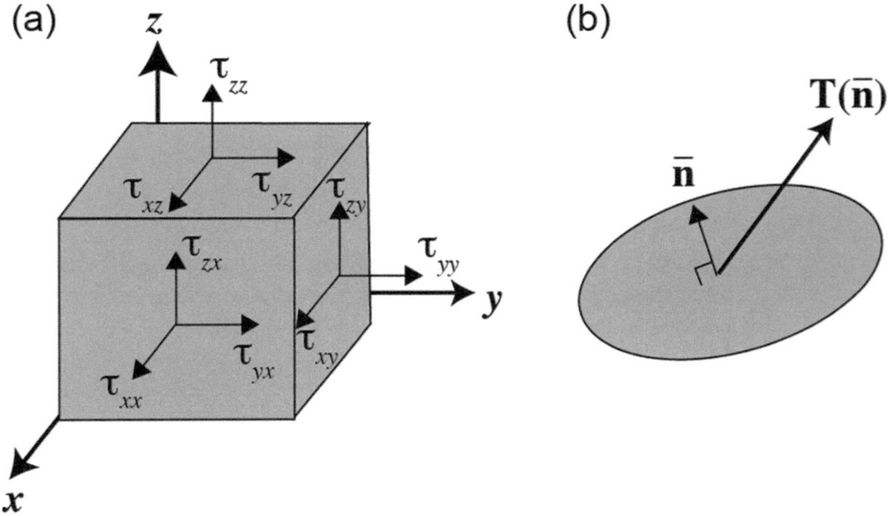

Fig. 3.1 (**a**) Each component of stress tensor τ_{ij} indicates the traction in the ith direction exerted on the plane of j. Traction is force per unit area exerted on a plane. (**b**) Traction $\mathbf{T}(\bar{\mathbf{n}})$ on a plane of which the normal vector is $\bar{\mathbf{n}}$. Traction has a dimension of force per area

Stress in a fluid medium is a special case of the stress in a continuum medium. In an ideal fluid, the shear stress is zero. The pressure P at a location in the fluid does not depend on the direction of the plane on which the pressure acts. The corresponding stress tensor is then represented as

$$\boldsymbol{\tau} = \begin{pmatrix} -P & 0 & 0 \\ 0 & -P & 0 \\ 0 & 0 & -P \end{pmatrix} \text{ or } \tau_{ij} = -P\delta_{ij}. \tag{3.3}$$

The negative sign in Eq. (3.3) is due to the definition of the difference in the signs of the stress and pressure. Positive stress represents tensional force, whereas positive pressure represents compressional force. Substituting Eq. (3.3) into Eq. (3.2), we obtain the traction as $\mathbf{T}(\bar{\mathbf{n}}) = -P\bar{\mathbf{n}}$. This indicates that traction is always perpendicular to the plane ($\mathbf{T}(\bar{\mathbf{n}}) \propto \bar{\mathbf{n}}$) without shear stress. The amplitude of the traction is given by P, which does not depend on the plane orientation $\bar{\mathbf{n}}$.

3.1.1.2 Strain

We then consider deformation. We suppose that a particle in the medium is displaced by the deformation where the particle moves from location $\mathbf{x}$ to $\mathbf{x}'$. Then, the displacement at $\mathbf{x}$ is defined as

$$\mathbf{u}(\mathbf{x}) = \mathbf{x}' - \mathbf{x}. \tag{3.4}$$

The deformation in a continuum medium is represented by the displacement vector field $\mathbf{u}(\mathbf{x})$. If we consider that the displacement field is uniform in space as $\mathbf{u}(\mathbf{x}) = \text{const.}$, the medium moves but does not show any deformation. Hence, the deformation should be defined by using the spatial variation of the displacement field (note that the spatial variation of the displacement does not always produce deformation. Rotation of a rigid body gives spatial variation of the displacement but does not give deformation). The deformation of the medium is represented by the strain tensor defined by the spatial derivatives of the displacement as

$$e_{ij} = \frac{1}{2}\left(\frac{\partial u_i}{\partial x_j} + \frac{\partial u_j}{\partial x_i}\right) \tag{3.5}$$

and

$$\mathbf{e} = \begin{pmatrix} e_{xx} & e_{xy} & e_{xz} \\ e_{yx} & e_{yy} & e_{yz} \\ e_{zx} & e_{zy} & e_{zz} \end{pmatrix}. \tag{3.6}$$

Let us examine a simple deformation/displacement field to introduce the strain tensor.

3.1.1.3 Extensional Strain

Suppose the displacement field given by

$$u_x = \alpha x + x_0, u_y = 0, u_z = 0. \tag{3.7}$$

The displacement field is illustrated in Fig. 3.2. The displacement occurs only in the x direction and becomes large at large values of x with a constant rate of α (>0). In Fig. 3.2, we suppose two points at x and $x + \Delta x$ separated by the distance Δx at the same z in a medium. Due to this displacement field, the two points move to $x + u_x(x)$ and $x + \Delta x + u_x(x + \Delta x)$, respectively. The distance between the two points after this deformation $\Delta x'$ is given by

$$\Delta x' = x + \Delta x + u_x(x + \Delta x) - (x + u_x(x)) = (1 + \alpha)\Delta x. \tag{3.8}$$

This means that the medium is extended in the x direction with a constant rate of α and the rate does not depend on the location x. The corresponding stress tensor is $e_{xx} = \alpha$. This strain field e_{xx} (>0) is referred to as extensional strain.

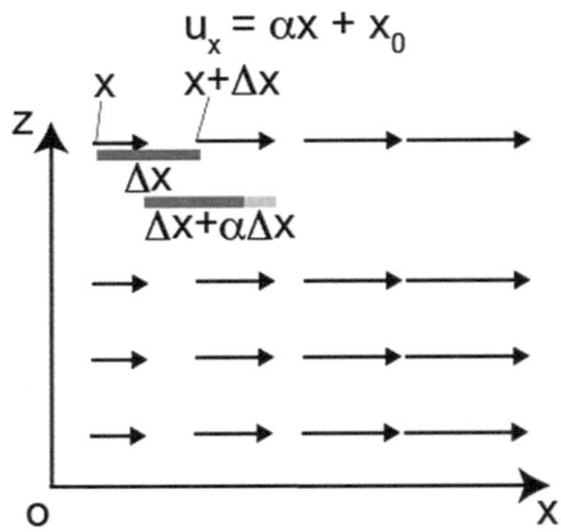

Fig. 3.2 An example of extensional strain represented by the displacement $u_x = \alpha x + x_0$ in the (x, z) plane. The distance Δx between two points at x and $x + \Delta x$ extends to $(1 + \alpha)\Delta x$

3.1.1.4 Dilatation

Suppose the deformation where extensional strain occurs in all three axes, i.e., the x-, y-, and z-axes, characterized by the strain tensor

$$\mathbf{e} = \begin{pmatrix} e_{xx} & 0 & 0 \\ 0 & e_{yy} & 0 \\ 0 & 0 & e_{zz} \end{pmatrix}. \tag{3.9}$$

When a cube whose sides are given by Δx, Δy, and Δz deforms due to the strain tensor of Eq. (3.9), the three sides Δx, Δy, and Δz change to $(1 + e_{xx})\Delta x$, $(1 + e_{yy})\Delta y$, and $(1 + e_{zz})\Delta z$, respectively. Hence, the volume of the cube changes from $V = \Delta x \Delta y \Delta z$ to $V + \Delta V = (1 + e_{xx})(1 + e_{yy})(1 + e_{zz})\Delta x \Delta y \Delta z$. The ratio of the volume change is given by

$$\begin{aligned} \frac{\Delta V}{V} &= \frac{(1+e_{xx})(1+e_{yy})(1+e_{zz})\Delta x \Delta y \Delta z - \Delta x \Delta y \Delta z}{\Delta x \Delta y \Delta z} \\ &\approx e_{xx} + e_{yy} + e_{zz} = \nabla \cdot \mathbf{u}. \end{aligned} \tag{3.10}$$

This value is referred to as dilatation, which is given by the divergence of the displacement field.

In tsunami modeling, seawater is usually assumed to show no dilatation. It is also referred to as incompressible fluid in fluid dynamics, and the condition is given by

$$\nabla \cdot \mathbf{u} = e_{xx} + e_{yy} + e_{zz} = 0. \tag{3.11}$$

3.1.1.5 Shear Strain

Suppose the displacement field given by

$$u_x = \alpha z. \tag{3.12}$$

The displacement field is described in Fig. 3.3. The rectangle OABC changes into a parallelogram OAB'C′ according to this displacement field. The stress tensor of the displacement field of Eq. (3.12) is given by

$$\mathbf{e} = \begin{pmatrix} 0 & 0 & \frac{1}{2}\alpha \\ 0 & 0 & 0 \\ \frac{1}{2}\alpha & 0 & 0 \end{pmatrix}. \tag{3.13}$$

This deformation is referred to as shear strain. The shear strain does not show the volume change as $\nabla \cdot \mathbf{u} = 0$.

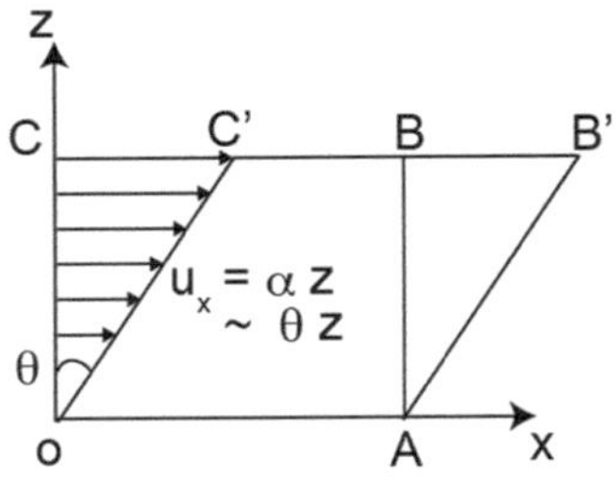

Fig. 3.3 An example of shear strain

Shear strain is an important factor in earthquakes because an earthquake can be considered a sudden release of the shear strain accumulated in elastic solid earth.

3.1.2 Equation of Motion and Constitutive Laws

3.1.2.1 Equation of Motion

When considering an infinitesimal cube in a medium and supposing an equilibrium of the total forces including the inertial force, traction, and the body force, we obtain an equation of motion as

$$\rho \frac{\partial^2 u_i(\mathbf{x}, t)}{\partial t^2} = \frac{\partial \tau_{ik}}{\partial x_k} + f_i, \tag{3.14}$$

where ρ is the density of the medium and f_i is the body force, that is, force applied per unit volume. For example, gravity works as the body force for the medium, which is given by $\mathbf{f} = (0, 0, -\rho g_0)$ where g_0 is gravitational acceleration and the z-axis is taken vertically upward. Also, the body force is often used as a source in seismic wave excitation because an earthquake fault, or shear dislocation source, has been proved to be represented by an equivalent body force distribution (see Chap. 4: *Earthquakes*). We use the Einstein summation convention: the summation is taken for a repeated index. For example, k in Eq. (3.14) is a repeated index. Equation (3.14) is then the same as

$$\begin{aligned} \rho \frac{\partial^2 u_i(\mathbf{x}, t)}{\partial t^2} &= \sum_{k=1}^{3} \frac{\partial \tau_{ik}}{\partial x_k} + f_i \\ &= \frac{\partial \tau_{ix}}{\partial x} + \frac{\partial \tau_{iy}}{\partial y} + \frac{\partial \tau_{iz}}{\partial z} + f_i, \qquad i = x, y, z. \end{aligned}$$

The equation of motion is one of the most fundamental equations for understanding tsunamis and earthquakes.

3.1.2.2 Laws of Medium Properties: Constitutive Law

The constitutive law, which determines the properties of the medium, is another fundamental equation. For an elastic medium, it describes the relation between the stress and the strain; in other words, it determines the elasticity of the medium: how soft or hard the material is. The linear relationship between the strain and stress is assumed to be

$$\tau_{ij} = c_{ijkl} e_{kl}, \tag{3.15}$$

where c_{ijkl} is referred to as the elastic tensor. This relation is referred to as generalized Hooke's law. If we assume that the elasticity of the medium is isotropic, the elastic tensor is represented as

$$c_{ijkl} = \lambda \delta_{ij} \delta_{kl} + \mu \left(\delta_{il} \delta_{jk} + \delta_{ik} \delta_{jl} \right) \tag{3.16}$$

where λ and μ are referred to as Lamé parameters. With Eqs. (3.15), (3.16), and (3.5), the stress tensor and strain tensor (or displacement) are related as

$$\tau_{ij} = \lambda \delta_{ij} e_{kk} + 2\mu e_{ij} = \lambda \delta_{ij} u_{k,k} + \mu \left(u_{i,j} + u_{j,i} \right), \tag{3.17}$$

where

$$u_{i,j} = \frac{\partial u_i}{\partial x_j} \text{ and } u_{k,k} = \frac{\partial u_x}{\partial x} + \frac{\partial u_y}{\partial y} + \frac{\partial u_z}{\partial z}.$$

The parameter μ is also referred to as the shear modulus or rigidity that relates shear strain and shear stress as

$$\tau_{ij} = 2\mu e_{ij} \text{ when } i \neq j$$

This represents how hard the material is for shearing. The average value of μ in the earth's crust is 30–40 GPa. On the other hand, the parameter λ by itself does not have a simple meaning. The constitutive law of Eq. (3.17) is often used for the elastic earth medium (e.g., Aki and Richards 2002; Shearer 2009).

We here introduce the bulk modulus. Supposing a sphere in a medium with pressure p, the sphere would shrink more with increasing pressure $p + \Delta p$. The bulk modulus K is then defined as

$$\Delta p = -K \frac{\Delta V}{V}, \tag{3.18}$$

where $\Delta V/V$ indicates the change in the volume with respect to the original volume. This means that the bulk modulus represents how hard the material is for compressing. Using Eq. (3.17), we calculate $\tau_{xx} + \tau_{yy} + \tau_{zz}$ as

$$\tau_{xx} + \tau_{yy} + \tau_{zz} = (3\lambda + 2\mu)\left(e_{xx} + e_{yy} + e_{zz}\right). \tag{3.19}$$

Since $e_{kk} + e_{yy} + e_{zz} = \Delta V/V$ and $\tau_{xx} + \tau_{yy} + \tau_{zz} = -3\Delta p$, Eq. (3.19) is rewritten as

$$\Delta p = -\left(\lambda + \frac{2}{3}\mu\right)\frac{\Delta V}{V}. \tag{3.20}$$

Comparing (3.20) with (3.18), we obtain

$$K = \lambda + \frac{2}{3}\mu. \tag{3.21}$$

3.1.2.3 Compressible Fluid

For seawater, the resistance of the material with respect to shear strain is zero. The resistance occurs only for the dilatation/compression. The corresponding constitutive relation is given by

$$\tau_{ij} = K\delta_{ij}e_{kk} = K\delta_{ij}u_{k,k}. \tag{3.22}$$

When modeling ocean acoustic waves, the seawater is usually approximated as compressible fluid.

3.1.2.4 Incompressible Fluid

In an incompressible fluid, there is no dilatation. The properties of the medium are represented as

$$\nabla \cdot \mathbf{u}(\mathbf{x}, t) = 0. \tag{3.23}$$

In such media, Eq. (3.23) describes the material properties instead of the constitutive law. When modeling tsunamis, the seawater is usually approximated as an incompressible fluid without elasticity.

3.1.2.5 Boundary Conditions

As a boundary condition of the earth surface, the traction at the surface is assumed to be zero as

$$\mathbf{T}(x, y, z = z_{\text{sur}}) = 0. \tag{3.24}$$

where $z = z_{\text{sur}}(x, y, t)$ represents the surface. This is referred to as free surface.

At the boundary between two different materials, both the traction vectors and the displacement vectors are assumed to be continuous. For seismic wave problems, we often encounter the boundary between seawater (fluid) and crust (solid). Since the shear stress is zero in the seawater, the shear stress on the boundary between the seawater and the crust is zero. Also, large shear strain is acceptable in the seawater adjacent to the boundary since the shear modulus is zero in the seawater. This gives that the tangential component of the displacements in the seawater and the crust can be discontinuous. Only the normal component of the displacement is continuous at the boundary between seawater and crust. For tsunami propagation problems, the crust is often assumed to be rigid body where the normal component of the displacement at the sea bottom is set as zero.

By using the equation of motion, the constitutive law, and boundary conditions, we describe the wave propagation of tsunamis and seismic waves. The equation of motion is common in tsunamis and seismic waves, whereas the constitutive laws are different for tsunamis and seismic waves. We investigate tsunami propagation in Sect. 3.2 *Tsunami: Dynamics of Incompressible Fluid* and seismic wave propagation in Sect. 3.3 *Seismic Waves: Elastic Dynamics*.

3.2 Tsunami: Dynamics of Incompressible Fluid

3.2.1 Fundamental Equations of Tsunami

Tsunamis can be represented as surface-wave motion in incompressible sea. The velocity in the water medium $\mathbf{v}(\mathbf{x}, t)$ is given by the time derivative of the displacement $\mathbf{u}(\mathbf{x}, t)$ as

$$\mathbf{v}(\mathbf{x}, t) = \frac{\partial \mathbf{u}(\mathbf{x}, t)}{\partial t}. \tag{3.25}$$

For tsunami modeling, the seawater is assumed to be an incompressible fluid, which means the medium does not show dilatation (Eq. 3.23):

$$\nabla \cdot \mathbf{v}(\mathbf{x}, t) = 0. \tag{3.26}$$

Also, we assume that the velocity field $\mathbf{v}(\mathbf{x}, t)$ is given by using the velocity potential $\phi(\mathbf{x}, t)$ as

$$\mathbf{v}(\mathbf{x}, t) = \nabla \phi(\mathbf{x}, t). \tag{3.27}$$

This means that the velocity field does not contain the rotation component (rot $\mathbf{v}(\mathbf{x}, t) = 0$). Substituting Eq. (3.27) into Eq. (3.26), we obtain the Laplace equation with respect to the velocity potential as

$$\Delta \phi(\mathbf{x}, t) = 0. \tag{3.28}$$

In seawater, there exists no shear stress. We represent the stress tensor as

$$\tau_{ij} = -p\delta_{ij}, \tag{3.29}$$

where p is the pressure. Substituting Eq. (3.29) into the equation of motion (Eq. 3.14), we obtain

$$\frac{\partial \mathbf{v}(\mathbf{x}, t)}{\partial t} = -\frac{1}{\rho_0} \nabla p(\mathbf{x}, t) + \mathbf{g}_0 \tag{3.30}$$

where ρ_0 is the seawater density. We consider gravity as the body force: $\mathbf{g}_0$ is the vector of gravitational acceleration. When we use the Cartesian coordinate where the z-axis is taken as positive in the vertically upward direction and (x, y) space is in the horizontal plane as shown in Fig. 3.4, the vector $\mathbf{g}_0$ is given by $\mathbf{g}_0 = -g_0 \boldsymbol{e}_z$ ($g_0 = 9.8$ m/s^2 and $\boldsymbol{e}_z$ is the unit base vector for the z-axis).

Let us consider a case when the system is in equilibrium. We describe it, assuming no time dependence in Eq. (3.30), as

$$0 = -\frac{1}{\rho_0} \nabla p_0(\mathbf{x}) + \mathbf{g}_0, \tag{3.31}$$

where $p_0(\mathbf{x})$ is the pressure distribution in the static state, referred to as hydrostatic pressure. We represent the total pressure $p(\mathbf{x}, t)$ in Eq. (3.30) as the sum of the hydrostatic pressure $p_0(\mathbf{x})$ and excess pressure $p_e(\mathbf{x}, t)$ as

$$p(\mathbf{x}, t) = p_0(\mathbf{x}) + p_e(\mathbf{x}, t). \tag{3.32}$$

Substituting Eq. (3.32) into Eq. (3.30) and using Eq. (3.31), we obtain

$$\frac{\partial \mathbf{v}(\mathbf{x}, t)}{\partial t} = -\frac{1}{\rho_0} \nabla p_e(\mathbf{x}, t).$$

Substituting Eq. (3.27) into this equation, we obtain

$$\nabla\left[\frac{\partial\phi}{\partial t}+\frac{1}{\rho_0}p_e(\mathbf{x},t)\right]=0.$$

This gives the excess pressure $p_e(\mathbf{x}, t)$ as

$$p_e(\mathbf{x},t)=-\rho_0\frac{\partial\phi(\mathbf{x},t)}{\partial t}+X(t).$$

Since $p_e(\mathbf{x}, t)$ should be zero when there is no velocity (ϕ = const.), we represent the excess pressure $p_e(\mathbf{x}, t)$ using the velocity potential as

$$p_e(\mathbf{x},t)=-\rho_0\frac{\partial\phi(\mathbf{x},t)}{\partial t}. \tag{3.33}$$

As shown in Fig. 3.4, the sea surface at rest is set at $z = 0$, and the sea-surface height distribution is represented as $z = \eta(x, y, t)$. We assume that the pressure is zero at the sea surface, using $p(x, y, z = \eta, t) = 0$ as the boundary condition. When $\eta(x, y, t)$ is small enough, we consider it a linear problem. We treat only the linear problem in this chapter, although some nonlinear problems are treated in Chap. 6. The pressure at the sea surface $z = \eta$ is represented as

$$\begin{aligned}
p(x,y,\eta,t) &\approx p(x,y,0,t)+\left.\tfrac{\partial p}{\partial z}\right|_{z=0}\eta\\
&=p_0(z=0)+\left.\tfrac{dp_0}{dz}\right|_{z=0}\eta+p_e(x,y,0,t)+\left.\tfrac{\partial p_e}{\partial z}\right|_{z=0}\eta\\
&=-\rho_0 g_0\eta+p_e(x,y,0,t)+\left.\tfrac{\partial p_e}{\partial z}\right|_{z=0}\eta\\
&\approx-\rho_0 g_0\eta+p_e(x,y,0,t).
\end{aligned}$$

We neglected the higher-order term $\partial_z p_e|_{z=0}\eta$. Then, the condition $p(x, y, \eta, t) = 0$ gives the boundary condition of p_e at $z = 0$ as

$$p_e(x,y,z=0,t)=\rho_0\, g_0\eta(x,y,t). \tag{3.34}$$

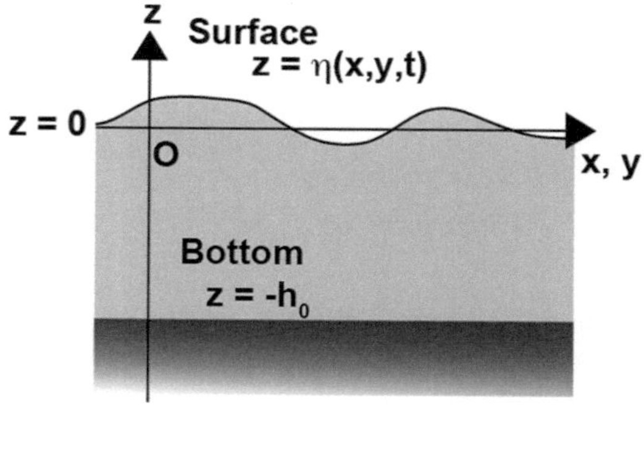

Fig. 3.4 Coordinates used in the formulation of water waves

Substituting Eq. (3.33) into Eq. (3.34), we obtain the boundary condition with respect to the velocity potential as

$$\left.\frac{\partial\phi(\mathbf{x},t)}{\partial t}\right|_{z=0} + g_0\eta(x,y,t) = 0. \tag{3.35}$$

Since this boundary condition determines the relation between the force and sea-surface motion, this is referred to as the dynamic boundary condition. On one hand, the sea surface is assumed to be continuous; in other words, wave breaking does not occur. Then, there exists a relation between the sea-surface height and the velocity at the sea surface as $\partial\eta/\partial t = v_z$ at the sea surface, as the first-order approximation (see also the second-order approximation as described in (6.11)). The relation is given by using the velocity potential as

$$\left.\frac{\partial\eta(x,y,t)}{\partial t}\right|_{z=0} = \left.\frac{\partial\phi(\mathbf{x},t)}{\partial z}\right|_{z=0}. \tag{3.36}$$

This is referred to as a kinematic boundary condition. By using Eqs. (3.35) and (3.36) and eliminating $\eta(x,y,t)$, we represent the boundary condition with respect to the velocity potential at $z = 0$ as

$$\left.\frac{\partial^2\phi(\mathbf{x},t)}{\partial t^2}\right|_{z=0} + g_0\left.\frac{\partial\phi(\mathbf{x},t)}{\partial z}\right|_{z=0} = 0. \tag{3.37}$$

The sea bottom is assumed to be a rigid boundary. Then, only horizontal flow exists, but vertical flow is not allowed at the boundary. Then, the boundary condition at the sea bottom ($z = -h_0$) is given by

$$\left.\frac{\partial\phi(\mathbf{x},t)}{\partial z}\right|_{z=-h_0} = 0. \tag{3.38}$$

As a result, the velocity potential $\phi(\mathbf{x},t)$ needs to satisfy the Laplace equation (Eq. 3.28) and the boundary condition at the sea surface (Eq. 3.37) and the sea bottom (Eq. 3.38). In other words, to find the velocity potential $\phi(\mathbf{x},t)$ that satisfies these equations is a mathematical problem in the wave theory of incompressible fluids.

3.2.2 *Tsunami*

Suppose a constant sea depth and the sea bottom are located at $z = -h_0$. Let us consider that the sea surface $\eta(x,y,t)$ propagates along the x-axis as a monochromatic plane wave with the angular frequency ω and the wavenumber k as

$$\eta(x, y, t) = \eta_0 e^{i(kx - \omega t)}. \tag{3.39}$$

We seek a solution of the velocity potential $\phi(\mathbf{x}, t)$ that satisfies the Laplace equation (Eq. 3.28) and the boundary conditions of Eqs. (3.37) and (3.38). In addition to these requirements, the sea-surface height η is represented as Eq. (3.39). This represents the sea surface $\eta(x, y, t)$ as complex number. Taking the real part of the solution, we can synthesize or visualize the wavefield that is comparable to actual records.

Since the wave propagation along the x-axis is represented as $\exp[i(kx - \omega t)]$, the velocity potential $\phi(\mathbf{x}, t)$ may be represented as

$$\phi(\mathbf{x}, t) = f(z) e^{i(kx - \omega t)}. \tag{3.40}$$

Substituting this into the Laplace equation of $\phi(\mathbf{x}, t)$ (Eq. 3.28) gives an ordinary differential equation with respect to z as

$$\frac{d^2 f(z)}{dz^2} = k^2 f(z). \tag{3.41}$$

The general solution of this equation is given by

$$f(z) = A\cosh(kz) + B\sinh(kz), \tag{3.42}$$

where A and B are the coefficients. These coefficients are to be determined so as to satisfy the boundary conditions. The velocity potential is then represented as

$$\phi(\mathbf{x}, t) = [A\cosh(kz) + B\sinh(kz)] e^{i(kx - \omega t)}. \tag{3.43}$$

Consider a boundary condition such that the vertical velocity is zero at the sea bottom (Eq. 3.38):

$$\begin{aligned} \left.\frac{\partial \phi(\mathbf{x}, t)}{\partial z}\right|_{z=-h_0} &= [Ak\sinh(kz) + Bk\cosh(kz)]_{z=-h_0} e^{i(kx - \omega t)} \\ &= [-Ak\sinh(kh_0) + Bk\cosh(kh_0)] e^{i(kx - \omega t)} \\ &= 0. \end{aligned} \tag{3.44}$$

Since this always needs to be satisfied for any values of x and t, we find

$$B = \frac{\sinh(kh_0)}{\cosh(kh_0)} A.$$

The velocity potential is then written as

$$\phi(\mathbf{x}, t) = A\left[\cosh(kz) + \frac{\sinh(kh_0)}{\cosh(kh_0)} \sinh(kz)\right] e^{i(kx - \omega t)}. \tag{3.45}$$

By substituting this into the boundary condition at the sea surface (Eq. 3.37), we obtain

$$-\omega^2 A\left[\cosh(kz)+\tfrac{\sinh(kh_0)}{\cosh(kh_0)}\sinh(kz)\right]_{z=0}e^{i(kx-\omega t)} + g_0 A\left[k\sinh(kz)+k\tfrac{\sinh(kh_0)}{\cosh(kh_0)}\cosh(kz)\right]_{z=0}e^{i(kx-\omega t)}=0. \tag{3.46}$$

This gives

$$\left[-\omega^2+g_0 k\frac{\sinh(kh_0)}{\cosh(kh_0)}\right]Ae^{i(kx-\omega t)}=0.$$

Hence, to satisfy this equation, we obtain the relation between the angular frequency ω and the wavenumber k as

$$\omega^2=g_0 k\tanh(kh_0). \tag{3.47}$$

This is the dispersion relation. Furthermore, by using the dynamic boundary condition at the sea surface (Eq. 3.35), the sea-surface height $\eta(x,y,t)$ is given by

$$\begin{aligned}\eta(x,y,t)&=-\tfrac{1}{g_0}\tfrac{\partial\phi(\mathbf{x},t)}{\partial t}\Big|_{z=0}\\&=\frac{i\omega A}{g_0}\left[\cosh(kz)+\frac{\sinh(kh_0)}{\cosh(kh_0)}\sinh(kz)\right]_{z=0}e^{i(kx-\omega t)}\\&=\frac{i\omega A}{g_0}e^{i(kx-\omega t)}\end{aligned} \tag{3.48}$$

Comparing this with $\eta(x,y,t)=\eta_0\exp[i(kx-\omega t)]$ (Eq. 3.39), we determine the coefficient A. Finally, we obtain the velocity potential as

$$\begin{aligned}\phi(\mathbf{x},t)&=\frac{g_0\eta_0}{i\omega}\left[\cosh(kz)+\frac{\sinh(kh_0)}{\cosh(kh_0)}\sinh(kz)\right]e^{i(kx-\omega t)}\\&=\frac{g_0\eta_0}{i\omega}\frac{\cosh[k(z+h_0)]}{\cosh(kh_0)}e^{i(kx-\omega t)}\end{aligned} \tag{3.49}$$

This velocity potential satisfies the Laplace equation (Eq. 3.28) and all the boundary conditions Eqs. (3.37), (3.38), and (3.39).

The velocity and pressure change from the hydrostatic pressure are represented by using the velocity potential as

$$\mathbf{v}(\mathbf{x},t)=\nabla\phi(\mathbf{x},t), \tag{3.27}$$

and

$$p_e(\mathbf{x},t)=-\rho_0\frac{\partial\phi(\mathbf{x},t)}{\partial t}, \tag{3.33}$$

respectively. Then, when the sea surface propagates along the x-axis as $\eta(x, y, t) = \eta_0 \exp[i(kx - \omega t)]$, $v_x(\mathbf{x}, t)$ is given by the gradient of the velocity potential (Eq. 3.27) as

$$\begin{aligned} v_x(\mathbf{x}, t) &= \frac{\partial \phi(\mathbf{x}, t)}{\partial x} \\ &= \frac{g_0 \eta_0 k}{\omega} \frac{\cosh[k(z + h_0)]}{\cosh(kh_0)} e^{i(kx - \omega t)} \\ &= \omega \eta_0 \frac{g_0 k}{\omega^2} \frac{\cosh[k(z + h_0)]}{\cosh(kh_0)} e^{i(kx - \omega t)} \end{aligned}$$

and by using the dispersion relation (Eq. 3.47)

$$v_x(\mathbf{x}, t) = \omega \eta_0 \frac{\cosh[k(z + h_0)]}{\sinh(kh_0)} e^{i(kx - \omega t)}. \tag{3.50}$$

Similarly, we obtain

$$v_y(\mathbf{x}, t) = 0, \tag{3.51}$$

and

$$v_z(\mathbf{x}, t) = -i\omega \eta_0 \frac{\sinh[k(z + h_0)]}{\sinh(kh_0)} e^{i(kx - \omega t)}. \tag{3.52}$$

An integration of the velocity with respect to time gives the displacement as

$$u_x(\mathbf{x}, t) = i\eta_0 \frac{\cosh[k(z + h_0)]}{\sinh(kh_0)} e^{i(kx - \omega t)}, \tag{3.53}$$

$$u_y(\mathbf{x}, t) = 0, \tag{3.54}$$

and

$$u_z(\mathbf{x}, t) = \eta_0 \frac{\sinh[k(z + h_0)]}{\sinh(kh_0)} e^{i(kx - \omega t)}. \tag{3.55}$$

Figure 3.5 which illustrates Eqs. (3.53) and (3.55) shows the spatial distributions of the sea-surface height $\eta(x, y, t) = \eta_0 \exp[i(kx - \omega t)]$ and the displacement vector (u_x, u_z). We made this figure by taking the real part of the sea-surface height $\eta(x, y, t)$ and the water particle displacement vector (u_x, u_z). The sea depth is set as $h_0 = 4$ km, and the wavelength is set as $\lambda = 40$ km. Due to the dispersion relation, $\omega^2 = g_0 k \tanh(kh_0)$ (Eq. 3.47), the angular frequency (or wave period T) $\omega = 2\pi/T$ is uniquely determined when the wavelength is given by $\lambda = 2\pi/k$. The wave period is

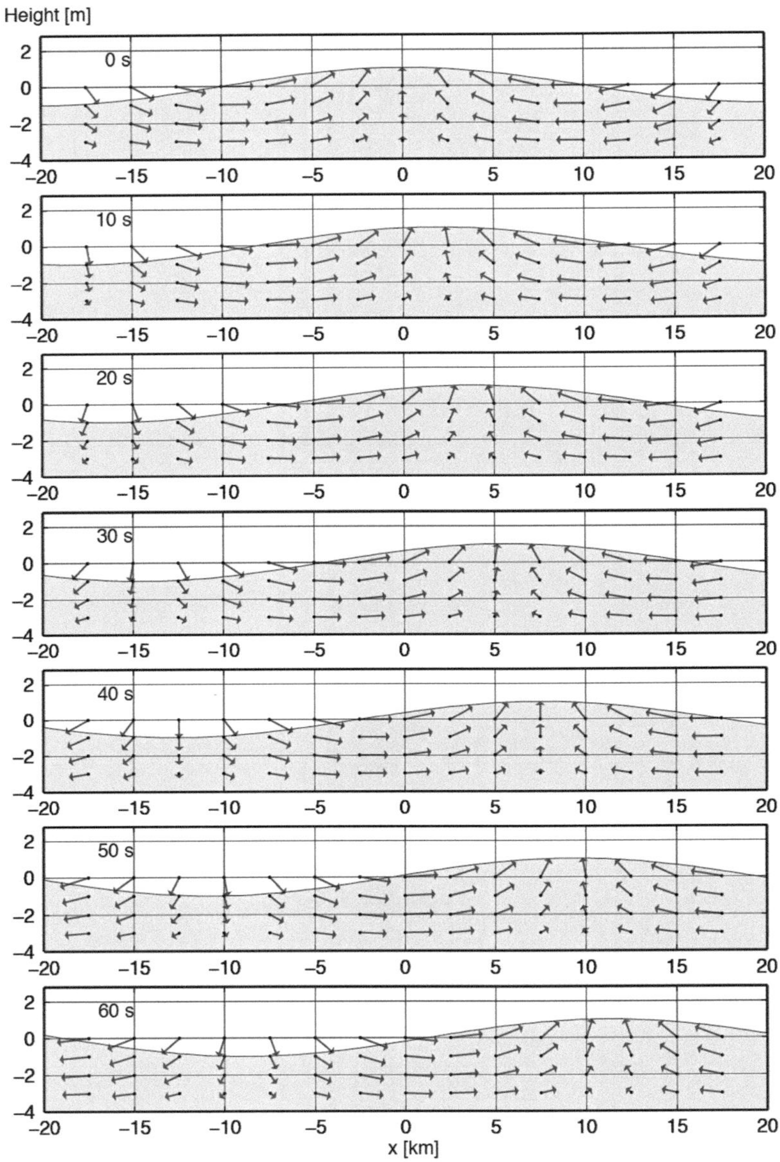

Fig. 3.5 Spatial distributions of the sea-surface height $\eta(x, y, t) = \eta_0 \exp[-i(\omega t - kx)]$ ($\eta_0 = 1\text{m}$) and the water particle displacement vector (u_x, u_z) for elapsed times of 0, 10, 20, 30, 40, 50, and 60 s. The sea depth is 4 km. The wavelength of the ocean wave is 40 km

T~210 s for the wavelength $\lambda = 40$ km at the sea depth of 4 km. This wave is long-wavelength wave or shallow-water wave.

At $t = 0$ s, the peak height of 1 m is located at $x = 0$ km. The peak moves rightward (positive x direction) with increasing time. At the time of $t = 60$ s, the peak

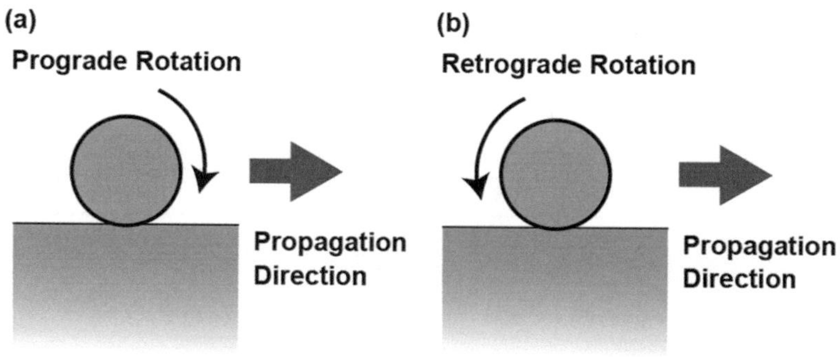

Fig. 3.6 Rotation directions. (**a**) When the rotation direction is the same as the direction of the rolling wheel in the propagation, the rotation direction is referred to as prograde rotation. (**b**) When the rotation direction is the opposite of that of the rolling wheel in the propagation, the rotation direction is referred to as retrograde rotation

arrives at the point x~12 km, which corresponds to a phase velocity of 0.2 km/s. If we confine our attention to a single water particle located at $(x, z) = (0 \text{ km}, 0 \text{ km})$, at the time of $t = 0$ s, the water particle is displaced vertically upward. Then, as time elapses, the displacement vector rotates in a clockwise direction. At the time of about 50–60 s, the vector points in the positive x direction. It takes T~210 s for one rotation. Because the wave propagates toward the positive x direction, this particle motion is referred to as prograde rotation (Figure 3.6).

Figure 3.7 displays the spatial distributions of the sea-surface height and the water particle displacement vector for the short-wavelength wave given by $\lambda = 10$ km. A peak sea-surface height of 1 m located at $x = 0$ km at the time of $t = 0$ s arrives at the point x~7.5 km at the elapsed time of $t = 60$ s. This corresponds to a phase velocity of 0.125 km/s. This phase velocity is slower than that of the long wavelength (0.2 km/s for $\lambda = 40$ km). The particle motion rotates in the clockwise direction as in the case of the long wavelength (prograde rotation), but the displacement is smaller. In particular, for the displacements in deep ocean (e.g., $z = -2$ and -3 km), there is large discrepancy between the long-wave and short-wave cases (Figs. 3.5 and 3.7). Figure 3.7 indicates that the particle motion is limited within the shallower part ($z = -2$ - 0 km) when the wavelength is short. This contrasts with the fact that the motion of the water particle extends the whole water layer in the case of the long wavelength.

3.2.2.1 Phase Velocity

The equations for u_x and u_z (Eqs 3.53 and 3.55) indicate that the wave propagates in the positive x direction according to $\exp[i(kx - \omega t)]$. The phase velocity c is given by $c = \omega/k$, where the dispersion relation is given by $\omega^2 = g_0 k \tanh(kh_0)$ (Eq. 3.47). Then, the phase velocity is

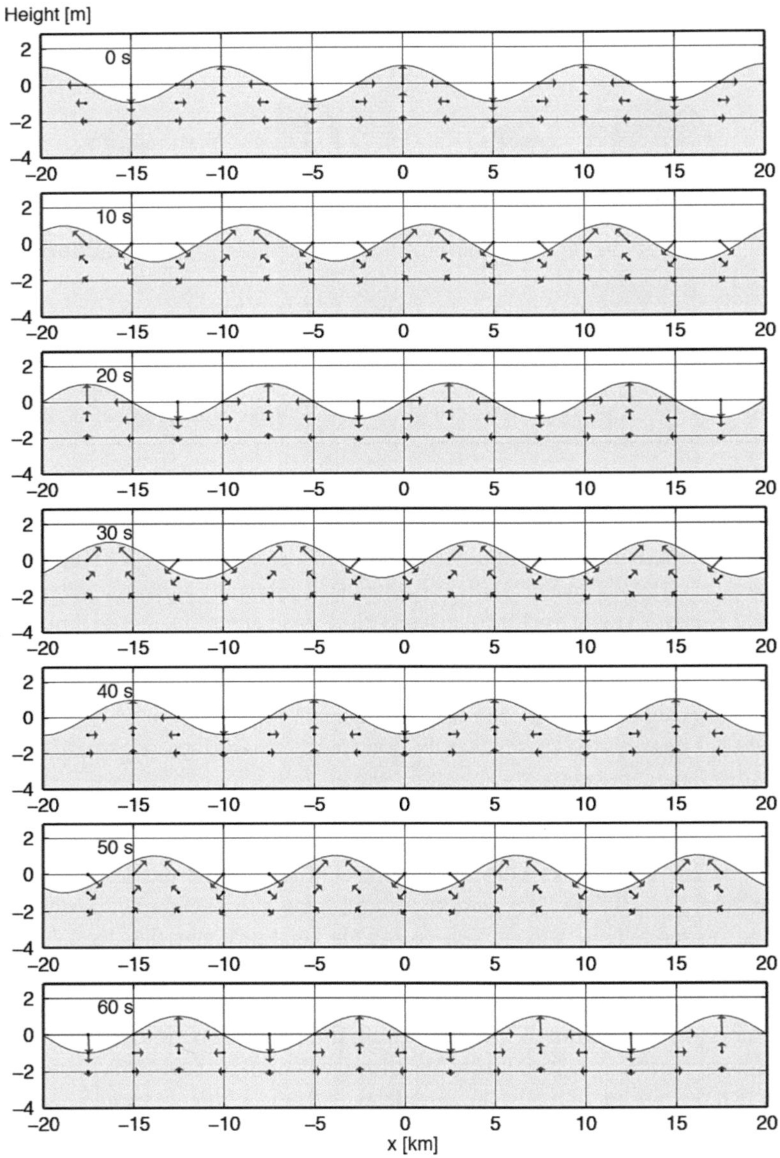

Fig. 3.7 Spatial distributions of the sea-surface height $\eta(x, y, t) = \eta_0 \exp[-i(\omega t - kx)]$ ($\eta_0 = 1$m) and the water particle displacement vector (u_x, u_z) for the elapsed times of 0, 10, 20, 30, 40, 50, and 60 s. The sea depth is 4 km. The wavelength of the ocean wave is 10 km

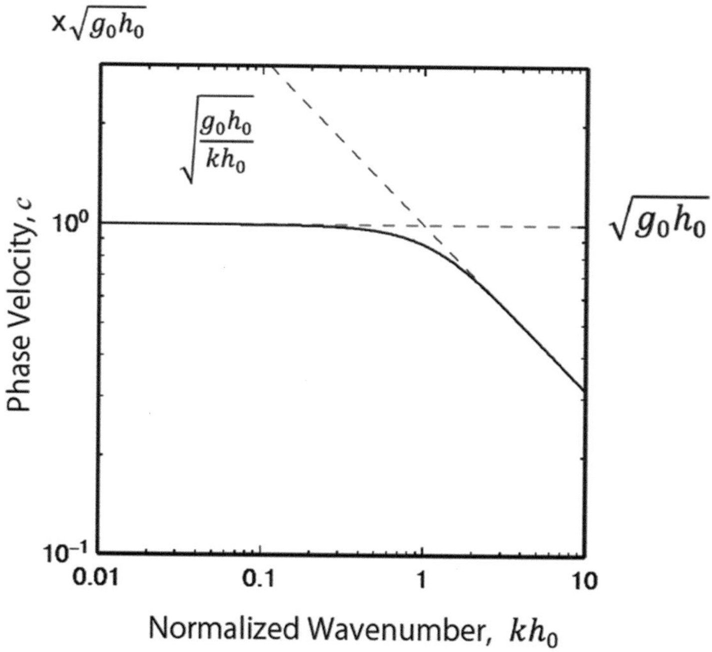

Fig. 3.8 Phase velocity as a function of kh_0, where k is the wavenumber of the ocean wave and h_0 is the sea depth

$$c = \frac{\omega}{k} = \sqrt{g_0 h_0}\sqrt{\frac{\tanh(kh_0)}{kh_0}} = \begin{cases} \sqrt{\frac{g_0}{k}} & \text{for } kh_0 \gg 1 \\ \sqrt{g_0 h_0} & \text{for } kh_0 \ll 1 \end{cases}$$

$$= \begin{cases} \sqrt{\frac{g_0 \lambda}{2\pi}} & \text{for } \lambda \ll h_0 \\ \sqrt{g_0 h_0} & \text{for } \lambda \gg h_0 \end{cases}. \tag{3.56}$$

Figure 3.8 plots the phase velocity c as a function of the wavenumber k normalized by the sea depth h_0. For a sea depth of 4 km, the wavelengths $\lambda = 40$ km and 10 km correspond to the normalized values of $kh_0 = 2\pi h_0/\lambda = 0.63$ and 2.5, respectively. Figure 3.8 indicates that the phase velocity is $c\tilde{}0.95\sqrt{g_0 h_0}$ (0.19 km/s) and $0.63\sqrt{g_0 h_0}$ (0.125 km/s) for $kh_0 = 2\pi h_0/\lambda = 0.63$ ($\lambda = 40$ km) and 2.5 ($\lambda = 10$ km), respectively. These estimated phase velocities were confirmed in Figs. 3.5 and 3.7.

3.2.2.2 Particle Orbit

Figures 3.9a and b shows the orbit of the particle motion at various depths of $z=$ 0, -1, -2, -3, -3.9 km for the wave of $\eta_0 = 1$ m.

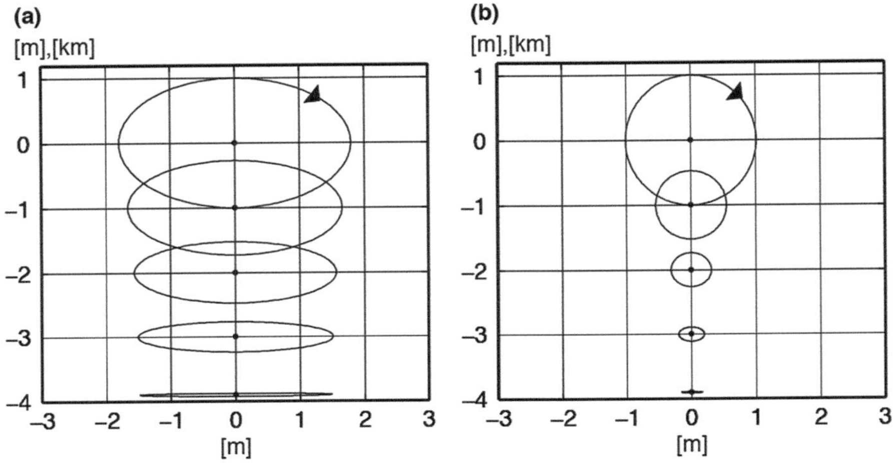

Fig. 3.9 Orbits of the particle motion at various depths of 0, 1, 2, 3, and 3.9 km for a sea depth of 4 km. (**a**) The wavelength λ is 40 km and (**b**) 10 km

First, we consider the case when the wavelength is relatively long given by $\lambda=$ 40 km in a sea depth of 4 km ($kh_0 = 0.63$). When the wavelength is relatively longer than the sea depth ($kh_0 < 1$) (and note that $k(h_0 + z) < kh_0 < 1$ because z is always negative), the real part of the water particle displacements is approximated as

$$\begin{aligned} u_x(\mathbf{x},t) &= \eta_0 \frac{\cosh[k(z+h_0)]}{\sinh(kh_0)} \sin(\omega t - kx) \\ &\approx \frac{\eta_0}{kh_0}\left[1+\frac{1}{2}k^2(z+h_0)^2\right] \sin(\omega t - kx) \quad \text{for } kh_0 \ll 1, \end{aligned} \tag{3.57}$$

and

$$\begin{aligned} u_z(\mathbf{x},t) &= \eta_0 \frac{\sinh[k(z+h_0)]}{\sinh(kh_0)} \cos(\omega t - kx) \\ &\approx \eta_0\left(1+\frac{z}{h_0}\right) \cos(\omega t - kx) \quad \text{for } kh_0 \ll 1. \end{aligned} \tag{3.58}$$

These waves are long waves or shallow-water waves. At the sea surface ($z = 0$), these are given by

$$\begin{aligned} u_x(x,y,z=0,t) &\approx \frac{\eta_0}{kh_0}\left[1+\frac{1}{2}(kh_0)^2\right] \sin(\omega t - kx) \\ &\approx \frac{\eta_0}{kh_0} \sin(\omega t - kx), \end{aligned} \tag{3.59}$$

and

$$u_z(x,y,z=0,t) \approx \eta_0 \cos(\omega t - kx) \quad \text{for } kh_0 \ll 1. \tag{3.60}$$

The orbit is an ellipse with a long horizontal axis of η_0/kh_0~1.6 m and a short vertical axis of $\eta_0 = 1$ m. The horizontal displacement is longer than the vertical displacement at any depth from the sea surface to the sea bottom. In the deeper portion, the vertical short axis decreases appreciably. It becomes zero at the sea bottom (z = -4 km). Since the horizontal and vertical displacements are represented using $\sin(\omega t - kx)$ and $\cos(\omega t - kx)$, respectively, the particle motion shows a clockwise rotation for all depths.

We then consider the case when the wavelength is short. Figure 3.9b shows the orbit of the particle motion for the wavelength $\lambda = 10$ km with a sea depth of 4 km (kh_0=2.5). When the wavelength is relatively shorter than the sea depth ($kh_0 \gg 1$), the real parts of the water particle displacements are approximated as

$$\begin{aligned} u_x(\mathbf{x}, t) &= \eta_0 \frac{\cosh[k(z + h_0)]}{\sinh(kh_0)} \sin(\omega t - kx) \\ &\approx \eta_0 e^{kz} \sin(\omega t - kx) \ \text{ for } \ kh_0 \gg 1, \end{aligned} \tag{3.61}$$

and

$$\begin{aligned} u_z(\mathbf{x}, t) &= \eta_0 \frac{\sinh[k(z + h_0)]}{\sinh(kh_0)} \cos(\omega t - kx) \\ &\approx \eta_0 e^{kz} \cos(\omega t - kx) \ \text{ for } \ kh_0 \gg 1. \end{aligned} \tag{3.62}$$

At the sea surface (z= 0 km), the orbit is a circle with a radius of η_0= 1 m. In the deeper portion, both the horizontal and vertical displacements rapidly decrease as exp(kz). This indicates that the wave motions are localized only within a shallow sea depth ($-1/k < z < 0$) for short wavelengths. These waves are short-wavelength waves or deep water waves.

3.2.2.3 Pressure Change

Ocean-bottom pressure gauges are among the most useful and common sensors for the detection of tsunamis. Therefore, it is important to derive the relation between the ocean wave height $\eta(x, y, t) = \eta_0 \exp[i(kx - \omega t)]$ and the ocean-bottom pressure change $p_e(x, y, z = -h_0, t)$. The pressure caused by the wave motion is represented using the velocity potential (Eq. 3.49), $p_e(\mathbf{x}, t) = -\rho_0 \partial\phi(\mathbf{x}, t)/\partial t$ (Eq. 3.33) as

$$p_e(\mathbf{x}, t) = \rho_0 g_0 \eta_0 \frac{\cosh[k(z + h_0)]}{\cosh(kh_0)} e^{i(kx - \omega t)}. \tag{3.63}$$

For the sea-bottom pressure change, substituting $z = -h_0$, we obtain the sea-bottom pressure change due to the ocean wave height (Eq. 3.39) as

$$p_e(x, y, -h_0, t) = \frac{\rho_0 g_0 \eta_0}{\cosh(kh_0)} e^{i(kx - \omega t)}. \tag{3.64}$$

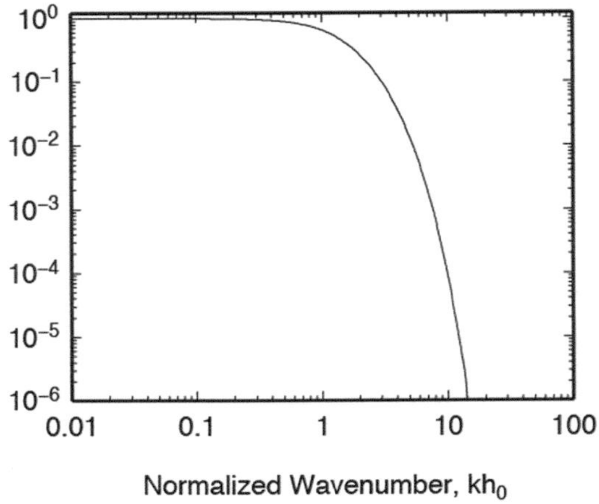

Fig. 3.10 The factor of $1/\cosh(kh_0)$ as a function of the wavenumber k normalized by the sea depth h_0

If the tsunami wavelength is much longer than the sea depth, this is approximated as

$$\begin{aligned} p_e(x, y, -h_0, t) &\approx \rho_0 g_0 \eta_0 e^{i(kx-\omega t)} \quad \text{for } kh_0 \ll 1 \\ &= \rho_0 g_0 \eta(x, y, t). \end{aligned} \tag{3.65}$$

The factor $1/\cosh(kh_0)$ (in Eq. 3.64) represents the effect of the water layer on the ocean-bottom pressure change. Figure 3.10 shows the factor $1/\cosh(kh_0)$ as a function of the wavelength normalized by the sea depth h_0. When the wavelength λ is ten times greater than the sea depth ($kh_0 = 2\pi h_0/\lambda \sim 0.6$), the pressure change at the sea bottom is about 83% of the value predicted by the simple relation $\rho_0 g_0 \eta$. For tsunamis, the wavelength is usually much longer than the sea depth, so that the hydrostatic relation of $p_e(x, y, t) \sim \rho_0 g_0 \eta(x, y, t)$ (Eq. 3.65) may be a good approximation. However, it should be noted that it would be necessary to use Eq. (3.64) instead of the simple hydrostatic relation (3.65) in analyzing short-wavelength tsunamis.

The pressure change described here is caused by the tsunami propagation only. In addition to this effect, a pressure change is also brought about by the tsunami generation. This effect is important when using pressure sensors installed inside the focal area. We treat this mechanism in Chap. 5: *Tsunami Generation.*

3.2.2.4 Tsunamis as Band-Limited Ocean Waves: Group Velocity

We have treated tsunami propagation as a monochromatic plane wave. Actually, a tsunami is a transient wave that is represented by the sum of monochromatic waves with different frequencies.

An idealistic impulsive wave is represented by the delta function as

$$f(t) = \delta(t). \tag{3.66}$$

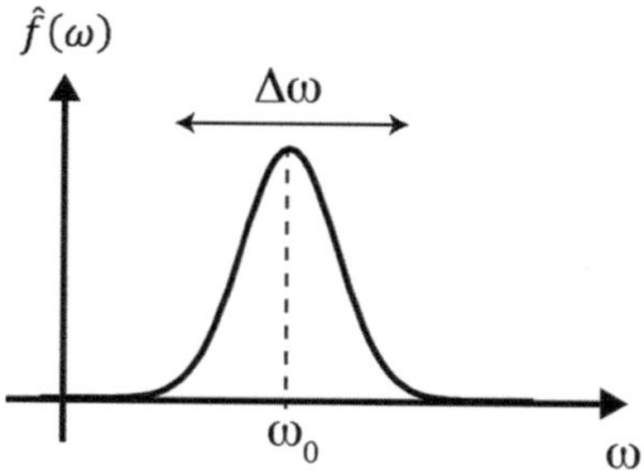

Fig. 3.11 Spectrum of a wave packet. It is characterized by the dominant angular frequency ω_0 with the bandwidth of $\Delta\omega$

The impulsive wave represented by the delta function is composed of the entire angular frequency band ranging from $\omega = -\infty$ to $\omega = \infty$ as

$$f(t) = \frac{1}{2\pi} \int_{-\infty}^{\infty} \widehat{f}(\omega) e^{-i\omega t} d\omega, \tag{3.67}$$

where

$$\widehat{f}(\omega) = 1 \text{ for all } \omega. \tag{3.68}$$

An actual tsunami is a transient wave and is considered as a wave packet with a limited band of angular frequency. As a simple model, we consider the wave packet for which the frequency content is characterized by the dominant angular frequency ω_0 with a bandwidth of $\Delta\omega$ as shown in Fig. 3.11.

The frequency content, for example, is given by the Gaussian function as

$$\widehat{f}(\omega) = e^{-\frac{(\omega-\omega_0)^2}{(\Delta\omega/4)^2}}. \tag{3.69}$$

A monochromatic ocean wave is represented by $\exp[i(kx - \omega t)]$ where the angular frequency ω and the wavenumber k need to satisfy the dispersion relation $\omega^2 = g_0 k \tanh(kh_0)$ (Eq. 3.47). Hence, the sea-surface height η which is composed of finite-band frequency contents is given by

$$f(x,t) = \frac{1}{2\pi} \int_{-\infty}^{\infty} \widehat{f}(\omega) e^{i(kx-\omega t)} d\omega, \tag{3.70}$$

where the wavenumber k is a function of ω according to the dispersion relation, $k = k(\omega)$ (Eq. 3.47). By substituting the frequency component $\widehat{f}(\omega)$ given by Eq. (3.69) into Eq. (3.70), we describe the ocean waves composed of a finite-frequency band. However, a rigorous integration with respect to ω in Eq. (3.70) is difficult, because k is a function of ω.

To gain a clear perspective on wave description, we use an approximation method instead of a rigorous integration. A function $\omega t - kx$ in the exponent in Eq. (3.70) is expanded around the dominant angular frequency $\omega = \omega_0$ by a Taylor series (e.g., Snieder and van Wijk 2015) as

$$\begin{aligned}\omega t - kx &= (\omega_0 t - k_0 x) + \frac{d(\omega t - kx)}{d\omega}\bigg|_{\omega=\omega_0} (\omega - \omega_0) + \mathrm{O}(\omega - \omega_0)^2 \\ &\approx (\omega_0 t - k_0 x) + \left(t - \frac{dk}{d\omega}\bigg|_{\omega=\omega_0} x\right)(\omega - \omega_0) \\ &= (\omega_0 t - k_0 x) + \left(t - \frac{x}{U_0}\right)(\omega - \omega_0)\end{aligned} \tag{3.71}$$

where we defined

$$U = \frac{1}{\frac{dk}{d\omega}}, \tag{3.72}$$

and $k_0 = k(\omega_0)$ and $U_0 = U(\omega_0)$. Substituting (3.71) into Eq. (3.70) gives

$$f(x,t) \approx \frac{1}{2\pi} e^{i(k_0 x - \omega_0 t)} \int_{-\infty}^{\infty} \widehat{f}(\omega) e^{-i\left(t - \frac{x}{U_0}\right)(\omega - \omega_0)} d\omega \tag{3.73}$$

where $\widehat{f}(\omega)$ is given by Eq. (3.69). Then, we calculate the following:

$$\begin{aligned}f(x,t) &\approx \frac{1}{2\pi} e^{i(k_0 x - \omega_0 t)} \int_{-\infty}^{\infty} e^{-\frac{(\omega-\omega_0)^2}{(\Delta\omega/4)^2}} e^{-i\left(t - \frac{x}{U_0}\right)(\omega - \omega_0)} d\omega \\ &= \frac{1}{2\pi} e^{i(k_0 x - \omega_0 t)} \int_{-\infty}^{\infty} e^{-\frac{\omega'^2}{(\Delta\omega/4)^2}} e^{-i\left(t - \frac{x}{U_0}\right)\omega'} d\omega' \\ &= \frac{\Delta\omega}{8\sqrt{\pi}} e^{i(k_0 x - \omega_0 t)} e^{-\frac{1}{4}\left(\frac{\Delta\omega}{4}\right)^2 \left(t - \frac{x}{U_0}\right)^2}.\end{aligned} \tag{3.74}$$

Figure 3.12 plots the waveforms calculated with Eq. (3.74) for the travel distances x= 400, 500, and 600 km across a sea of 4 km in depth using black lines. The frequency content $\widehat{f}(\omega)$ is set with $\omega_0 = 0.029\ \mathrm{s}^{-1}$ and $\Delta\omega = 0.0071\ \mathrm{s}^{-1}$. The dominant angular frequency corresponds to the wavelength $\lambda = 40$ km in a sea of 4 km in depth. The dominant period of the waves is given by $T_0 = 2\pi/\omega_0 \sim 215$ s.

Eq. (3.74) indicates that the envelope of the waveform is given by

$$g(x,t) = \frac{\Delta\omega}{8\sqrt{\pi}} e^{-\left(\frac{\Delta\omega}{8}\right)^2 \left(t - \frac{x}{U_0}\right)^2} = \frac{\Delta\omega}{8\sqrt{\pi}} e^{-\left(\frac{1}{4}\frac{32}{\Delta\omega}\right)^{-2} \left(t - \frac{x}{U_0}\right)^2}. \tag{3.75}$$

The envelopes are also plotted using gray lines in Fig. 3.12. Equation (3.75) indicates that the envelope duration is determined by the bandwidth of $\Delta\omega$; the duration is given by $\sim 32/\Delta\omega$. The envelope peak propagates with the velocity of U_0

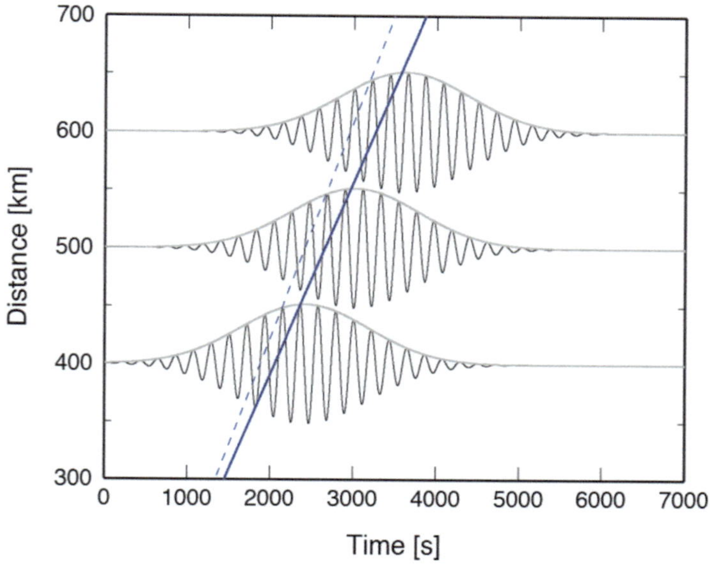

Fig. 3.12 Waveforms at the distance of $x = 400$, 500, and 600 km. Each waveform is composed of the frequency contents characterized by the dominant angular frequency of $\omega_0 = 0.029\ \mathrm{s}^{-1}$ and the bandwidth of $\Delta\omega = 0.0071\ \mathrm{s}^{-1}$ in Eq. (3.69). The blue line and blue dashed line indicate the group velocity of $U = 0.166$ km/s and the phase velocity of $c = 0.19$ km/s, respectively

where $U_0 = U(\omega_0)$ is derived by Eq. (3.72) with the dispersion relation (Eq. 3.47). The velocity U is referred to as the group velocity. Differentiating the dispersion relation (Eq. 3.47) with respect to ω gives

$$\begin{aligned} 2\omega &= \frac{dk}{d\omega}\frac{d}{dk}[g_0 k \tanh(kh_0)] \\ &= \frac{dk}{d\omega}\left[g_0 \tanh(kh_0) + \frac{g_0 kh_0}{\cosh^2(kh_0)}\right] \\ &= \frac{dk}{d\omega}\left[\frac{\omega^2}{k} + \frac{h_0\omega^2}{\cosh(kh_0)\sinh(kh_0)}\right], \end{aligned}$$

and

$$2 = \frac{dk}{d\omega}\left[\frac{\omega}{k} + \frac{2h_0\omega}{\sinh(2kh_0)}\right]. \tag{3.76}$$

Thus, by using Eqs. (3.72) and (3.76), we represent the group velocity as

$$U = \left(\frac{dk}{d\omega}\right)^{-1} = \frac{1}{2}\frac{\omega}{k}\left[1 + \frac{2kh_0}{\sinh(2kh_0)}\right] = \frac{c}{2}\left[1 + \frac{2kh_0}{\sinh(2kh_0)}\right], \tag{3.77}$$

where c is the phase velocity given by Eq. (3.56).

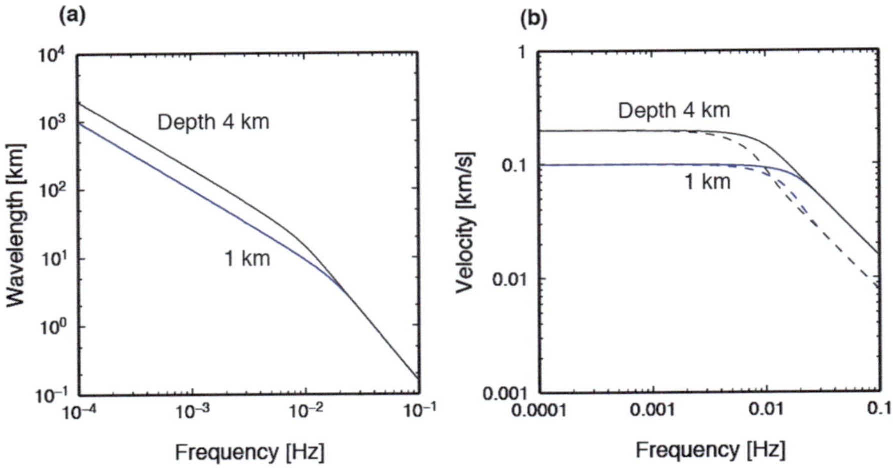

Fig. 3.13 (**a**) Wavelength as a function of wave frequency for the sea depths of 1 km (blue line) and 4 km (black line) derived by the dispersion relation (Eq. 3.47). (**b**) Phase velocity for the sea depths of 1 km (blue solid line) and 4 km (black solid line) and group velocity for the sea depths of 1 km (blue dashed line) and 4 km (black dashed line)

The phase velocity is plotted as a function of the normalized wavenumber in Fig. 3.8. In practice, we observe tsunami waveforms as functions of time. Hence, treating the phase velocity and the group velocity as functions of frequency is useful. The dispersion relation (Eq. 3.47) is approximated as Eq. (3.78)

$$\omega^2 = g_0 k \tanh(kh_0) = \begin{cases} g_0 k(kh_0) & \text{for } kh_0 \ll 1 \\ g_0 k & \text{for } kh_0 \gg 1. \end{cases} \tag{3.78}$$

Figure 3.13a plots the relation between the wavelength and the wave frequency based on the dispersion relation. When the wave frequency is very low, the relation between the frequency and the wavenumber is given by $\omega^2 \approx g_0 k(kh_0)$ or $\omega \approx \sqrt{g_0 h_0} k$. Then, the wavelength is approximately given by $\lambda = 2\pi/k \approx 2\pi\sqrt{g_0 h_0}/\omega = \sqrt{g_0 h_0}/f$. On the other hand, when the wave frequency is very high, the dispersion relation becomes $\omega^2 \approx g_0 k$. The wavelength is then given by $\lambda \approx g_0/(2\pi f^2)$, indicating that the wavelength does not depend on the sea depth.

The phase velocity and group velocity are given by

$$c = \frac{\omega}{k} = \sqrt{g_0 h_0}\sqrt{\frac{\tanh(kh_0)}{kh_0}} = \begin{cases} \sqrt{g_0 h_0} & \text{for } kh_0 \ll 1 \\ \sqrt{g_0/k} & \text{for } kh_0 \gg 1 \end{cases} \tag{3.79}$$

and

$$U = \left(\frac{dk}{d\omega}\right)^{-1} = \frac{c}{2}\left[1 + \frac{2kh_0}{\sinh(2kh_0)}\right] = \begin{cases} \sqrt{g_0 h_0} & \text{for } kh_0 \ll 1 \\ \frac{1}{2}\sqrt{\frac{g_0}{k}} & \text{for } kh_0 \ll 1 \end{cases}. \tag{3.80}$$

In Figure 3.13b, the group velocity (Eq. 3.80) is plotted by dashed lines. At a low frequency or a long wavelength ($kh_0 \ll 1$), the group velocity becomes the same as the phase velocity, which does not depend on the frequency. At a high frequency (or a short wavelength, $kh_0 \gg 1$), the group velocity decreases with increasing wave frequency and is independent of the sea depth. The group velocity is generally smaller than the phase velocity and half of the phase velocity when the frequency is very high.

It is interesting to consider why the tsunami propagation velocity depends on the frequency in high-frequency waves and depends on the sea depth in low-frequency waves (see Eqs. (3.79) and (3.80)). We may intuitively interpret it from the viewpoint of the restoring force. The restoring force of the ocean waves is gravity. When the wavelength is short enough compared to the sea depth ($kh_0 \gg 1$), the particle motions of the ocean waves are confined to near the sea surface given by $-1/k < z < 0$, because of the exponential decaying factors of $\exp(kz)$ (Eqs. 3.61 and 3.62). In this case, the waves do not detect the sea bottom (Figure 3.14a). As a result, the phase and group velocities are independent of the sea depth h_0 and depend on the wavenumber or the wavelength. When considering the ocean wave oscillation, the water volume of $\sim \lambda/2\pi \cdot S$ is oscillating (Figure 3.14a), where S is the area in the horizontal plane. The force due to the gravity on this region is proportional to the volume $\lambda/2\pi \cdot S$. This means that when the wavelength is longer, the restoring force becomes large. The large restoring force results in wave propagation at a higher velocity. If you suppose elastic wave propagation (which we also treat in Sect. 3.3 *Seismic Waves: Elastic Dynamics*), you may consider the role of this oscillating volume $\lambda/2\pi \cdot S$ in ocean wave propagation as the role of the elastic coefficient. A large-volume oscillating body in the ocean waves or a hard elastic material in elastic waves gives a high restoring force, which results in fast wave propagation.

When the wavelength becomes much longer than the sea depth ($kh_0 \ll 1$), the particle motions of the ocean waves are distributed from the sea bottom to the sea surface ($-h_0 \leq z \leq 0$). Since the sea depth is finite, the oscillating body has a limitation even if the wavelength increases (Figure 3.14b). As a result, the phase and group velocities become independent of the wavelength when the wavelength is much longer than the sea depth but depend on the sea depth h_0.

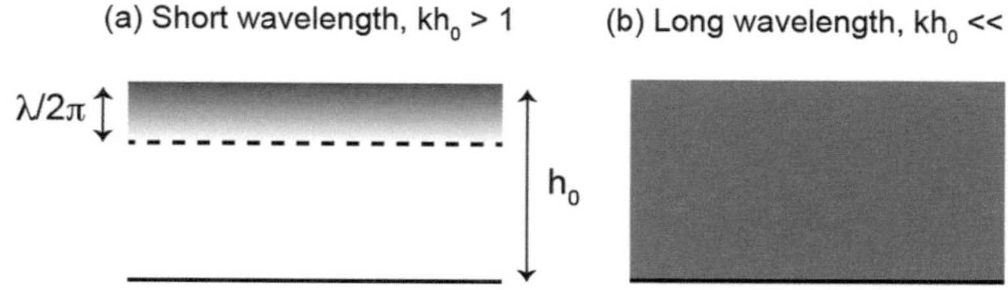

Fig. 3.14 (**a**) When the wavelength of the ocean waves is short ($kh_0 \gg 1$), the shallow part of the water layer ($\sim -\lambda/2\pi < z < 0$) is in motion, but the ocean wave does not reach deeper part. The portion in motion increases with increasing wavelength. (**b**) When the wavelength of the ocean waves is long ($kh_0 \ll 1$)), the whole water layer ($-h_0 < z < 0$) is in motion

3.2.3 Energy Density and Energy Flux Density for Incompressible Fluid

It is fundamentally important to introduce energy density and energy flux density in wave theory. These parameters satisfy a continuity equation in an incompressible fluid medium.

Substituting Eq. (3.32) into Eq. (3.30) and using Eq. (3.31), we rewrite the equation of motion for the ith component as

$$\frac{\partial v_i(\mathbf{x},t)}{\partial t} = -\frac{1}{\rho_0}\frac{\partial p_e(\mathbf{x},t)}{\partial x_i}. \tag{3.81}$$

We consider complex number for the velocity and the pressure in this formulation as in Eq. (3.39). Multiplying $v_i^*(\mathbf{x},t)$ by Eq. (3.81) (* indicates complex conjugate),

$$v_i^*\frac{\partial v_i(\mathbf{x},t)}{\partial t} = -\frac{1}{\rho_0}v_i^*\frac{\partial p_e(\mathbf{x},t)}{\partial x_i},$$

and adding its complex conjugate gives

$$v_i^*\frac{\partial v_i}{\partial t} + v_i\frac{\partial v_i^*}{\partial t} = -\frac{1}{\rho_0}\left(v_i^*\frac{\partial p_e}{\partial x_i} + v_i\frac{\partial p_e^*}{\partial x_i}\right).$$

This is calculated as Eq. (3.82)

$$\begin{aligned}\rho_0\frac{\partial}{\partial t}(v_iv_i^*) &= -\left(\frac{\partial p_e}{\partial x_i}v_i^* + \frac{\partial p_e^*}{\partial x_i}v_i\right)\\ &= -\left(\frac{\partial}{\partial x_i}(p_ev_i^*) - p_e\frac{\partial v_i^*}{\partial x_i} + \frac{\partial}{\partial x_i}(p_e^*v_i) - p_e^*\frac{\partial v_i}{\partial x_i}\right)\\ &= -\frac{\partial}{\partial x_i}(p_ev_i^* + p_e^*v_i),\end{aligned} \tag{3.82}$$

where we assumed that the seawater is incompressible $(\partial v_k)/(\partial x_k) = 0$. We here define energy density $E(\mathbf{x},t)$ and energy flux density vector $\mathbf{J}(\mathbf{x},t)$ as

$$E(\mathbf{x},t) = \frac{1}{2}\rho_0 v_i v_i^*, \tag{3.83}$$

and

$$J_i(\mathbf{x},t) = \frac{1}{2}(p_ev_i^* + p_e^*v_i), \tag{3.84}$$

respectively. Energy density $E(\mathbf{x},t)$ represents the energy per unit volume (e.g., J/m^3), and the energy flux density vector $\mathbf{J}(\mathbf{x},t)$ represents the amount of energy passing through a unit area perpendicular to the propagation direction per unit time (e.g., J/m^2/s). Substituting Eqs. (3.83) and (3.84) into Eq. (3.82), we obtain

$$\frac{\partial E(\mathbf{x},t)}{\partial t} + \text{div}\,\mathbf{J}(\mathbf{x},t) = 0. \tag{3.85}$$

This equation represents a continuity equation of energy.

Equation (3.85) is integrated in space over volume V_0 as

$$\begin{aligned}\frac{\partial}{\partial t}\iiint_{V_0} E(\mathbf{x},t)dV(\mathbf{x}) &= -\iiint_{V_0} \text{div}\,\mathbf{J}(\mathbf{x},t)dV(\mathbf{x}) \\ &= -\iint_{S_0} \mathbf{J}(\mathbf{x},t)\cdot\bar{\mathbf{n}}\,dS(\mathbf{x})\end{aligned} \tag{3.86}$$

in which Gauss's theorem (the divergence theorem) was used. This equation means that the increasing rate of the total energy inside the volume V_0 (the left-hand side) is given by the total energy flowing into the volume through its surface S_0 (the right-hand side) (Figure 3.15).

Equation (3.86) represents the conservation of energy in infinite three-dimensional space. Because tsunami propagation involves horizontal propagation but does not propagate along the vertical direction, it is useful to derive the energy conservation in the two-dimensional (horizontal) space in tsunami problems. In the Cartesian coordinates shown in Fig. 3.4, Eq. (3.85) is integrated over the vertical direction from the sea bottom to the sea surface as

$$\begin{aligned}0 &= \frac{\partial}{\partial t}\int_{-h_0}^{0} E(\mathbf{x},t)dz + \int_{-h_0}^{0} \text{div}\,\mathbf{J}(\mathbf{x},t)dz \\ &= \frac{\partial}{\partial t}\int_{-h_0}^{0} E(\mathbf{x},t)dz + \int_{-h_0}^{0}\left(\frac{\partial J_x(\mathbf{x},t)}{\partial x} + \frac{\partial J_y(\mathbf{x},t)}{\partial y}\right)dz \\ &\quad + J_z(x,y,0,t) - J_z(x,y,-h_0,t)\end{aligned} \tag{3.87}$$

From the definition of the energy flux density in 3-D space (Eq. 3.84) and the boundary conditions at $z = 0$ (Eqs. 3.35 and 3.36), the energy flux density in the z component $J_z(x, y, z, t)$ at the surface $z = 0$ is calculated as

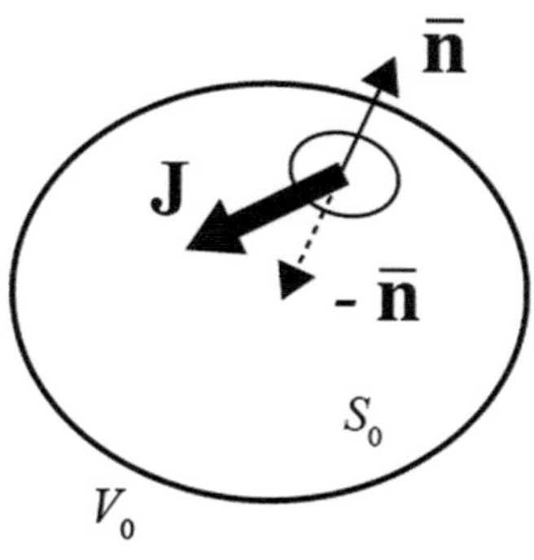

Fig. 3.15 Energy flux density $\mathbf{J}(\mathbf{x}, t)$ flows into a region of volume V_0 with the surface S_0. The unit vector normal to the surface is given by $\bar{\mathbf{n}}$. The positive sign of $\bar{\mathbf{n}}$ means that the vector is directed outward from the volume. Then, the flux that flows into the volume through the small surface dS_0 is given by $-\mathbf{J}(\mathbf{x}, t)\cdot\bar{\mathbf{n}}\,dS_0$

$$\begin{aligned} J_z(x,y,0,t) &= \frac{1}{2}\left(p_e v_z^* + p_e^* v_z\right)\Big|_{z=0} = \frac{1}{2}\left(\rho_0 g_0 \eta \frac{\partial \phi^*}{\partial z} + \rho_0 g_0 \eta^* \frac{\partial \phi}{\partial z}\right)\Big|_{z=0} \\ &= \frac{1}{2}\rho_0 g_0 \left(\eta \frac{\partial \eta^*}{\partial t} + \eta^* \frac{\partial \eta}{\partial t}\right) = \frac{1}{2}\rho_0 g_0 \frac{\partial}{\partial t}(\eta\eta^*). \end{aligned} \tag{3.88}$$

The energy flux density in the z component at the surface $z = 0$ is represented using the sea-surface height η for incompressible fluid. Equation (3.87) is then rewritten as Eq. (3.89)

$$\begin{aligned} 0 &= \frac{\partial}{\partial t}\int_{-h_0}^{0} E(\mathbf{x},t)dz + \int_{-h_0}^{0}\left(\frac{\partial J_x(\mathbf{x},t)}{\partial x} + \frac{\partial J_y(\mathbf{x},t)}{\partial y}\right)dz \\ &\quad + \frac{1}{2}\rho_0 g_0 \frac{\partial}{\partial t}(\eta\eta^*) - J_z(x,y,-h_0,t) \\ &= \frac{\partial}{\partial t}\left[\int_{-h_0}^{0} E(\mathbf{x},t)dz + \frac{1}{2}\rho_0 g_0 \eta\eta^*\right] + \frac{\partial}{\partial x}\int_{-h_0}^{0} J_x(\mathbf{x},t)dz + \frac{\partial}{\partial y}\int_{-h_0}^{0} J_y(\mathbf{x},t)dz \end{aligned} \tag{3.89}$$

where we assumed that the vertical velocity is zero at the sea bottom and set it as $J_z(x, y, -h_0, t) = 0$. We then define the energy density $E_{2D}(x, y, t)$ and energy flux density $\mathbf{J}_{2D}(x, y, t)$ in the two-dimensional horizontal space as Eq. (3.90)

$$\begin{aligned} E_{2D}(x,y,t) &= \int_{-h_0}^{0} E(\mathbf{x},t)dz + \frac{1}{2}\rho_0 g_0 |\eta|^2 \\ &= \frac{1}{2}\rho_0 \int_{-h_0}^{0} \left(|v_x|^2 + |v_y|^2 + |v_z|^2\right)dz + \frac{1}{2}\rho_0 g_0 |\eta|^2, \end{aligned} \tag{3.90}$$

and

$$\begin{aligned} J_{2D,x}(x,y,t) &= \frac{1}{2}\int_{-h_0}^{0} \left(p_e v_x^* + p_e^* v_x\right)dz, \\ J_{2D,y}(x,y,t) &= \frac{1}{2}\int_{-h_0}^{0} \left(p_e v_y^* + p_e^* v_y\right)dz. \end{aligned} \tag{3.91}$$

Then, Eq. (3.89) represents a continuity equation of energy in 2D form as

$$\frac{\partial E_{2D}(x,y,t)}{\partial t} + \nabla_{xy} \cdot \mathbf{J}_{2D}(x,y,t) = 0, \tag{3.92}$$

where $\nabla_{xy} \cdot$ is the divergence in the two dimensional (x, y) space. The energy flux density $\mathbf{J}_{2D}$ in the two-dimensional horizontal space (Eq. 3.91) may be considered a straightforward extension from the three-dimensional space (Eq. 3.84). On the other hand, the energy density E_{2D} in the two-dimensional horizontal space includes the additional term $\rho_0 g_0 |\eta|^2/2$ compared to the three-dimensional space (Eq. 3.83). We may consider the term $\rho_0 g_0 |\eta|^2/2$ as potential energy density stored in the sea layer, whereas the first term in Eq. (3.90) can be considered the kinetic energy density in the sea layer. The sum of potential and kinetic energy gives the energy density in the two-dimensional horizontal space.

3.2.3.1 Tsunami Energy and Energy Flux

As derived in the previous Sect. 3.2.2, for a monochromatic tsunami propagating along the positive x-axis, the wavefield is represented by the following solutions as

$$\eta(x, y, t) = \eta_0 e^{i(kx-\omega t)}, \tag{3.39}$$

$$v_x(x, y, z, t) = \omega\eta_0 \frac{\cosh[k(z + h_0)]}{\sinh(kh_0)} e^{i(kx-\omega t)}, \tag{3.50}$$

$$v_z(x, y, z, t) = -i\omega\eta_0 \frac{\sinh[k(z + h_0)]}{\sinh(kh_0)} e^{i(kx-\omega t)}, \tag{3.52}$$

and

$$p_e(x, y, z, t) = \rho_0 g_0 \eta_0 \frac{\cosh[k(z + h_0)]}{\cosh(kh_0)} e^{i(kx-\omega t)}. \tag{3.63}$$

Substituting the solutions of Eqs. (3.39), (3.50), and (3.52) into Eq. (3.90), the energy density for a plane wave is obtained as

$$\begin{aligned} E_{2D}(x, y, t) &= \frac{\rho_0}{2} \frac{\omega^2 \eta_0^2}{\sinh^2(kh_0)} \int_{-h_0}^{0} \{\cosh^2[k(z + h_0)] + \sinh^2[k(z + h_0)]\} dz \\ &\quad + \frac{1}{2}\rho_0 g_0 |\eta|^2 \\ &= \frac{\rho_0}{2} \frac{\omega^2 \eta_0^2}{\sinh^2(kh_0)} \int_{-h_0}^{0} \cosh[2k(z + h_0)] dz + \frac{1}{2}\rho_0 g_0 |\eta|^2 \\ &= \frac{\rho_0}{4} \frac{\omega^2 \eta_0^2}{k} \frac{\sinh(2kh_0)}{\sinh^2(kh_0)} + \frac{1}{2}\rho_0 g_0 |\eta|^2 \\ &= \frac{\rho_0}{4} \frac{\omega^2 \eta_0^2}{k} \frac{2 \sinh kh_0 \cosh kh_0}{\sinh^2(kh_0)} + \frac{1}{2}\rho_0 g_0 |\eta|^2 \\ &= \frac{\rho_0 \eta_0^2}{2} \frac{\omega^2}{k \tanh(kh_0)} + \frac{1}{2}\rho_0 g_0 |\eta|^2 \\ &= \frac{\rho_0 g_0 \eta_0^2}{2} + \frac{1}{2}\rho_0 g_0 |\eta|^2 \\ &= \rho_0 g_0 |\eta|^2, \end{aligned} \tag{3.93}$$

where the dispersion relation $\omega^2 = g_0 k \tanh(kh_0)$ (Eq. 3.47) and an identity $\sinh 2kh_0 = 2 \sinh kh_0 \cosh kh_0$ are used. The first term of Eq. (3.93) ($\rho_0/2 \cdot \int_{-h_0}^{0} \left(|v_x|^2 + |v_y|^2 + |v_z|^2\right) dz$) represents the kinetic energy, and the second term ($\rho_0/2 \cdot g_0 |\eta|^2$) is the potential energy. This calculation indicates that for a propagating plane ocean wave, the kinetic energy density is identical to the potential energy density. The total energy density that is the sum of the kinetic and potential energy density is given by $\rho_0 g_0 |\eta|^2$. The total energy density is proportional to the squared amplitude of the surface height.

Substituting the solutions of Eqs. (3.50) and (3.63) into Eq. (3.91) gives the energy flux density as

$$
\begin{aligned}
J_{2D,x}(x,y,t) &= \frac{1}{2}\int_{-h_0}^{0} \left(p_e v_x^* + p_e^* v_x\right) dz \\
&= \int_{-h_0}^{0} \rho_0 g_0 \eta_0 \frac{\cosh k(z+h_0)}{\cosh kh_0} \omega \eta_0 \frac{\cosh k(z+h_0)}{\sinh kh_0} dz \\
&= \frac{\rho_0 g_0 \eta_0^2 \omega}{\cosh(kh_0)\sinh(kh_0)} \int_{-h_0}^{0} \cosh^2 k(z+h_0) dz \\
&= \frac{\rho_0 g_0 \eta_0^2 \omega}{2\cosh(kh_0)\sinh(kh_0)} \left[h_0 + \frac{1}{2k}\sinh 2kh_0\right] \\
&= \frac{\rho_0 g_0 \eta_0^2 \omega}{\sinh(2kh_0)} \left[h_0 + \frac{1}{2k}\sinh 2kh_0\right] \\
&= \rho_0 g_0 \eta_0^2 \frac{1\omega}{2k}\left[1 + \frac{2kh_0}{\sinh(2kh_0)}\right].
\end{aligned}
\tag{3.94}
$$

We should remember that the group velocity of the ocean wave is given by

$$
U = \frac{1}{2}\frac{\omega}{k}\left[1 + \frac{2kh_0}{\sinh(2kh_0)}\right]. \tag{3.80}
$$

Then, the energy flux density of Eq. (3.94) is represented by the product of the energy density and the group velocity as

$$
J_{2D,x}(x,y,t) = UE_{2D}. \tag{3.95}
$$

Hence, we may consider that the group velocity is the velocity of the energy propagation.

Considering the case when the wavelength is much longer than the sea depth, $kh_0 \ll 1$, Eqs. (3.50) and (3.63) indicate that the velocity and pressure fields become independent of z. The velocity is approximated as

$$
v_x(x,y,z,t) \approx v_x^{\mathrm{av}}(x,y,t) = \frac{\omega}{kh_0}\eta(x,y,t) \quad \text{for } kh_0 \ll 1 \tag{3.96}
$$

where $v_x^{\mathrm{av}}(x, y, t)$ denotes the average horizontal velocity that is independent of the depth z. The excess pressure is then approximated as

$$p_e(x, y, z, t) \approx \rho_0 g_0 \eta(x, y, t) \quad \text{for} \quad kh_0 \ll 1. \tag{3.97}$$

Equation (3.91) is then rewritten as

$$\begin{aligned} J_{2D,x}(x, y, t) &\approx \frac{1}{2}\rho_0 g_0 h_0 \left(v_x^{\mathrm{av}*} \eta + \eta^* v_x^{\mathrm{av}} \right) \quad \text{for } kh_0 \ll 1 \\ &= \rho_0 g_0 h_0 \mathrm{Re}\left(v_x^{\mathrm{av}*} \eta \right) \\ &= \rho_0 g_0 \sqrt{g_0 h_0} |\eta|^2. \end{aligned} \tag{3.98}$$

We have considered a plane wave propagation along the x-axis. It is straightforward to extend this to a plane wave propagation in any direction in the (x, y) plane. The energy flux density $\mathbf{J}_{2D}$ in the (x, y) plane for 2-D horizontal particle velocity vector $\mathbf{v} = (v_x, v_y)$ and tsunami height η is given by

$$\mathbf{J}_{2D}(x, y, t) \approx \rho_0 g_0 h_0 \eta \mathbf{v} \quad \text{for } kh_0 \ll 1. \tag{3.99}$$

By using this equation, the energy flux density of the tsunami is visualized based on the simulation. Figure 3.16 shows the energy flux density for the tsunami (average flux density during the first hour after the first tsunami arrival) in the 2011 Tohoku-Oki earthquake (Fine et al. 2013). The energy flux density is directed toward the southeast and decreases with travel distance. The decrease occurs due to the geometrical spreading, the backward scattering of waves, and the intrinsic energy absorption. Fine et al. (2013) estimated that the total energy flux density decreases with distance r according to $\exp[-r/4,700\ \mathrm{km}]$.

The energy flux density is an important idea from the viewpoint of conservation law. When we assume that the reflected tsunami does not appear during tsunami propagation, in other words, when we consider multiple forward scattering but neglect backward scattering, the energy flux density J (Eq. 3.98) should be constant irrespective of the location and the sea depth. This leads to

$$\rho_0 g_0 |\eta(x_1)|^2 \sqrt{g_0 h(x_1)} = \rho_0 g_0 |\eta(x_2)|^2 \sqrt{g_0 h(x_2)},$$

and

$$|\eta(x_2)| = \left| \frac{h(x_1)}{h(x_2)} \right|^{\frac{1}{4}} |\eta(x_1)|, \tag{3.100}$$

where $\eta(x)$ and $h(x)$ are the tsunami height and the sea depth at the location x, respectively. When the point x_2 is located near the coast and the sea depth is shallower than the point x_1, $h(x_2) < h(x_1)$, the tsunami height $\eta(x_2)$ at location x_2 is then larger than $\eta(x_1)$ at location x_1. This is called as Green's law. Note that this does

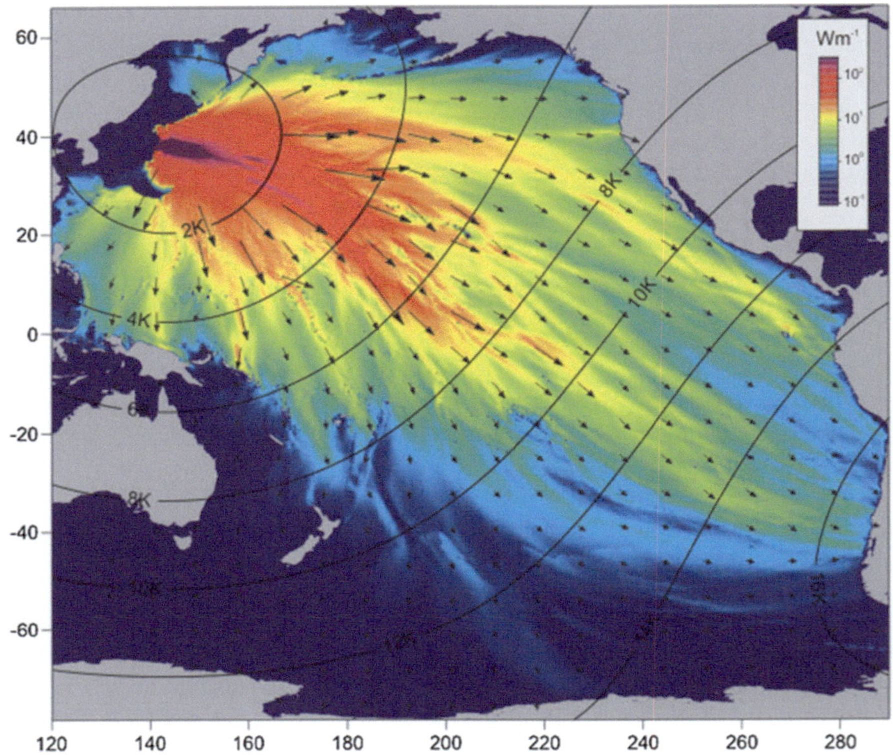

Fig. 3.16 Energy flux averaged over 1 h during the passage of the front of the 2011 Tohoku-Oki earthquake tsunami. Black contours mark the distance (in km) from the tsunami source (Fine et al. 2013, copyright by Springer)

not consider reflected waves, whereas reflected tsunamis usually appear in observations. Nevertheless, the Green's law provides a good perspective of the tsunami amplification mechanisms. Hayashi (2010) investigated the relationship between the tsunami heights obtained at offshore and coastal stations by introducing an additional parameter into Green's law. Understanding the relation between offshore and coastal tsunamis would be useful for rapidly forecasting the tsunami height on the coast using the tsunami height recorded at offshore stations.

3.3 Seismic Waves: Elastic Dynamics

Section 3.2 investigated tsunami or water waves with seawater assumed to be incompressible. This means that the elasticity of the medium (or stress due to strain) was neglected. However, the earth medium also works as an elastic body in which compressional and shear stress are excited by strain. The stress works as a restoring force for seismic waves. Furthermore, a surface Rayleigh wave is excited due to a stress-free boundary. These elastic wave excitations and propagations have been extensively described in books on seismology (e.g., Aki and Richards (2002), Shearer (2009), Saito (2009a, b), Sato et al. (2012)). Hence, this section does not thoroughly describe elastic waves in general but focuses on the points closely related to tsunami generation and propagation. For example, the Love wave, which is accompanied by displacement only in its horizontal components but not in its vertical component, is not treated here.

3.3.1 P, S, *and Rayleigh Waves*

By substituting the stress-displacement relation in an isotropic medium (Eq. 3.17) into the equation of motion (Eq. 3.14), we rewrite the equation of motion as a system of equations with respect to the displacement u_i:

$$\rho \frac{\partial^2 u_i(x,t)}{\partial t^2} = \frac{\partial}{\partial x_k}[\lambda \delta_{ik} u_{l,l} + \mu(u_{i,k} + u_{k,i})] + f_i. \tag{3.101}$$

An earthquake fault is represented by the body force f_i as a source (Chap. 4: *Earthquakes*). This section considers the wave propagation in a homogeneous medium but does not consider the wave excitation by setting the body force to zero ($f_i = 0$). When the medium is homogeneous, Eq. (3.101) is then rewritten as

$$\rho \frac{\partial^2 \mathbf{u}(x,t)}{\partial t^2} = (\lambda + \mu)\nabla(\nabla \cdot \mathbf{u}) + \mu \nabla^2 \mathbf{u}. \tag{3.102}$$

Here, note that $\nabla^2 \mathbf{u}$ is given by the identity:

$$\nabla^2 \mathbf{u} = \nabla(\nabla \cdot \mathbf{u}) - \nabla \times (\nabla \times \mathbf{u}). \tag{3.103}$$

With this equation, the equation of motion in an isotropic homogeneous medium is given by the differential equation:

$$\rho \frac{\partial^2 \mathbf{u}(x,t)}{\partial t^2} = (\lambda + 2\mu)\nabla(\nabla \cdot \mathbf{u}) - \mu \nabla \times (\nabla \times \mathbf{u}), \tag{3.104}$$

with respect to the displacement vector $\mathbf{u}$.

Taking the divergence of the equations of motion (Eq. 3.104), we calculate

$$\begin{aligned}\rho \frac{\partial^2}{\partial t^2}(\nabla \cdot \mathbf{u}) &= (\lambda + 2\mu)\nabla \cdot \nabla(\nabla \cdot \mathbf{u}) - \mu \nabla \cdot \nabla \times (\nabla \times \mathbf{u}) \\ &= (\lambda + 2\mu)\nabla^2(\nabla \cdot \mathbf{u})\end{aligned}$$

and obtain a wave equation

$$\frac{1}{V_P^2}\frac{\partial^2}{\partial t^2}(\nabla\cdot\mathbf{u}) = \nabla^2(\nabla\cdot\mathbf{u}), \tag{3.105}$$

where

$$V_P = \sqrt{\frac{\lambda+2\mu}{\rho}}. \tag{3.106}$$

Because the value of $\nabla\cdot\mathbf{u}$ is a scalar quantity representing dilatation or the ratio of the volume change ($\Delta V/V$, Eq. (3.10)), Eq. (3.105) indicates a wave equation of the dilatation. The volume change propagates as a wave with the velocity of V_P. The wave is referred to as the longitudinal wave, compressional wave, and *P* wave.

On one hand, when taking the rotation of the equations of motion (Eq. 3.104), we calculate

$$\begin{aligned}\rho\frac{\partial^2}{\partial t^2}(\nabla\times\mathbf{u}) &= (\lambda+2\mu)\nabla\times\nabla(\nabla\cdot\mathbf{u}) - \mu\nabla\times\nabla\times(\nabla\times\mathbf{u})\\ &= -\mu\nabla\times\nabla\times(\nabla\times\mathbf{u})\\ &= -\mu\nabla[\nabla\cdot(\nabla\times\mathbf{u})] + \mu\nabla^2(\nabla\times\mathbf{u})\\ &= \mu\nabla^2(\nabla\times\mathbf{u})\end{aligned}$$

and obtain a wave equation

$$\frac{1}{V_S^2}\frac{\partial^2}{\partial t^2}(\nabla\times\mathbf{u}) = \nabla^2(\nabla\times\mathbf{u}) \tag{3.107}$$

where

$$V_S = \sqrt{\frac{\mu}{\rho}}. \tag{3.108}$$

The value of $\nabla\times\mathbf{u}$ is a vector representing the rotation and does not show the volume change. It propagates with the velocity of V_S. The wave is referred to as the transverse wave, shear wave, or *S* wave.

3.3.1.1 Plane Wave Propagation

The *P* wave and *S* wave propagate through an isotropic homogeneous medium. These are classified as body wave. Moreover, a stress-free surface is intrinsically important for seismic wave propagation, which enables the waves to propagate along the surface. This is called the surface wave. A coupling of plane *P* and *S* waves with

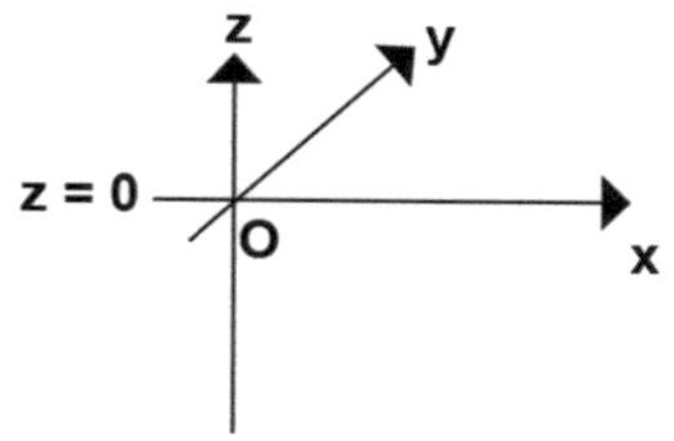

Fig. 3.17 Coordinates used in the formulation. The x- and y-axes are in the horizontal plane and the z-axis is in the vertical direction (upward positive). The vector normal to the wave front is in the (x, z) plane

the free surface results in a surface wave referred to as the Rayleigh wave. A plane wave propagation is considered in order to investigate surface waves. The Cartesian coordinates (x, y, z) shown in Fig. 3.17 are used where the x- and y-axes are in the horizontal plane and the z-axis is in the vertical direction. When the vector normal to the wave front is in the (x, z) plane, variation with respect to the y-axis does not occur in the propagation. The equation of motion (Eq. 3.104) is then simplified as

$$\rho\frac{\partial^2 u_x}{\partial t^2} = (\lambda+\mu)\frac{\partial}{\partial x}\left(\frac{\partial u_x}{\partial x}+\frac{\partial u_z}{\partial z}\right)+\mu\left(\frac{\partial^2 u_x}{\partial x^2}+\frac{\partial^2 u_x}{\partial z^2}\right), \tag{3.109}$$

$$\rho\frac{\partial^2 u_y}{\partial t^2} = \mu\left(\frac{\partial^2 u_y}{\partial x^2}+\frac{\partial^2 u_y}{\partial z^2}\right), \tag{3.110}$$

and

$$\rho\frac{\partial^2 u_z}{\partial t^2} = (\lambda+\mu)\frac{\partial}{\partial z}\left(\frac{\partial u_x}{\partial x}+\frac{\partial u_z}{\partial z}\right)+\mu\left(\frac{\partial^2 u_z}{\partial x^2}+\frac{\partial^2 u_z}{\partial z^2}\right). \tag{3.111}$$

3.3.1.2 SH Wave

The displacement in the y-axis, u_y, appears only in Eq. (3.110) which is rewritten as

$$\frac{1}{V_S^2}\frac{\partial^2 u_y}{\partial t^2} = \frac{\partial^2 u_y}{\partial x^2}+\frac{\partial^2 u_y}{\partial z^2} \tag{3.112}$$

where $V_S = \sqrt{\mu/\rho}$. This is a wave equation for u_y. We consider a plane wave propagation with the horizontal slowness p and the vertical slowness η. The wave is described as

$$u_y = \exp[-i\omega(t - px - \eta z)]. \tag{3.113}$$

We consider the case in which the horizontal slowness p is real and $p > 0$ (the wave propagates toward and along the positive x-axis). If p is not real but purely imaginary, the wave does not propagate horizontally (along the x-axis) as a wave. Substituting Eq. (3.113) into the wave equation (Eq. 3.112), we obtain the following relation:

$$\frac{1}{V_S^2} = p^2 + \eta^2. \tag{3.114}$$

We consider the case in which the value of η is real when $0 < p < 1/V_S$. The horizontal slowness p is represented as

$$p = \frac{\sin\phi}{V_S} \tag{3.115}$$

where ϕ is the angle between the z-axis and the wave propagation direction (Fig. 3.18).

Then, the vertical slowness η is represented as

$$\eta = \frac{\cos\phi}{V_S}. \tag{3.116}$$

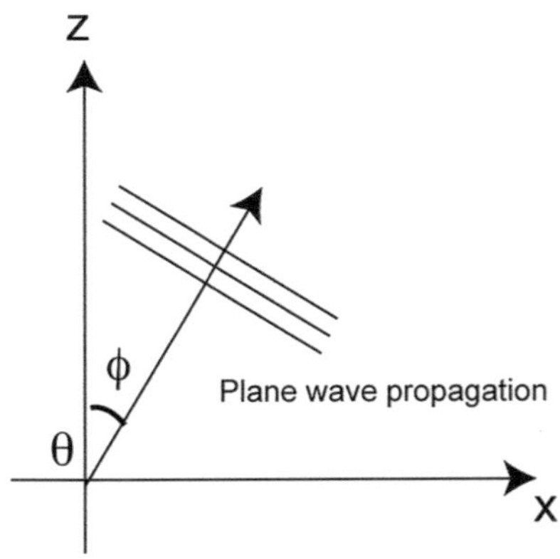

Fig. 3.18 A plane wave propagation. The propagation direction lies in the (x, z) plane. The angle between the direction of the propagating wave and the z-axis is ϕ

3.3.1.3 P-SV Wave

Unlike the case of u_y, the displacements of u_x and u_z appear in the two equations (Eqs. 3.109 and 3.111). Hence, the components of u_x and u_z are coupled, and we need to simultaneously solve two equations:

$$\rho\frac{\partial^2 u_x}{\partial t^2} = (\lambda+\mu)\frac{\partial}{\partial x}\left(\frac{\partial u_x}{\partial x}+\frac{\partial u_z}{\partial z}\right)+\mu\left(\frac{\partial^2 u_x}{\partial x^2}+\frac{\partial^2 u_x}{\partial z^2}\right), \tag{3.109}$$

and

$$\rho\frac{\partial^2 u_z}{\partial t^2} = (\lambda+\mu)\frac{\partial}{\partial z}\left(\frac{\partial u_x}{\partial x}+\frac{\partial u_z}{\partial z}\right)+\mu\left(\frac{\partial^2 u_z}{\partial x^2}+\frac{\partial^2 u_z}{\partial z^2}\right). \tag{3.111}$$

We intend to find a solution for plane wave propagation with a horizontal slowness of p (or an apparent horizontal velocity of $1/p$) along the x direction as

$$\begin{aligned} u_x(x,z,t) &= u_0\exp[-i\omega(t-px-qz)], \\ u_z(x,z,t) &= \exp[-i\omega(t-px-qz)]. \end{aligned} \tag{3.117}$$

Substituting Eq. (3.117) into Eqs. (3.109) and (3.111), the parameter q, or vertical slowness, is given by

$$q = \pm\xi,\ \pm\eta, \tag{3.118}$$

where

$$\xi = \sqrt{\frac{1}{V_P^2}-p^2} \text{ and } \eta = \sqrt{\frac{1}{V_S^2}-p^2}. \tag{3.119}$$

The parameter u_0 in Eq. (3.117) is given by

$$u_0 = \pm\frac{p}{\xi},\ \mp\frac{\eta}{p}, \tag{3.120}$$

corresponding to each value of q in Eq. (3.118). Then, the solution of Eqs. (3.109) and (3.111) is given by a linear combination of the four waves as

$$\begin{aligned} u_x(x,z,t) &= \left[p\left(Ae^{i\omega\xi z}-Be^{-i\omega\xi z}\right)+\eta\left(-Ce^{i\omega\eta z}+De^{-i\omega\eta z}\right)\right]e^{-i\omega(t-px)}, \\ u_z(x,z,t) &= \left[\xi\left(Ae^{i\omega\xi z}+Be^{-i\omega\xi z}\right)+p\left(Ce^{i\omega\eta z}+De^{-i\omega\eta z}\right)\right]e^{-i\omega(t-px)}, \end{aligned} \tag{3.121}$$

where A, B, C, and D are constants. In Eq. (3.121), the constants A, B, C, and D were multiplied by ξ or p for a simple representation.

As shown in Eq. (3.119), the variables ξ and η correspond to the vertical slownesses of the P wave and S wave, respectively. Therefore, Eq. (3.121) means that the wavefield is represented as a sum of P waves and S waves.

3.3.1.4 P Wave

Let us consider the wave represented by the constants A and B by setting $C = 0$ and $D = 0$ in Eq. (3.121):

$$\begin{aligned} u_x(x,z,t) &= p\left(Ae^{i\omega\xi z} - Be^{-i\omega\xi z}\right)e^{-i\omega(t-px)}, \\ u_z(x,z,t) &= \xi\left(Ae^{i\omega\xi z} + Be^{-i\omega\xi z}\right)e^{-i\omega(t-px)}, \end{aligned} \tag{3.122}$$

where

$$\xi = \sqrt{\frac{1}{V_P^2} - p^2} \text{ and } V_P = \sqrt{\frac{\lambda + 2\mu}{\rho}}. \tag{3.123}$$

This wave represents a P wave. Consider the case when ξ is real. The first terms of Eq. (3.122), which include the term $\exp[-i\omega(t - px - \xi z)]$, indicate a plane P wave propagating in the direction (p, ξ) with a velocity of V_P. The second term indicates the plane P wave propagating in the direction $(p, -\xi)$. The horizontal slowness p is given by $p = \sin\theta/V_P$ when the angle between the wave propagation direction and the z-axis is θ. Then, the slowness along the z-axis is $\xi = \cos\theta/V_P$.

As shown in Fig. 3.19, consider the plane P wave with a positive vertical slowness by setting $B = 0$:

$$\begin{aligned} u_x(x,z,t) &= pAe^{-i\omega(t-px-\xi z)}, \\ u_z(x,z,t) &= \xi Ae^{-i\omega(t-px-\xi z)}. \end{aligned} \tag{3.124}$$

The displacement along the propagation direction (radial component) W and the displacement perpendicular to the propagation direction (transverse component) U are calculated as

$$\begin{aligned} W &= u_x \sin\theta + u_z \cos\theta = A(p\sin\theta + \xi\cos\theta) = A/V_P \\ U &= u_x \cos\theta - u_z \sin\theta = A(p\cos\theta - \xi\sin\theta) = 0 \end{aligned} \tag{3.125}$$

where $p = \sin\theta/V_P$ and $\xi = \cos\theta/V_P$ are used. The P wave in an isotropic homogeneous medium has displacement only in the propagation direction.

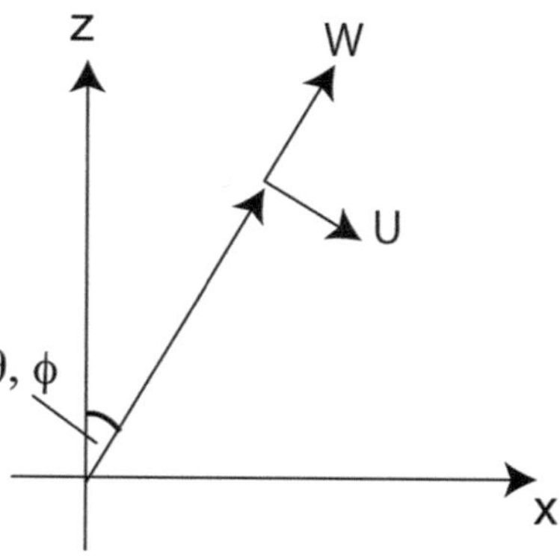

Fig. 3.19 A plane wave propagation. The propagation direction lies in the (x, z) plane. The angle between the direction of the propagating wave and the z-axis is θ or ϕ. The radial and transverse components of the displacement are W and U, respectively

3.3.1.5 SV Wave

Let us consider the wave represented by the constants C and D by setting $A = 0$ and $B = 0$ in Eq. (3.121). This wave is the S wave.

$$\begin{aligned} u_x(x, z, t) &= \eta\left(-Ce^{i\omega\eta z} + De^{-i\omega\eta z}\right)e^{-i\omega(t-px)}, \\ u_z(x, z, t) &= p\left(Ce^{i\omega\eta z} + De^{-i\omega\eta z}\right)e^{-i\omega(t-px)}, \end{aligned} \tag{3.126}$$

where

$$\eta = \sqrt{\frac{1}{V_S^2} - p^2} \quad \text{and} \quad V_S = \sqrt{\frac{\mu}{\rho}}. \tag{3.127}$$

Consider the case when η is real. The first terms of Eq. (3.126), which include the term $\exp[-i\omega(t - px - \eta z)]$, denote a plane S wave propagating in the direction (p, η) with a velocity of V_S. The second term indicates a plane S wave propagating in the direction $(p, -\eta)$. The horizontal slowness p is given by $p = \sin\phi/V_S$ when the angle between the wave propagation direction and the z-axis is ϕ. Then, the slowness along the z-axis is $\eta = \cos\phi/V_S$.

Consider the plane S wave with positive vertical slowness $(D = 0)$:

$$\begin{aligned} u_x(x, z, t) &= -\eta Ce^{-i\omega(t-px-\eta z)} \\ u_z(x, z, t) &= pCe^{-i\omega(t-px-\eta z)} \end{aligned} \tag{3.128}$$

The displacement along the propagation direction W and the displacement perpendicular to the propagation direction U (Fig. 3.19) are calculated as

$$\begin{aligned} W &= u_x \sin\phi + u_z \cos\phi = C(-\eta\sin\phi + p\cos\phi) = 0 \\ U &= u_x \cos\phi - u_z \sin\phi = C(-\eta\cos\phi - p\sin\phi) = -C/V_S \end{aligned} \tag{3.129}$$

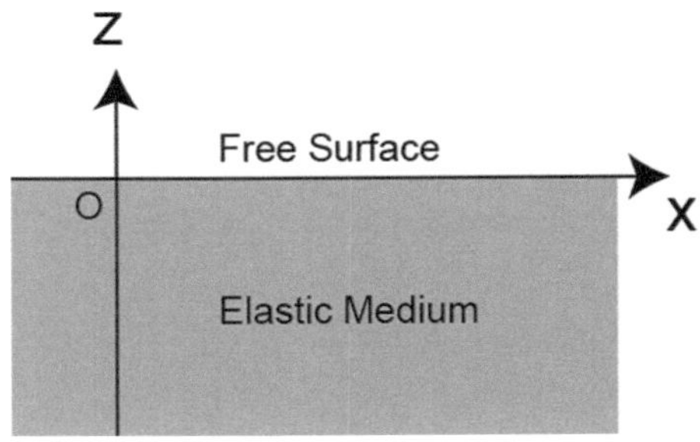

Fig. 3.20 The coordinates (x, z). The free surface is located at $z = 0$

where $p = \ \sin\phi/V_S$ and $\eta = \ \cos\phi/V_S$ are used. The S wave in an isotropic homogeneous medium has displacement only in the direction perpendicular to the propagation direction.

3.3.1.6 Rayleigh Wave

In an infinite elastic medium, the P and S waves propagate through the medium. Now, setting a free surface (i.e., the traction is zero on the surface), we consider a wave propagating along the surface (surface wave). The free surface is one of the most important inhomogeneities of the earth medium.

A free surface is set at $z = 0$, and the homogenous half medium exists in the region $z < 0$ as shown in Fig. 3.20. Although in the above section we considered the case in which the vertical slownesses ξ and η are real for P and S waves, respectively, we here consider that ξ and η are imaginary in Eq. (3.121):

$$\begin{aligned} u_x(x,z,t) &= \left[p\left(Ae^{i\omega\xi z} - Be^{-i\omega\xi z}\right) + \eta\left(-Ce^{i\omega\eta z} + De^{-i\omega\eta z}\right)\right]e^{i\omega(px-t)}, \\ u_z(x,z,t) &= \left[\xi\left(Ae^{i\omega\xi z} + Be^{-i\omega\xi z}\right) + p\left(Ce^{i\omega\eta z} + De^{-i\omega\eta z}\right)\right]e^{i\omega(px-t)}. \end{aligned} \tag{3.121}$$

When waves do not propagate vertically (along the z-axis) but propagate horizontally (along the x-axis), the vertical slowness is imaginary. The vertical slowness is represented as imaginary as follows:

$$\xi = i\hat{\xi},\ \hat{\xi} = \sqrt{p^2 - \frac{1}{V_P^2}},\ \eta = i\hat{\eta},\ \hat{\eta} = \sqrt{p^2 - \frac{1}{V_S^2}}, \tag{3.130}$$

where $\hat{\xi} > 0$ and $\hat{\eta} > 0$. Substituting Eq. (3.130) into Eq. (3.121), we will obtain a solution for the Rayleigh wave. Constants A and C need to be zero so that the displacement becomes zero when $z = -\infty$. Then, Eq. (3.121) is calculated as

$$\begin{aligned} u_x(x,z,t) &= \left[ipB'e^{\omega\hat{\xi}z} + i\hat{\eta}De^{\omega\hat{\eta}z}\right]e^{i\omega(px-t)}, \\ u_z(x,z,t) &= \left[\hat{\xi}B'e^{\omega\hat{\xi}z} + pDe^{\omega\hat{\eta}z}\right]e^{i\omega(px-t)}. \end{aligned} \tag{3.131}$$

By using the constitutive law and the displacement given by Eq. (3.131), the stress τ_{xz} is calculated as

$$\begin{aligned}\tau_{xz}(x,z,t) &= \mu\left(\frac{\partial u_x}{\partial z}+\frac{\partial u_z}{\partial x}\right)\\ &= \mu\left[2i\omega p\hat{\xi}B' e^{\omega\hat{\xi}z} + i\omega\left(\hat{\eta}^2+p^2\right)De^{\omega\hat{\eta}z}\right]e^{i\omega(px-t)}.\end{aligned}$$

By using $\mu=\rho V_S^2$ (Eq. 3.108) and Eq. (3.130) and introducing a parameter γ as

$$\gamma = 2V_S^2p^2 \tag{3.132}$$

we obtain

$$\tau_{xz}(x,z,t) = \left[2i\omega\rho V_S^2p\hat{\xi}B' e^{\omega\hat{\xi}z} - i\omega\rho(1-\gamma)De^{\omega\hat{\eta}z}\right]e^{i\omega(px-t)}. \tag{3.133}$$

Similarly, using $\lambda=\rho\left(V_P^2-2V_S^2\right)$, we calculate

$$\begin{aligned}\tau_{zz}(x,z,t) &= \lambda\frac{\partial u_x}{\partial x}+(\lambda+2\mu)\frac{\partial u_z}{\partial z}\\ &= \left[\left(-\lambda\omega p^2+(\lambda+2\mu)\omega\hat{\xi}^2\right)B' e^{\omega\hat{\xi}z} + (-\lambda\omega p\hat{\eta}+(\lambda+2\mu)\omega p\hat{\eta})De^{\omega\hat{\eta}z}\right]e^{i\omega(px-t)}\end{aligned}$$

and obtain

$$\tau_{zz}(x,z,t) = \left[-\omega\rho(1-\gamma)B' e^{\omega\hat{\xi}z} + 2\omega\rho V_S^2p\hat{\eta}De^{\omega\hat{\eta}z}\right]e^{i\omega(px-t)}. \tag{3.134}$$

On the surface ($z=0$), the traction is set as zero as follows: $\tau_{xz}(x,0,t)=0$ and $\tau_{zz}(x,0,t)=0$. Substituting the stress tensor of Eqs. (3.133) and (3.134) into this free surface condition, we obtain

$$\begin{aligned}2V_S^2p\hat{\xi}B' - (1-\gamma)D &= 0,\\ -(1-\gamma)B' + 2V_S^2p\hat{\eta}D &= 0.\end{aligned} \tag{3.135}$$

These are homogeneous equations. These have a nontrivial solution (a solution other than $B'=0$ and $D=0$) only when the determinant of the simultaneous equation of Eq. (135) is zero:

$$\Delta_R(p) = (1-\gamma)^2 - 4V_S^4p^2\hat{\xi}\hat{\eta} = (1-\gamma)^2 - 2V_S^2\gamma\hat{\xi}\hat{\eta} = 0. \tag{3.136}$$

This is referred to as the Rayleigh function. Because $\hat{\xi}$ is a function of V_P and $\hat{\eta}$ is a function of V_S, the Rayleigh function gives the horizontal slowness p as a function of V_P and V_S. It is difficult to represent an analytical solution for p, but we can numerically estimate the slowness p. The corresponding phase velocity (along the surface), in other words, the phase velocity of the Rayleigh wave, is given by the inverse of the horizontal slowness as $c_R = 1/p$. When $\lambda = \mu$ or $V_P = \sqrt{3}V_S$ (Poisson solid), the phase velocity of the Rayleigh wave is given by $c_R = 0.92V_S$.

Representing the coefficient D by B' using Eq. (3.135) and $c_R = 1/p$, Eq. (3.131) is represented as

$$\begin{aligned} u_x(x,z,t) &= ipB'\left(e^{\omega\hat{\xi}z} + \frac{1-\gamma}{\gamma}e^{\omega\hat{\eta}z}\right)\exp\left[-i\omega\left(t-\frac{x}{c_R}\right)\right], \\ u_z(x,z,t) &= \hat{\xi}B'\left(e^{\omega\hat{\xi}z} + \frac{\gamma}{1-\gamma}e^{\omega\hat{\eta}z}\right)\exp\left[-i\omega\left(t-\frac{x}{c_R}\right)\right]. \end{aligned} \tag{3.137}$$

In order to display the wave propagation, only the real parts of Eq. (3.137) are taken as

$$\begin{aligned} u_x(x,z,t) &= pB'\left(e^{\omega\hat{\xi}z} + \frac{1-\gamma}{\gamma}e^{\omega\hat{\eta}z}\right)\sin\omega\left(t-\frac{x}{c_R}\right) \\ u_z(x,z,t) &= \hat{\xi}B'\left(e^{\omega\hat{\xi}z} + \frac{\gamma}{1-\gamma}e^{\omega\hat{\eta}z}\right)\cos\omega\left(t-\frac{x}{c_R}\right). \end{aligned} \tag{3.138}$$

By using Eq. (3.138), the wavefield of the Rayleigh wave is visualized in Fig. 3.21. The elastic medium was characterized by V_S= 4.0 km/s and V_P= 6.93 km/s, and the angular frequency is set as $\omega = 2\pi/T$= 0.63 s^{-1} (T= 10 s). By taking a look at the shape of the surface, we notice that the wave propagates rightward. The minimum height is located at x= 0 km at the time of 0 s. The minimum arrives at $x\sim$ 22.5 km at the time of 6 s. Hence, the (horizontal) velocity is roughly estimated to be 3.7 km/s. This estimation is reasonable because the phase velocity should be $c_R = 0.92V_S = 3.68$ km/s. Then, we expect a wavelength of 37 km (= 3.7 km/s $\times$ 10 s) from the phase velocity of 3.7 km/s and the wave period of 10 s. The corresponding wavelength is confirmed in Fig. 3.21. The particle motion at the surface ($z = 0$) indicates counterclockwise rotation. Since the wave now propagates rightward, this is retrograde motion (see Figure 3.6). On the other hand, the particle motion at the deeper part (e.g., z = 15 km) becomes prograde motion. Figure 3.22 shows the particle orbit for various depths. It is interesting that the particle motions of Rayleigh wave and tsunami are opposite to each other near the surface $z = 0$ km, although they both propagate along a free surface. The particle motion of a tsunami does not depend on depth and always shows prograde motion (Figure 3.5).

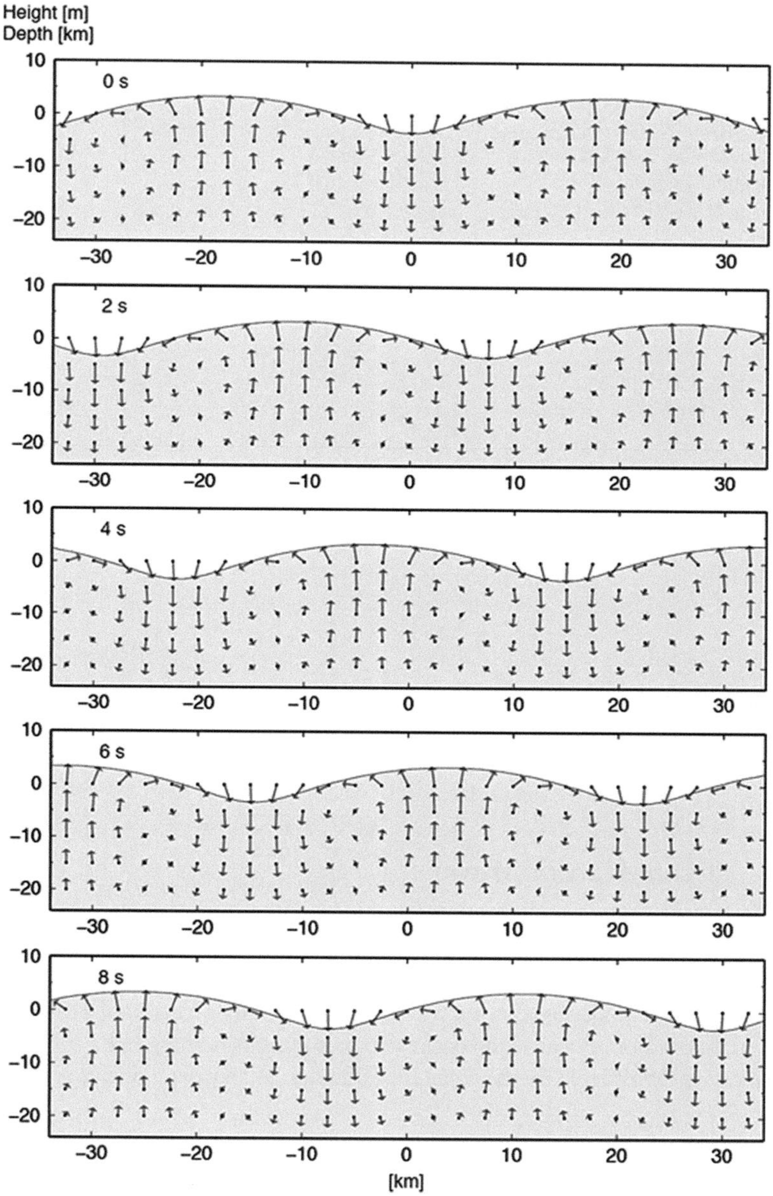

Fig. 3.21 Spatial distribution of the particle displacement caused by the Rayleigh wave for elapsed times of 0, 2, 4, 6, and 8 s. The medium is characterized by V_S= 4.0 km/s and V_P= 6.93 km/s. The angular frequency is set as $\omega = 2\pi/T$= 0.63 s^{-1} (T= 10 s)

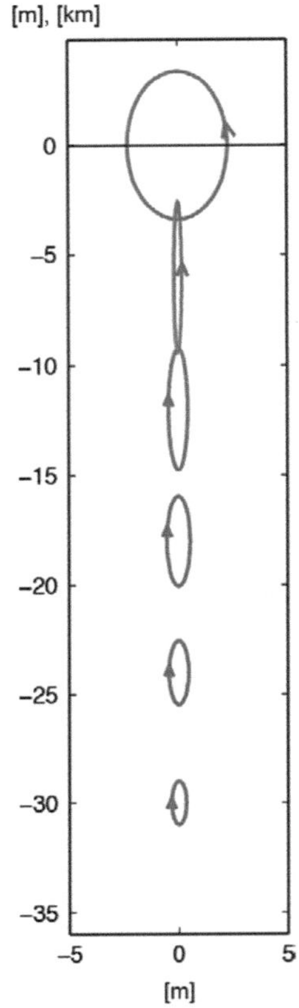

Fig. 3.22 Orbits of the particle displacement caused by the Rayleigh wave for various depths. The medium is characterized by V_S= 4.0 km/s and V_P= 6.93 km/s. The angular frequency is set as $\omega = 2\pi/T$= 0.63 s^{-1} (T= 10 s)

3.3.2 *Ocean Acoustic Waves*

When an earthquake occurs in an offshore region, a P wave that propagates in a sea layer is excited in addition to P, S, and Rayleigh waves in solid earth. The P wave in the sea layer is called an ocean acoustic wave. We derive the properties of ocean acoustic waves in this section. Because the ocean acoustic wave is nothing but a P wave in the sea layer, it is represented by the equation of motion in an elastic medium.

3.3.2.1 Acoustic Waves in Seawater

The equation of motion in an elastic medium is given by

$$\rho_0 \frac{\partial^2 u_i(x,t)}{\partial t^2} = \frac{\partial}{\partial x_k} \tau_{ik}, \tag{3.139}$$

where we set ρ_0 as the density of seawater. This is assumed to be constant. We do not consider the source term since we focus on the propagation. The constitutive law of seawater (3.22) is

$$\tau_{ij} = K\delta_{ij}u_{k,k}, \tag{3.140}$$

where K is the bulk modulus of seawater. An equivalent equation can be obtained if the rigidity is set as zero $\mu = 0$ in the constitutive law of an elastic medium (Eq. 3.17). As shown by Eq. (3.140), in the seawater, the shear stress is always zero $\tau_{ij} = 0$ $(i \neq j)$ even when there exists a large shear strain e_{ij} $(i \neq j)$. Substituting Eq. (3.140) into Eq. (3.139) and taking the divergence, we obtain

$$\Delta\,(\nabla\cdot\mathbf{u}) - \frac{1}{c_0^2}\frac{\partial^2(\nabla\cdot\mathbf{u})}{\partial t^2} = 0, \tag{3.141}$$

where a constant c_0 is defined as

$$c_0 = \sqrt{\frac{K}{\rho_0}}. \tag{3.142}$$

Equation (3.141) is a wave equation with respect to dilatation $\nabla\cdot\mathbf{u}$ with a phase velocity of c_0.

We introduce the displacement potential $\phi(\mathbf{x}, t)$ as

$$\mathbf{u} = \nabla\phi \quad \text{or} \quad u_i = \partial\phi/\partial x_i. \tag{3.143}$$

In seismology, $\phi(\mathbf{x}, t)$ is often referred to as the P wave scalar potential (e.g., Shearer 2009). By using the displacement potential, the wave equation (Eq. 3.141) is written as

$$\left(\Delta - \frac{1}{c_0^2}\frac{\partial^2}{\partial t^2}\right)\phi = 0. \tag{3.144}$$

We introduce p_e as the pressure change caused by the acoustic wave. The relation between the pressure change and the stress change is given by $\tau_{ij} = -p_e\delta_{ij}$. Then, Eq. (3.140) is represented by using the potential as

$$p_e = -K\Delta\phi. \tag{3.145}$$

By using (3.144) and (3.145), the pressure change is also represented as

$$p_e = -\rho_0\frac{\partial^2\phi}{\partial t^2}. \tag{3.146}$$

We here consider a plane wave propagating along the z-axis with the angular frequency ω and the wavenumber k, which is represented as

$$\phi(x, y, z, t) = A_0 e^{i(kz-\omega t)}. \tag{3.147}$$

Using Eq. (3.146), the pressure change accompanying the plane wave is rewritten as

$$\begin{aligned} p_e &= -\rho_0 \frac{\partial^2 \phi}{\partial t^2} \\ &= \rho_0 \omega^2 A_0 e^{i(kz-\omega t)} = \frac{\rho_0 \omega}{k}(-i\omega) ik A_0 e^{i(kz-\omega t)} \\ &= \rho_0 c_0 \frac{\partial}{\partial t}\frac{\partial \phi}{\partial z} = \rho_0 c_0 \frac{\partial u_z}{\partial t} \\ &= \rho c_0 v_z, \end{aligned} \tag{3.148}$$

where $c_0 = \omega/k$ is used. This indicates that the pressure change due to the ocean acoustic waves is proportional to the particle velocity $v_z = \partial u_z/\partial t$.

3.3.2.2 Ocean Acoustic Waves: P Waves in a Sea Layer

We then suppose a constant water layer lying on a rigid crust. Note that the actual sea bottom is not rigid but shows elasticity, but here we suppose a rigid crust for simplicity. The Cartesian coordinates shown in Fig. 3.23 are used: the z-axis runs vertically upward from the sea surface at $z = 0$, the sea depth is at $z = -h_0$, and the (x, y) axes are in the horizontal plane. In addition to the wave equation (Eq. 3.144), the displacement potential ϕ satisfies the boundary conditions at the sea surface and the sea bottom. At the sea surface, the traction is free or the pressure is zero:

$$p_e(x, y, z = 0, t) = 0. \tag{3.149}$$

At the sea bottom, the vertical displacement is zero:

$$u_z(x, y, z = -h_0, t) = 0, \tag{3.150}$$

when the crust is assumed to be rigid. These boundary conditions are rewritten using the displacement potential as

$$\left.\frac{\partial^2 \phi}{\partial t^2}\right|_{z=0} = 0, \tag{3.151}$$

and

$$\left.\frac{\partial \phi}{\partial z}\right|_{z=-h_0} = 0. \tag{3.152}$$

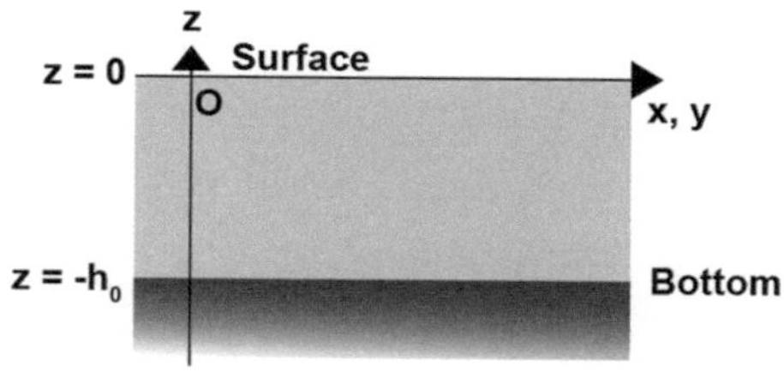

Fig. 3.23 Coordinates for propagating ocean acoustic wave

We then seek a solution of ϕ that satisfies Eqs. (3.144), (3.151), and (3.152). Suppose a monochromatic plane wave that propagates horizontally with the wavenumber vector (k_x, k_y) in the (x, y) coordinate plane as

$$\phi(x, y, z, t) = \widehat{\phi}\left(k_x, k_y, z\right)e^{i\left(k_x x + k_y y - \omega t\right)}. \tag{3.153}$$

Using Eq. (3.153), the wave Eq. (3.144) is rewritten as

$$\frac{d^2}{dz^2}\widehat{\phi} + \left(\frac{\omega^2}{c_0^2} - k^2\right)\widehat{\phi} = 0 \tag{3.154}$$

where $k = \sqrt{k_x^2 + k_y^2}$ is the amplitude of the horizontal wavenumber. Introducing κ as

$$\kappa^2 = \frac{\omega^2}{c_0^2} - k^2, \tag{3.155}$$

we rewrite (3.154) as

$$\frac{d^2}{dz^2}\widehat{\phi} + \kappa^2\widehat{\phi} = 0. \tag{3.156}$$

At first, we assume that κ^2 is positive. The general solution of Eq. (3.156) is then given by

$$\widehat{\phi} = C_1 \cos(\kappa z) + C_2 \sin(\kappa z). \tag{3.157}$$

We estimate the coefficients C_1 and C_2 so that the solution (3.157) satisfies the boundary conditions (Eqs. 3.151 and 3.152). The boundary condition at the sea surface (Eq. 3.151) gives $C_1 = 0$. The boundary condition at the sea bottom (Eq. 3.152) gives

$$\kappa C_2 \cos(\kappa h_0) = 0. \tag{3.158}$$

In order to obtain a nontrivial solution ($\kappa \neq 0$) that satisfies Eq. (3.158), we obtain

$$\kappa h_0 = \pi\left(n - \frac{1}{2}\right), \qquad n = 1, 2, \cdots. \tag{3.159}$$

Using Eqs. (3.155) and (3.159), we get the dispersion relation

$$\frac{\omega^2}{c_0^2} - k^2 = \left[\frac{\pi}{h_0}\left(n - \frac{1}{2}\right)\right]^2, n = 1, 2, \cdots. \tag{3.160}$$

We then assume that κ^2 is negative, i.e., $\kappa^2 = -\kappa'^2$ where κ' is real. In this case, the general solution is given by

$$\widehat{\phi} = D_1 \cosh(\kappa' z) + D_2 \sinh(\kappa' z).$$

We cannot find a nontrivial solution that satisfies the boundary conditions of Eqs. (3.151) and (3.152) in this case.

Equation (3.160) represents a dispersion relation ($k - \omega$ relation) for acoustic waves in the sea layer. In order for the acoustic waves to propagate horizontally, the horizontal wavenumber k needs to be real. Otherwise, the waves are trapped within the source region due to the exponential terms in Eq. (3.153). Hence, we obtain

$$k^2 = \frac{\omega^2}{c_0^2} - \left[\frac{\pi}{h_0}\left(n - \frac{1}{2}\right)\right]^2 > 0, \ \ n = 1, 2, \cdots. \tag{3.161}$$

Equation (3.161) limits the range of the angular frequency ω (>0) as follows:

$$\omega > \pi\left(n - \frac{1}{2}\right)\frac{c_0}{h_0}, n = 1, 2, \cdots. \tag{3.162}$$

This indicates that the minimum angular frequency and minimum frequency for horizontally propagating ocean acoustic waves are given by

$$\omega_{\min} = \frac{\pi\, c_0}{2\, h_0}, \ \ \text{and} \ f_{\min} = \frac{c_0}{4 h_0}, \tag{3.163}$$

respectively. In a sea layer with finite depth and a rigid sea bottom, ocean acoustic waves can propagate horizontally for a long distance only when the frequency is higher than $f_{\min}$. This critical frequency is often used in the analysis of ocean-bottom pressure records (e.g., Nosov and Kolesov 2007; Matsumoto et al. 2012; Saito 2017). Using Eq. (3.161), the phase velocity of the plane wave in the horizontal space $c = \omega/k$ is given by

$$\frac{c}{c_0} = \left[1 - \frac{1}{4}\left(n - \frac{1}{2}\right)^2 \left(\frac{c_0/h_0}{f}\right)^2\right]^{-1/2}$$
$$= \begin{cases} \infty & \text{for } f/(c_0/h_0) \to 1/2(n - 1/2) \\ 1 & \text{for } f/(c_0/h_0) \to +\infty. \end{cases} \quad (3.164)$$

Figure 3.24 shows the phase velocity of the horizontally propagating ocean acoustic wave in the case of $n = 1, 2, 3,$ and 4. This indicates that for a fixed wave frequency, there exist different and discrete phase velocities. The discrete set of phase velocities originates from the rigid sea-bottom boundary condition. Equation (3.162) indicates that the frequency f needs to be larger than $f > (c_0/h_0)\ (1/2)\ (n - 1/2)$. When the frequency f is slightly larger than $(c_0/h_0)\ (1/2)\ (n - 1/2)$, the phase velocity is infinitely larger. Setting $f = (c_0/h_0)[\ (1/2)\ (n - 1/2) + \epsilon]\ (\epsilon \ll 1)$, we obtain

$$\frac{c}{c_0} \approx \left(\frac{4\epsilon}{n - 1/2}\right)^{-1/2} \quad \text{for} \quad f/(c_0/h_0) = 1/2(n - 1/2) + \epsilon. \quad (3.165)$$

As the frequency f increases, the phase velocity decreases. It converges to the phase velocity of the ocean acoustic wave c_0 when f is much larger than c_0/h_0.

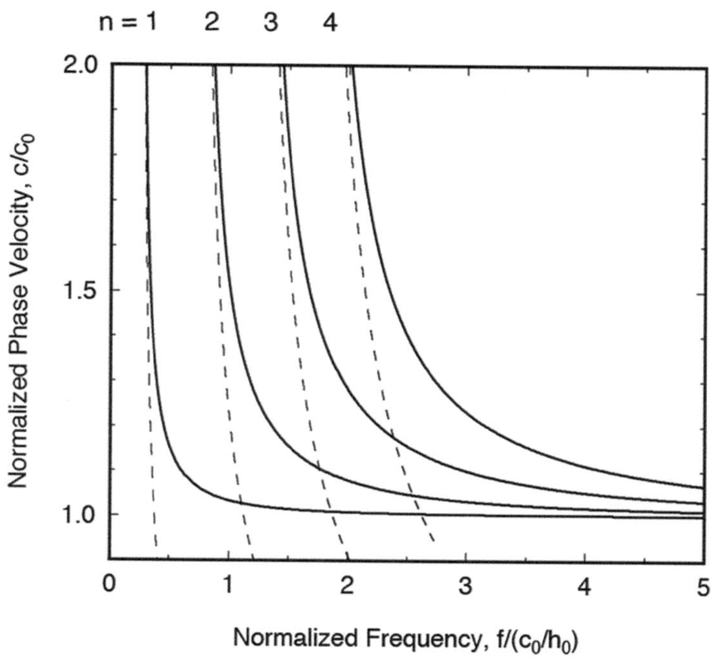

Fig. 3.24 Phase velocity of the horizontally propagating ocean acoustic wave with a sea depth of h_0 for the case of $n = 1, 2, 3,$ and 4, where $c_0 = \sqrt{K/\rho_0}$ (K is the bulk modulus and ρ_0 is the density of seawater). The asymptotic equation $c/c_0 \sim (4\epsilon/(n - 1/2))^{-1/2}$ is plotted by dashed lines

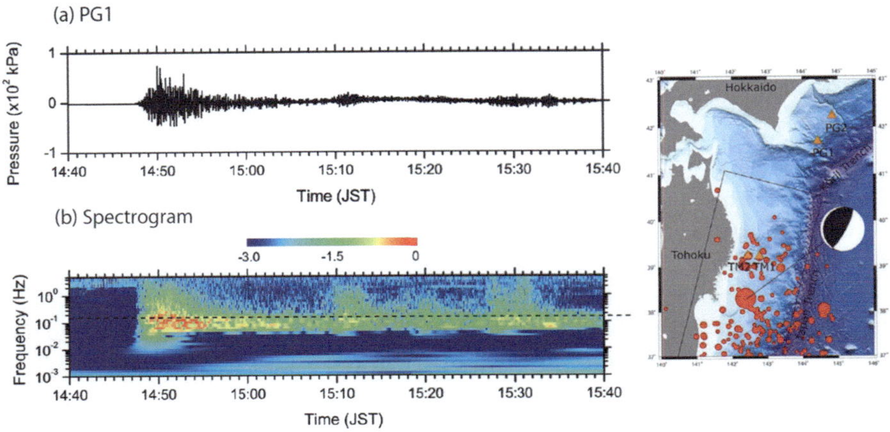

Fig. 3.25 (**a**) Ocean-bottom pressure record observed at station PG1 located off Hokkaido for the 2011 Tohoku-Oki earthquake. (**b**) Spectrogram of the ocean-bottom pressure record. The dashed line indicates the frequency predicted by $f = c_0/4h_0$ where c_0 is the phase velocity of the ocean acoustic wave and h_0 is the sea depth. (Matsumoto et al. 2012, copyright by Fuji Technology Press Ltd)

Ocean acoustic waves are usually a dominant signal in ocean-bottom pressure records of a large earthquake. Figure 3.25a shows the ocean-bottom pressure records observed at PG1 located off Hokkaido for the 2011 Tohoku-Oki earthquake (Matsumoto et al. 2012). Figure 3.25b shows a spectrogram of the pressure record. It shows a dominant wave energy around 0.1 Hz and a clear shadow of low-wave energy below 0.05 Hz. This suggests the minimum frequency for the ocean acoustic wave excitation. Matsumoto et al. (2012) noticed that the observed minimum frequency is slightly lower than the theoretical prediction of Eq. (3.163). The real sea bottom should be modeled as soft sediment rather than a rigid body, which predicts a lower minimum frequency (Nosov and Kolesov 2007).

3.3.3 *Energy Density and Energy Flux Density for Elastic Medium*

We define the energy density and the energy flux density of the elastic waves. The equation of motion is written with the velocity v_i as follows:

$$\rho \frac{\partial v_i(\mathbf{x}, t)}{\partial t} = \frac{\partial \tau_{ik}(\mathbf{x}, t)}{\partial x_k} \tag{3.166}$$

After multiplying both sides of the equations of motion by v_i

$$\rho v_i \frac{\partial v_i}{\partial t} = v_i \frac{\partial \tau_{ik}(\mathbf{x}, t)}{\partial x_k} \tag{3.167}$$

we calculate the equation as Eq. (3.168)

$$\begin{aligned}
\frac{\partial}{\partial t}\left(\frac{1}{2}\rho v_i v_i\right) &= \frac{\partial}{\partial x_j}(v_i \tau_{ij}) - \frac{\partial v_i}{\partial x_j}\tau_{ij} = \frac{\partial}{\partial x_j}(v_i \tau_{ij}) - \frac{\partial e_{ij}}{\partial t}\tau_{ij} \\
&= \frac{\partial}{\partial x_j}(v_i \tau_{ij}) - \frac{\partial e_{ij}}{\partial t} c_{ijkl} e_{kl} \\
&= \frac{\partial}{\partial x_j}(v_i \tau_{ij}) - \frac{1}{2}\left(c_{ijkl}\frac{\partial e_{ij}}{\partial t} e_{kl} + c_{ijkl}\frac{\partial e_{ij}}{\partial t} e_{kl}\right) \\
&= \frac{\partial}{\partial x_j}(v_i \tau_{ij}) - \frac{1}{2}\frac{\partial}{\partial t}(c_{ijkl} e_{ij} e_{kl}) \\
&= \frac{\partial}{\partial x_j}(v_i \tau_{ij}) - \frac{1}{2}\frac{\partial}{\partial t}(e_{ij} \tau_{ij}),
\end{aligned} \tag{3.168}$$

where the generalized Hooke's law $\tau_{ij} = c_{ijkl} e_{kl}$ (Eq. 3.15) was used. We finally obtain

$$\frac{\partial}{\partial t}\left(\frac{1}{2}\rho v_i^2 + \frac{1}{2}\tau_{ij} e_{ij}\right) = \frac{\partial}{\partial x_j}(v_i \tau_{ij}) \tag{3.169}$$

By defining the energy density as

$$E = \frac{1}{2}\rho v_i^2 + \frac{1}{2} e_{ij} \tau_{ij} \tag{3.170}$$

and the energy flux density as

$$J_j = -v_i \tau_{ij}, \tag{3.171}$$

we obtain an equation for the balance of the energy density and energy flux density as

$$\frac{\partial E(\mathbf{x}, t)}{\partial t} + \nabla \cdot \mathbf{J} = 0. \tag{3.172}$$

3.4 Waves in Continuum Medium

We have explained the waves that can exist in the continuum earth medium. As summarized in Fig. 3.26, these include *P* waves, *S* waves, Rayleigh waves, ocean acoustic waves, and tsunami. In addition to these waves, there also exist other kinds

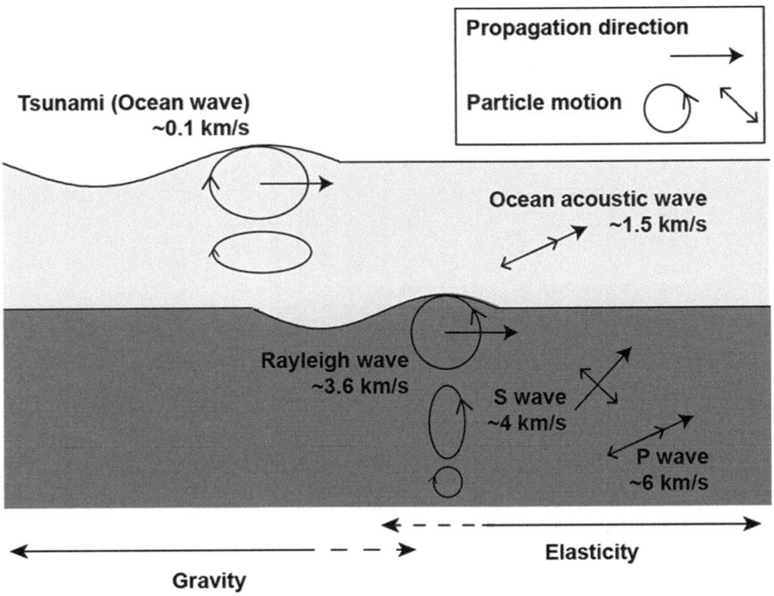

Fig. 3.26 Waves in a continuum medium treated in this chapter. The characteristics of each wave are briefly summarized as follows. The *P* wave in the crust has longitudinal particle motion with a propagation velocity ~6 km/s. The *S* wave in the crust has transverse motion with a phase velocity of ~4 km/s. The Rayleigh wave is composed of *P* and *S* waves with existence of the free surface. It shows retrograde motion at the surface with a horizontal phase velocity of ~3.6 km/s. The ocean acoustic wave is a *P* wave in the sea layer. The phase velocity is ~1.6 km/s. Horizontally propagating modes exist in the high-frequency range $f > c_0/4h_0$. The restoring force of the *P*, *S*, Rayleigh, and ocean acoustic waves is the elasticity of the medium. The tsunami shows prograde motion. The phase velocity depends on the sea depth, which is approximately given by $c = \sqrt{g_0 h}$. The restoring force of the tsunami is gravity

of waves such as Love waves in elastic earth and Kelvin waves in the ocean. Numerous excellent books have explained the details of various waves (e.g., Aki and Richards 2002; Pedlosky 2003). This book treats only the minimum set of waves that are related with tsunami generation due to an earthquake.

For the *P* wave, the restoring force is the stress generated mainly by the dilatation (volume change) of the elastic medium. The propagation velocity is represented by using Lamé parameters as $V_P = \sqrt{(\lambda + 2\mu)/\rho}$. The *P* wave propagates through the earth's crust at ~6 km/s. The particle motion of the *P* wave is in the same direction as the propagation direction. The *P* wave is also referred to as the longitudinal wave.

For the *S* wave, the restoring force is the shear stress of the elastic medium. Hence, the *S* wave does not propagate through the sea layer where the shear stress vanishes. The *S* wave does not accompany the dilatation, unlike the *P* wave. The *S* wave velocity is represented by using shear modulus as $V_S = \sqrt{\mu/\rho}$. The *S* wave propagates through the earth's crust at ~4 km/s. The particle motion of the *S* wave is perpendicular to the propagation direction. The S wave is also referred to as the transverse wave.

The Rayleigh wave is a seismic surface wave. When a free surface exists, the P and S waves with the same horizontal velocity cause a wave propagating along the surface as the Rayleigh wave. The Rayleigh wave propagates at ~3.6 km/s along the surface of the earth's crust. The particle motion of the Rayleigh wave shows retrograde motion at the surface. The amplitude of the motion decreases with increasing depth, and the particle motion changes to prograde motion in the deeper part.

The ocean acoustic wave is considered to be a P wave in the sea layer. The restoring force of the ocean acoustic wave is the compressibility of seawater. It propagates with a phase velocity of $c_0 = \sim 1.6$ km/s. Considering a sea layer with a finite depth, discrete sets of horizontally propagating modes exist. The minimum frequency of the ocean acoustic wave is given by $f_{\min} = c_0/(4h_0)$. If the wave frequency is low (i.e., the wavelength is long), a significant portion of the energy would penetrate into the crust, which should be classified as a P wave because the restoring force of this long-period wave would come from the crust rather than the sea layer.

Tsunamis propagate by gravity as the restoring force. They show prograde motion from the sea surface to the sea bottom. The phase velocity depends on the sea depth, which is approximately given by $c = \sqrt{g_0 h}$ when the wavelength is much longer than the sea depth, while it depends on the wavelength given by $c = \sqrt{g_0 h_0} \sqrt{\tanh(kh_0)/(kh_0)}$ in general. A longer-wavelength tsunami propagates faster than a shorter-wavelength tsunami.

In many practical applications, we may treat tsunami (waves due to gravity) and seismic waves (waves due to elasticity) separately because the dominant wave frequency/period is considerably different between them. This book, hence, treats the tsunami without considering the elasticity and treats seismic waves without considering the gravity. However, the elasticity of the crust and seawater compressibility affect tsunami propagation, although the effect is usually very small (e.g., Nakamura 1961; Ward 1980; Okal 1982). Recent observations proved that the elastic crust makes the tsunami propagation velocity slightly smaller than that theoretically predicted from a rigid crust (e.g., Watada et al. 2014). Also, we need to take into account gravity in addition to elasticity to synthesize long-period seismic waves (e.g., Aki and Richards 2002). In such cases, we need to employ a theory that includes gravity and elasticity simultaneously.

References

Aki K, Richards P (2002) Quantitative seismology. University Science Books, Mill Valley

Fine IV, Kulikov EA, Cherniawsky JY (2013) Japan's 2011 tsunami: characteristics of wave propagation from observations and numerical modelling. Pure Appl Geophys 170(6–8):1295–1307. https://doi.org/10.1007/s00024-012-0555-8

Hayashi Y (2010) Empirical relationship of tsunami height between offshore and coastal stations. Earth Planets Space 62(3):269–275. https://doi.org/10.5047/eps.2009.11.006

Matsumoto H, Inoue S, Ohmachi T (2012) Dynamic response of bottom water pressure due to the 2011 Tohoku Earthquake. J Disaster Res 7(7):468–475. https://doi.org/10.20965/jdr.2012.p0468

Nakamura K (1961) Velocity of long gravity waves in the ocean, Science Reports of Tohoku University, Ser. 5. Geophysics 13(3):164–173

Nosov MA, Kolesov SV (2007) Elastic oscillations of water column in the 2003 Tokachi-oki tsunami source: in-situ measurements and 3-D numerical modelling. Nat Hazards Earth Syst Sci 7:243–249

Okal EA (1982) Mode-wave equivalence and other asymptotic problems in tsunami theory. Phys Earth Planet Inter 30(1):1–11. https://doi.org/10.1016/0031-9201(82)90123-6

Pedlosky J (2003) Waves in the ocean and atmosphere: introduction to wave dynamics. Springer, Berlin

Saito M (2009a) The theory of seismic wave propagation. (in Japanese. University of Tokyo Press, Tokyo

Saito M (2009b) The theory of seismic wave propagation. TERRAPUB, Tokyo. http://www.terrapub.co.jp

Saito T (2017) Tsunami generation: validity and limitations of conventional theories. Geophys J Int 210(3):1888–1900. https://doi.org/10.1093/gji/ggx275

Sato H, Fehler MC, Maeda T (2012) Seismic wave propagation and scattering in the heterogeneous Earth: second edition. Springer, Berlin

Shearer PM (2009) Introduction to seismology. Cambridge University Press, Cambridge

Snieder R, van Wijk K (2015) A guided tour of mathematical methods for the physical sciences. Cambridge University Press, Cambridge

Ward SN (1980) Relationships of tsunami generation and an earthquake source. J Phys Earth 28:441–474

Watada S, Kusumoto S, Satake K (2014) Travel time delay and initial phase reversal of distant tsunamis coupled with the self-gravitating elastic Earth. J Geophys Res Solid Earth 119(5):4287–4310. https://doi.org/10.1002/2013JB010841

Chapter 4
Earthquakes

Abstract Describing earthquake fault motion is indispensable to understanding the mechanism of tsunami generation. Moreover, seismic waves, excited by the fault motion, are analyzed in order to estimate the magnitude and location of earthquakes. The information is used to perform rapid tsunami calculations and predictions. At the same time, we should note that seismic waves sometimes function as noise among tsunami signals. This chapter introduces earthquake seismology, which is closely related to tsunami phenomena, and illustrates a practical method of seismic wave simulation. Section 4.1 explains a mathematical representation of an earthquake fault as a point source in order to quantitatively describe the relation between the fault motion and seismic waves. Section 4.2 explains an empirical scaling law representing the fault size from small to large earthquakes. We also introduce the idea of earthquake stress change (stress drop) as a mechanism behind the scaling law. Section 4.3 illustrates the finite difference method as a practical method of seismic wave simulation. By using this numerical method, we investigate seismic waves, ocean acoustic waves, and the permanent displacement caused by an earthquake. The simulation results can be used in the simulation of tsunami propagation.

Keywords Shear dislocation · Moment tensor · Equivalent body force · Scaling law · Finite difference method

4.1 A Point Source and Seismic Wave Excitation

4.1.1 A Point Shear Dislocation Source: Equivalent Body Force and Moment Tensor

Prior to seeking a representation of the displacement field due to an earthquake fault, we consider the displacement field due to a point source. Suppose an impulsive unit body force in the n direction is applied at $\mathbf{x} = \boldsymbol{\eta}$ at the time $t = \tau$ in a homogeneous medium. This body force is represented as $\delta_{in}\delta(\mathbf{x} - \boldsymbol{\eta})\delta(t - \tau)$. Body force means force per unit volume. The displacement $G_{in}(\mathbf{x}, t; \boldsymbol{\eta}, \tau)$ in the ith direction at $\mathbf{x}$ at the

T. Saito, *Tsunami Generation and Propagation*, Springer Geophysics,
https://doi.org/10.1007/978-4-431-56850-6_4

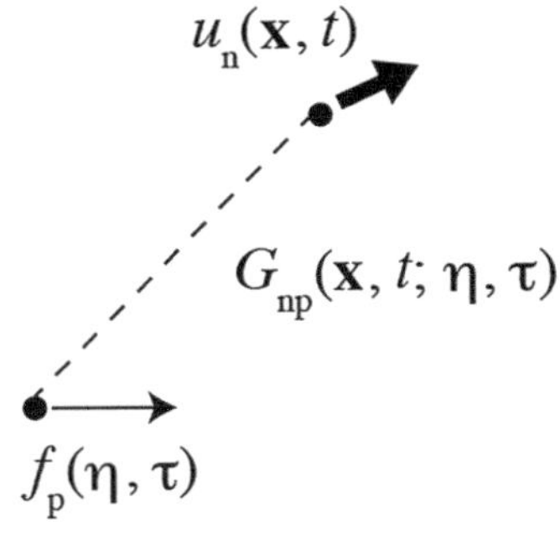

Fig. 4.1 The displacement from a point body force

time t from the impulsive unit body force satisfies the following equation derived by the equation of motion (Eq. (3.14)) and Hooke's law (Eq.(3.15)):

$$\rho\frac{\partial^2}{\partial t^2}G_{in}(\mathbf{x}, t; \boldsymbol{\eta}, \tau) - c_{ijkl}\frac{\partial^2}{\partial x_j \partial x_l}G_{kn}(\mathbf{x}, t; \boldsymbol{\eta}, \tau) = \delta_{in}\delta(\mathbf{x} - \boldsymbol{\eta})\delta(t - \tau), \tag{4.1}$$

where c_{ijkl} is elastic tensor (Eq. (3.16)). The displacement field $G_{in}(\mathbf{x}, t; \boldsymbol{\eta}, \tau)$ for the unit body force is referred to as the Green function. As shown in Fig. 4.1, for an infinite medium where the body force is distributed as $f_p(\boldsymbol{\eta}, \tau)$, the displacement field caused by the body force distribution is represented with the convolution of the source and the Green function based on the superposition principle as

$$u_n(\mathbf{x}, t) = \int_{-\infty}^{\infty} d\tau \iiint_{-\infty}^{\infty} G_{np}(\mathbf{x}, t; \boldsymbol{\eta}, \tau) f_p(\boldsymbol{\eta}, \tau) d\boldsymbol{\eta}. \tag{4.2}$$

Note that we here suppose an infinite medium but do not suppose a surface boundary of the volume.

We then consider how an earthquake fault is represented. An earthquake fault is represented by displacement discontinuity on a plane or a surface as shown in Fig. 4.2. A surface Σ has two sides Σ^+ and Σ^-, where the normal vector $\boldsymbol{\nu}$ of the surface Σ is pointing from Σ^- to Σ^+. The displacement distributions on the Σ^+ and Σ^- planes are denoted by $\mathbf{u}(\boldsymbol{\xi}, \tau)|_{\Sigma^+}$ and $\mathbf{u}(\boldsymbol{\xi}, \tau)|_{\Sigma^-}$, respectively. The displacement distributions of $\mathbf{u}(\boldsymbol{\xi}, \tau)|_{\Sigma^+}$ and $\mathbf{u}(\boldsymbol{\xi}, \tau)|_{\Sigma^-}$ are not equal, but there exists displacement discontinuity across the plane. The displacement discontinuity is given by $[u_i(\boldsymbol{\xi}, \tau)] \equiv u_i(\boldsymbol{\xi}, \tau)|_{\Sigma^+} - u_i(\boldsymbol{\xi}, \tau)|_{\Sigma^-}$. An earthquake fault is represented as the displacement discontinuity $[u_i(\boldsymbol{\xi}, \tau)]$ on a plane Σ.

Considering the displacement discontinuity on a plane as a boundary condition, one can derive the relation between the displacement discontinuity on a plane $[u_i(\boldsymbol{\xi}, \tau)]$ and the displacement at the location $\mathbf{x}$, $\mathbf{u}(\mathbf{x}, t)$. The displacement field $\mathbf{u}(\mathbf{x}, t)$ due to the displacement discontinuity on a surface Σ is given by using the Green function as

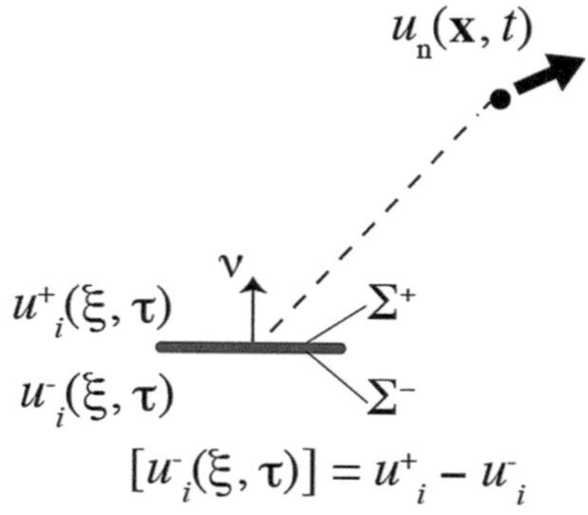

Fig. 4.2 The displacement from the displacement discontinuity on a plane. The displacement discontinuity represents an earthquake fault

$$u_n(\mathbf{x}, t) = \int_{-\infty}^{\infty} d\tau \iint_{\Sigma} [u_i(\boldsymbol{\xi}, \tau)] c_{ijpq} \nu_j \frac{\partial}{\partial \xi_q} G_{np}(\mathbf{x}, t; \boldsymbol{\xi}, \tau) d\Sigma(\boldsymbol{\xi}). \tag{4.3}$$

This derivation is not short and is not explained in this book. We use the resultant equation. Seismology textbooks (e.g., Aki and Richards 2002) illustrate the details of the derivation of Eq. (4.3). We may consider that the second tensor quantity $[u_i(\boldsymbol{\xi}, \tau)] c_{ijpq} \nu_j$ represents a source, and the derivative of the Green function with respect to the source location $\partial G_{np}/\partial \xi_q$ represents the propagation process from the source to the receiver. We may notice that $\partial G_{np}/\partial \xi_q$ appears in Eq. (4.3), while the propagation process is usually simply given by $G_{np}(\mathbf{x}, t; \boldsymbol{\xi}, \tau)$ (e.g., as in Eq. (4.2)). This implies that the displacement discontinuity (Eq. 4.3) is not simply represented by a single force vector. An earthquake fault source can be effectively given by force couples (e.g., Aki and Richards 2002).

As a source, the moment density tensor m_{pq} is introduced as

$$m_{pq}(\boldsymbol{\xi}, \tau) = [u_i(\boldsymbol{\xi}, \tau)] c_{ijpq} \nu_j. \tag{4.4}$$

Equation (4.3) is then rewritten as a convolution of the source term of m_{pq} and the propagation term $\partial G_{np}/\partial \xi_q$ as

$$u_n(\mathbf{x}, t) = \int_{-\infty}^{\infty} d\tau \iint_{\Sigma} \left[\frac{\partial}{\partial \xi_q} G_{np}(\mathbf{x}, t; \boldsymbol{\xi}, \tau) \right] m_{pq}(\boldsymbol{\xi}, \tau) d\Sigma(\boldsymbol{\xi}). \tag{4.5}$$

This equation represents the displacement field due to the moment density tensor.

When the elastic medium is isotropic and homogeneous, the elastic constant c_{ijpq} is represented with Lamé parameters as $c_{ijpq} = \lambda \delta_{ij} \delta_{pq} + \mu(\delta_{ip} \delta_{jq} + \delta_{iq} \delta_{jp})$ (Eq. (3.16)). The moment density tensor (Eq. (4.4)) is then given by

$$m_{pq}(\boldsymbol{\xi}, \tau) = \lambda \nu_k [u_k(\boldsymbol{\xi}, \tau)] \delta_{pq} + \mu \{ \nu_p [u_q(\boldsymbol{\xi}, \tau)] + \nu_q [u_p(\boldsymbol{\xi}, \tau)] \}. \tag{4.6}$$

As a model of a typical earthquake, the direction of the displacement discontinuity, or the direction of the slip on the fault, is parallel to the fault plane. This type of displacement discontinuity is referred to as shear dislocation. In that case, the slip vector $[\mathbf{u}]$ is perpendicular to the fault normal vector $\boldsymbol{\nu}$. The scalar product $\boldsymbol{\nu} \cdot [\mathbf{u}] = \nu_k \cdot [u_k]$ of the first term in Eq. (4.6) then becomes zero. The moment density tensor of shear dislocation in an isotropic homogeneous medium is given by

$$m_{pq}(\boldsymbol{\xi}, \tau) = \mu\{\nu_p[u_q(\boldsymbol{\xi}, \tau)] + \nu_q[u_p(\boldsymbol{\xi}, \tau)]\}. \tag{4.7}$$

In Eq. (4.5), the displacement $u_n(\mathbf{x}, t)$ is represented by the integration over an area Σ. By using the delta function, the areal integration is changed to a volume integration as

$$u_n(\mathbf{x}, t) = \int_{-\infty}^{\infty} d\tau \iiint_{-\infty}^{\infty} d\boldsymbol{\eta} \left[\frac{\partial}{\partial \eta_q} G_{np}(\mathbf{x}, t; \boldsymbol{\eta}, \tau)\right] \iint_{\Sigma} m_{pq}(\boldsymbol{\xi}, \tau)\delta(\boldsymbol{\eta} - \boldsymbol{\xi})d\Sigma(\boldsymbol{\xi}). \tag{4.8}$$

By integrating Eq. (4.8) with respect to $\boldsymbol{\eta}$ in part, we obtain the equation

$$u_n(\mathbf{x}, t) = -\int_{-\infty}^{\infty} d\tau \iiint_{-\infty}^{\infty} d\boldsymbol{\eta} G_{np}(\mathbf{x}, t; \boldsymbol{\eta}, \tau) \iint_{\Sigma} m_{pq}(\boldsymbol{\xi}, \tau)\frac{\partial}{\partial \eta_q}\delta(\boldsymbol{\eta} - \boldsymbol{\xi})d\Sigma(\boldsymbol{\xi}). \tag{4.9}$$

Comparing Eqs. (4.9) and (4.2), we found that the term

$$f_p(\boldsymbol{\eta}, t) = -\iint_{\Sigma} m_{pq}(\boldsymbol{\xi}, t)\frac{\partial}{\partial \eta_q}\delta(\boldsymbol{\eta} - \boldsymbol{\xi})d\Sigma(\boldsymbol{\xi}) \tag{4.10}$$

functions as the body force distributed in space and time. This indicates that the displacement discontinuity is equivalent to the body force distribution defined in Eq. (4.10). The force is referred to as the equivalent body force (Maruyama 1963; Burridge and Knopoff 1964).

We consider that the characteristic length of the fault plane L is small. One may consider L as the fault length or radius, for example. When the fault scale L is much smaller than the wavelength λ of the seismic wave ($L \ll \lambda$), the seismic waves radiating from each subfault $d\Sigma$ arrive at an observation point with the same phase (in phase). In this case, we may consider the finite size of the earthquake fault plane as a point source by neglecting the variation of $\partial G_{np}/\partial \xi_q$ with respect to the location on the fault $\boldsymbol{\xi}$. Then, we simplify Eq. (4.3) by neglecting the integration of $\partial G_{np}/\partial \xi_q$ over the fault plane Σ:

$$
\begin{aligned}
u_n(\mathbf{x},t) &\approx \int_{-\infty}^{\infty} d\tau \left[\frac{\partial}{\partial \xi_q} G_{np}(\mathbf{x},t;\boldsymbol{\xi},\tau)\right]_{\boldsymbol{\xi}=\boldsymbol{\xi}_e} \iint_{\Sigma} m_{pq}(\boldsymbol{\xi},\tau)d\Sigma(\boldsymbol{\xi}) \\
&= \int_{-\infty}^{\infty} d\tau \left[\frac{\partial}{\partial \xi_q} G_{np}(\mathbf{x},t;\boldsymbol{\xi}_e,\tau)\right] M_{pq}(\boldsymbol{\xi}_e,\tau) \\
&= \int_{-\infty}^{\infty} d\tau \left[\frac{\partial}{\partial \xi_q} G_{np}(\mathbf{x},t-\tau;\boldsymbol{\xi}_e,0)\right] M_{pq}(\boldsymbol{\xi}_e,\tau) \\
&= M_{pq}(\boldsymbol{\xi}_e,t) * \frac{\partial}{\partial \xi_q} G_{np}(\mathbf{x},t;\boldsymbol{\xi}_e,\tau),
\end{aligned}
\tag{4.11}
$$

where $\boldsymbol{\xi}_e$ represents the centroid of the fault, and moment tensor M_{pq} is introduced as an integration of the moment density tensor on the fault plane:

$$
\begin{aligned}
M_{pq}(\boldsymbol{\xi}_e,\tau) &= \iint_{\Sigma} m_{pq}(\boldsymbol{\xi},\tau)d\Sigma(\boldsymbol{\xi}) \\
&= \mu \iint_{\Sigma} \{\nu_p[u_q(\boldsymbol{\xi},\tau)] + \nu_q[u_p(\boldsymbol{\xi},\tau)]\}d\Sigma(\boldsymbol{\xi}).
\end{aligned}
\tag{4.12}
$$

Thus, the displacement field $\mathbf{u}$ is then simply given by the convolution of the moment tensor and the propagation process $\partial G_{np}/\partial \xi_q$ as

$$
u_n(\mathbf{x},t) = M_{pq}(\boldsymbol{\xi}_e,t) * \frac{\partial}{\partial \xi_q} G_{np}(\mathbf{x},t;\boldsymbol{\xi}_e,0). \tag{4.13}
$$

The average slip over the fault is given by

$$
\overline{[u_i(t)]} \equiv \frac{\int_{\Sigma}[u_i(\xi,\tau)]d\Sigma(\boldsymbol{\xi})}{S}, \tag{4.14}
$$

where S is the area of the fault plane Σ. Then the moment tensor (Eq. 4.12) is calculated as

$$
M_{pq}(t) = \mu S\left\{\nu_p\overline{[u_q(t)]} + \nu_q\overline{[u_p(t)]}\right\}. \tag{4.15}
$$

For example, when the plane Σ lies in the plane $z = 0$ with a normal vector of $\boldsymbol{\nu} = (0,0,1)$ and the slip occurs only in the x direction ($\overline{[\mathbf{u}(t)]} = (\bar{u}(t),0,0)$), the moment tensor becomes

$$
\mathbf{M} = \begin{pmatrix} 0 & 0 & \mu\bar{u}(t)S \\ 0 & 0 & 0 \\ \mu\bar{u}(t)S & 0 & 0 \end{pmatrix} = \begin{pmatrix} 0 & 0 & M(t) \\ 0 & 0 & 0 \\ M(t) & 0 & 0 \end{pmatrix}. \tag{4.16}
$$

The quantity appearing in the component of the moment tensor $M(t) = \mu\bar{u}(t)S$ is referred to as the moment time function. In particular, when the final displacement discontinuity, or total slip, is given by $D = \bar{u}(t = \infty)$, the corresponding moment time function becomes

$$M_0 = M(t = \infty) = \mu DS. \tag{4.17}$$

This is referred to as the seismic moment. The seismic moment represents the magnitude of an earthquake. This is one of the most important quantities used to characterize earthquakes. Actually, the seismic moment is related to the moment magnitude M_W as follows:

$$\log M_0\ [\mathrm{N \cdot m}] = 1.5M_\mathrm{W} + 9.1. \tag{4.18}$$

In the next section (Sect. 4.1.2) below, we show an analytical solution for the displacement field $u_n(\mathbf{x}, t)$ from a point shear dislocation source. On the other hand, we will consider the numerical simulation of seismic wave propagation from a point source in *4.3. Seismic Wave Simulation*. For the numerical simulation, it is useful to represent a point source as an equivalent body force. Equation (4.11) is calculated as

$$\begin{aligned} u_n(\mathbf{x}, t) &= \int_{-\infty}^{\infty} d\tau \left[\frac{\partial}{\partial \xi_q} G_{np}(\mathbf{x}, t - \tau; \boldsymbol{\xi}, 0)\right] M_{pq}(\boldsymbol{\xi}, \tau) \\ &= \int_{-\infty}^{\infty} d\tau \iiint_{-\infty}^{\infty} d\boldsymbol{\eta} \left[\frac{\partial}{\partial \eta_q} G_{np}(\mathbf{x}, t - \tau; \boldsymbol{\eta}, 0)\right] M_{pq}(\boldsymbol{\xi}, \tau)\delta(\boldsymbol{\eta} - \boldsymbol{\xi}) \\ &= -\int_{-\infty}^{\infty} d\tau \iiint_{-\infty}^{\infty} d\boldsymbol{\eta} G_{np}(\mathbf{x}, t - \tau; \boldsymbol{\eta}, 0) \frac{\partial}{\partial \eta_q} \left[M_{pq}(\boldsymbol{\xi}, \tau)\delta(\boldsymbol{\eta} - \boldsymbol{\xi})\right]. \end{aligned} \tag{4.19}$$

Then, comparing this with Eq. (4.2), we obtain the body force equivalent to the point moment tensor source as

$$f_p(\mathbf{x}, t) = -\frac{\partial}{\partial x_q} \left[M_{pq}(\boldsymbol{\xi}, t)\delta(\mathbf{x} - \boldsymbol{\xi})\right], \tag{4.20}$$

Note that Eq. (4.10) is for a finite fault size, but Eq. (4.20) is for a point source. We will use Eq. (4.20) in *4.3. Seismic Wave Simulation*.

4.1.2 Displacement Field from a Point Source

When the elastic constant is given by Lamé parameters as $c_{ijpq} = \lambda\delta_{ij}\delta_{pq} + \mu(\delta_{ip}\delta_{jq} + \delta_{iq}\delta_{jp})$, the Green function in a homogeneous, isotropic, infinite elastic medium that satisfies Eq. (4.1) is given by

$$G_{jk}(\mathbf{x}, t; \boldsymbol{\xi}, \tau) = \frac{\gamma_j \gamma_k}{4\pi\rho V_P^2 r} \delta\left(t - \tau - \frac{r}{V_P}\right) + \frac{\delta_{jk} - \gamma_j \gamma_k}{4\pi\rho V_S^2 r} \delta\left(t - \tau - \frac{r}{V_S}\right) + \frac{3\gamma_j \gamma_k - \delta_{jk}}{4\pi\rho r^3} (t - \tau) \left[H\left(t - \tau - \frac{r}{V_P}\right) - H\left(t - \tau - \frac{r}{V_S}\right)\right], \tag{4.21}$$

and

$$V_P = \sqrt{\frac{\lambda + 2\mu}{\rho}} \text{ and } V_S = \sqrt{\frac{\mu}{\rho}}, \tag{4.22}$$

where r is the distance from the source to the receiver as $r = |\mathbf{x} - \boldsymbol{\xi}|$ and γ_j is the jth component of the direction cosine for the vector $\mathbf{x} - \boldsymbol{\xi}$:

$$\gamma_j = \frac{x_j - \xi_j}{r} \tag{4.23}$$

and $H(t)$ is a step function defined as $H(t) = 1$ when $t > 0$ and $H(t) = 0$ when $t < 0$. Equation (4.21) is obtained if we set $X_0(t)$ as $X_0(t) = \delta(t)$ in Eq. (4.23) of Aki and Richards (2002) using a formal equation:

$$\begin{aligned} f(t) &= \int_{r/V_P}^{r/V_S} \tau\delta(t - \tau) d\tau \\ &= \begin{cases} t \text{ for } r/V_P \le t \le r/V_S \\ 0 \text{ for others} \end{cases} \\ &= t\left[H\left(t - \frac{r}{V_P}\right) - H\left(t - \frac{r}{V_S}\right)\right]. \end{aligned}$$

When the slip on the fault is $[u_i] = \bar{u}(t)s_i$ (s_i is a unit vector representing slip direction), substituting the Green function of (4.21) and the moment tensor (4.15) into (4.13), we obtain displacement field $u_n(\mathbf{x}, t)$ as

$$\begin{aligned} u_n = &\frac{2(\boldsymbol{\gamma} \cdot \mathbf{s})(\boldsymbol{\gamma} \cdot \boldsymbol{\nu})\gamma_n}{4\pi\rho V_P^3 r} \mu S \dot{\bar{u}}\left(t - \frac{r}{V_P}\right) \\ &+ \frac{-2(\boldsymbol{\gamma} \cdot \mathbf{s})(\boldsymbol{\gamma} \cdot \boldsymbol{\nu})\gamma_n + (\boldsymbol{\gamma} \cdot \mathbf{s})\nu_n + (\boldsymbol{\gamma} \cdot \boldsymbol{\nu})s_n}{4\pi\rho V_S^3 r} \mu S \dot{\bar{u}}\left(t - \frac{r}{V_S}\right) \\ &+ \frac{12(\boldsymbol{\gamma} \cdot \mathbf{s})(\boldsymbol{\gamma} \cdot \boldsymbol{\nu})\gamma_n - 2(\boldsymbol{\gamma} \cdot \mathbf{s})\nu_n - 2(\boldsymbol{\gamma} \cdot \boldsymbol{\nu})s_n}{4\pi\rho V_P^2 r^2} \mu S \bar{u}\left(t - \frac{r}{V_P}\right) \\ &+ \frac{-12(\boldsymbol{\gamma} \cdot \mathbf{s})(\boldsymbol{\gamma} \cdot \boldsymbol{\nu})\gamma_n + 3(\boldsymbol{\gamma} \cdot \mathbf{s})\nu_n + 3(\boldsymbol{\gamma} \cdot \boldsymbol{\nu})s_n}{4\pi\rho V_P^2 r^2} \mu S \bar{u}\left(t - \frac{r}{V_S}\right) \\ &+ \frac{30(\boldsymbol{\gamma} \cdot \mathbf{s})(\boldsymbol{\gamma} \cdot \boldsymbol{\nu})\gamma_n - 6(\boldsymbol{\gamma} \cdot \mathbf{s})\nu_n - 6(\boldsymbol{\gamma} \cdot \boldsymbol{\nu})s_n}{4\pi\rho r^4} \int_{r/V_P}^{r/V_S} \tau \mu S \bar{u}(t - \tau) d\tau. \end{aligned} \tag{4.24}$$

Using the vector representation and the moment time function:

$$M(t) = \mu\bar{u}(t)S, \tag{4.25}$$

Eq. (4.24) is rewritten as

$$\mathbf{u}(\mathbf{x},t) = \frac{\mathbf{A}^{\mathbf{FP}}}{4\pi\rho V_P^3 r}\dot{M}\left(t-\frac{r}{V_P}\right) + \frac{\mathbf{A}^{\mathbf{FS}}}{4\pi\rho V_S^3 r}\dot{M}\left(t-\frac{r}{V_S}\right) + \frac{\mathbf{A}^{\mathbf{IP}}}{4\pi\rho V_P^2 r^2}M\left(t-\frac{r}{V_P}\right) + \frac{\mathbf{A}^{\mathbf{IS}}}{4\pi\rho V_S^2 r^2}M\left(t-\frac{r}{V_S}\right) + \frac{\mathbf{A}^{\mathbf{N}}}{4\pi\rho r^4}\int_{r/V_P}^{r/V_S}\tau M(t-\tau)d\tau, \tag{4.26}$$

where the vector coefficients are

$$\begin{aligned}
\mathbf{A}^{\mathbf{FP}} &= 2(\boldsymbol{\gamma}\cdot\mathbf{s})(\boldsymbol{\gamma}\cdot\boldsymbol{\nu})\boldsymbol{\gamma} \\
\mathbf{A}^{\mathbf{FS}} &= -2(\boldsymbol{\gamma}\cdot\mathbf{s})(\boldsymbol{\gamma}\cdot\boldsymbol{\nu})\boldsymbol{\gamma} + (\boldsymbol{\gamma}\cdot\mathbf{s})\boldsymbol{\nu} + (\boldsymbol{\gamma}\cdot\boldsymbol{\nu})\mathbf{s} \\
\mathbf{A}^{\mathbf{IP}} &= 12(\boldsymbol{\gamma}\cdot\mathbf{s})(\boldsymbol{\gamma}\cdot\boldsymbol{\nu})\boldsymbol{\gamma} - 2(\boldsymbol{\gamma}\cdot\mathbf{s})\boldsymbol{\nu} - 2(\boldsymbol{\gamma}\cdot\boldsymbol{\nu})\mathbf{s} \\
\mathbf{A}^{\mathbf{IS}} &= -12(\boldsymbol{\gamma}\cdot\mathbf{s})(\boldsymbol{\gamma}\cdot\boldsymbol{\nu})\boldsymbol{\gamma} + 3(\boldsymbol{\gamma}\cdot\mathbf{s})\boldsymbol{\nu} + 3(\boldsymbol{\gamma}\cdot\boldsymbol{\nu})\mathbf{s} \\
\mathbf{A}^{\mathbf{N}} &= 30(\boldsymbol{\gamma}\cdot\mathbf{s})(\boldsymbol{\gamma}\cdot\boldsymbol{\nu})\boldsymbol{\gamma} - 6(\boldsymbol{\gamma}\cdot\mathbf{s})\boldsymbol{\nu} - 6(\boldsymbol{\gamma}\cdot\boldsymbol{\nu})\mathbf{s}.
\end{aligned} \tag{4.27}$$

Equation (4.26) represents the displacement field excited by a point shear dislocation source. Each term has each vector coefficient, such as $\mathbf{A}^{\mathbf{FP}}$ and $\mathbf{A}^{\mathbf{FS}}$. The terms including $\mathbf{A}^{\mathbf{FP}}$ and $\mathbf{A}^{\mathbf{FS}}$ are called far-field terms (far-field P and S, respectively) because Eq. (4.26) shows that those terms decay slowly as $1/r$ and become dominant far from the source. The terms including $\mathbf{A}^{\mathbf{IP}}$ and $\mathbf{A}^{\mathbf{IS}}$ are called intermediate terms. The terms including $\mathbf{A}^{\mathbf{N}}$ are called near-field terms. These coefficients are given by using the unit vectors of $\boldsymbol{\nu}$, $\boldsymbol{\gamma}$, and $\mathbf{s}$ representing the fault normal direction, source-to-receiver direction, and slip direction, respectively. Therefore, these vector coefficients represent a radiation pattern from a source or functions depending on the direction. Also, each term represents different characteristics of the displacement field.

4.1.2.1 Permanent Displacement Field

The displacement field due to a shear dislocation motion is constituted by various terms (Eq. (4.26)). We first focus on the displacement that persists after enough time elapses. By substituting $t = \infty$ into Eq. (4.26), we obtain

$$
\begin{aligned}
\mathbf{u}(\mathbf{x}, t=\infty) &= \frac{\mathbf{A}^{\mathrm{IP}}}{4\pi\rho V_P^2 r^2} M(\infty) + \frac{\mathbf{A}^{\mathrm{IS}}}{4\pi\rho V_S^2 r^2} M(\infty) + \frac{\mathbf{A}^{\mathrm{N}}}{4\pi\rho r^4} \int_{r/V_P}^{r/V_S} \tau M(\infty) d\tau \\
&= \frac{\mathbf{A}^{\mathrm{IP}}}{4\pi\rho V_P^2 r^2} M(\infty) + \frac{\mathbf{A}^{\mathrm{IS}}}{4\pi\rho V_S^2 r^2} M(\infty) \\
&\qquad + \frac{\mathbf{A}^{\mathrm{N}}}{4\pi\rho r^4} \frac{M(\infty)}{2} \left[\left(\frac{r}{V_S} \right)^2 - \left(\frac{r}{V_P} \right)^2 \right] \\
&= \frac{M(\infty)}{4\pi\rho r^2} \left[\frac{\mathbf{A}^{\mathrm{IP}}}{V_P^2} + \frac{\mathbf{A}^{\mathrm{IS}}}{V_S^2} + \frac{\mathbf{A}^{\mathrm{N}}}{2} \left(\frac{1}{V_S^2} - \frac{1}{V_P^2} \right) \right] \\
&\propto \frac{M_0}{r^2}.
\end{aligned}
\tag{4.28}
$$

We may consider that the moment time function becomes a non-zero constant $M(\infty) = M_0$ and the time derivative of the moment time function becomes zero $\dot{M}(\infty)$ when enough time elapses. The displacement field of Eq. (4.28) is referred to as permanent displacement. Equation (4.28) indicates that the permanent displacement field $\mathbf{u}(\mathbf{x}, t=\infty)$ is proportional to the seismic moment M_0. Since it decreases rapidly with an increasing hypocentral distance according to r^{-2}, the permanent displacement appears only near the hypocenter. We can rewrite (4.28) as

$$
\begin{aligned}
\mathbf{u}(\mathbf{x}, t=\infty) &= \frac{M(\infty)}{4\pi\rho r^2} \left[\frac{\mathbf{A}^{\mathrm{IP}}}{V_P^2} + \frac{\mathbf{A}^{\mathrm{IS}}}{V_S^2} + \frac{\mathbf{A}^{\mathrm{N}}}{2} \left(\frac{1}{V_S^2} - \frac{1}{V_P^2} \right) \right] \\
&= \frac{DS}{4\pi r^2} \left[\frac{V_S^2}{V_P^2} \mathbf{A}^{\mathrm{IP}} + \mathbf{A}^{\mathrm{IS}} + \frac{\mathbf{A}^{\mathrm{N}}}{2} \left(1 - \frac{V_S^2}{V_P^2} \right) \right],
\end{aligned}
\tag{4.29}
$$

where D is the total slip and S is the area of the fault plane. The medium parameter of V_P/V_S appears in the permanent displacement.

4.1.2.2 Propagating Seismic Waves: Far-Field Terms

Equation (4.26) indicates that the first two terms decrease with an increasing hypocentral distance according to r^{-1} more slowly than other terms. The first two terms, namely,

$$
\mathbf{u}^{\mathrm{Far}}(\mathbf{x}, t) = \frac{\mathbf{A}^{\mathrm{FP}}}{4\pi\rho V_P^3 r} \dot{M}\left(t - \frac{r}{V_P} \right) + \frac{\mathbf{A}^{\mathrm{FS}}}{4\pi\rho V_S^3 r} \dot{M}\left(t - \frac{r}{V_S} \right) \tag{4.30}
$$

become more dominant as the hypocentral distance r gets longer. These terms are referred to as far-field terms. The first and second terms in Eq. (4.30) correspond to the P and S waves, respectively, and represent the propagation with velocities of V_P and V_S. Equation (4.30) indicates that each far-field term is proportional to the time derivative of the moment time function:

$$\dot{M}(t)$$

This is called the moment rate function. Both the P and S waves show the same function shape as the moment rated function $\dot{M}(t)$.

Equation (4.30) indicates that the P and S waves have radiation patterns represented by $\mathbf{A}^{\mathrm{FP}}$ and $\mathbf{A}^{\mathrm{FS}}$, respectively, as defined in Eq. (4.27). Consider the case in which the fault normal is given by $\boldsymbol{\nu} = (1, 0, 0)$ and the slip direction is $\mathbf{s} = (0, 1, 0)$. If we use the spherical coordinates (r, θ, ϕ) as shown in Fig. 4.3.

$$\begin{aligned}\mathbf{e}_r &= \sin\theta\cos\phi\,\mathbf{e}_x + \sin\theta\sin\phi\,\mathbf{e}_y + \cos\theta\,\mathbf{e}_z = \boldsymbol{\gamma},\\ \mathbf{e}_\theta &= \cos\theta\cos\phi\,\mathbf{e}_x + \cos\theta\sin\phi\,\mathbf{e}_y - \sin\theta\,\mathbf{e}_z,\\ \mathbf{e}_\phi &= -\sin\phi\,\mathbf{e}_x + \cos\phi\,\mathbf{e}_y,\end{aligned} \tag{4.31}$$

the radiation patterns are represented as

$$\begin{aligned}\mathbf{A}^{FP} &= 2\mathbf{e}_r \sin\theta\sin\phi\sin\theta\cos\phi,\\ \mathbf{A}^{\mathrm{FS}} &= \frac{1}{2}\sin 2\theta\sin 2\phi\,\mathbf{e}_\theta + \sin\theta\cos 2\phi\,\mathbf{e}_\phi.\end{aligned} \tag{4.32}$$

Figure 4.4 plots the radiation pattern of the P and S waves from the shear dislocation source (Eq. 4.32). The P wave shows a four-lobe pattern. The $\mathbf{e}_\phi$ component of the S wave also shows a four-lobe pattern, and the $\mathbf{e}_\theta$ component shows eight lobes. The radiation pattern of both the P and S waves is a highly symmetric pattern. Actually, even if we take the normal vector $\boldsymbol{\nu} = (0, 1, 0)$ and the slip direction vector $\mathbf{s} = (1, 0, 0)$ instead of $\boldsymbol{\nu} = (1, 0, 0)$ and $\mathbf{s} = (0, 1, 0)$, we obtain the identical radiation patterns $\mathbf{A}^{\mathrm{FP}}$ and $\mathbf{A}^{\mathrm{FS}}$ for the P and S waves, respectively. This means that it is impossible to determine the fault plane uniquely from the two fault planes by performing far-field P and S wave analysis.

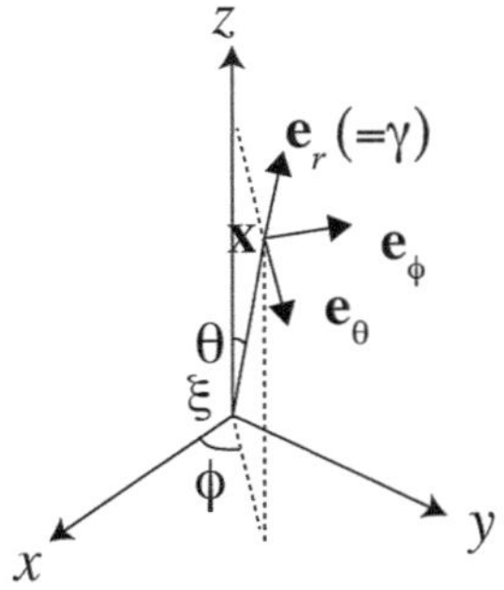

Fig. 4.3 The location of the point source $\boldsymbol{\xi}$ is set as the origin. The spherical coordinates (r, θ, ϕ) are introduced. $(\mathbf{e}_r, \mathbf{e}_\theta, \mathbf{e}_\phi)$ are unit base vectors

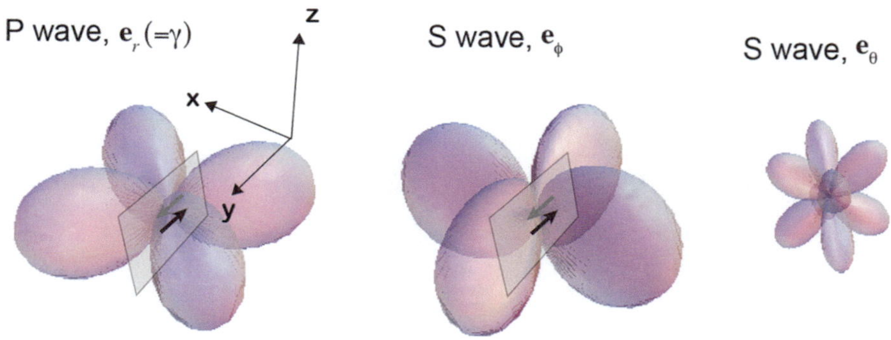

Fig. 4.4 Radiation patterns of P and S waves from a point shear dislocation source. The amplitudes of the coefficients of $\mathbf{e}_r$, $\boldsymbol{e}_\theta$, and $\mathbf{e}_\phi$ in Eq. (4.32) are plotted

The ratio of the S wave amplitude to the P wave amplitude is given by $(V_P/V_S)^3$~5 if a Poisson solid is assumed, $V_P/V_S = \sqrt{3}$, indicating that the S wave is dominant over the P wave for a shear dislocation source. We also note that the far-field terms become zero when enough time has elapsed:

$$\mathbf{u}^{\mathrm{Far}}(\mathbf{x}, \infty) = 0, \tag{4.33}$$

because the moment time function becomes constant $\dot{M}(t \to \infty) = 0$ when the earthquake rupture ends.

4.1.2.3 Centroid Moment Tensor (CMT) Solution

As explained above, far-field P and S waves each have specific radiation patterns depending on the earthquake fault direction (strike, dip, rake), and the displacement waveforms are proportional to the moment rate function of the fault motion. Therefore, if we analyze the observed seismograms based on a point shear dislocation model, we can estimate the parameters of a point shear dislocation source. Because the location of the point source corresponds to a centroid of the earthquake fault motion, the point source is referred to as the centroid moment tensor (CMT) solution. Aki (1966) estimated the seismic moment of a large earthquake by analyzing the seismograms of the 1964 Niigata earthquake, Japan. His study proved the usefulness of modeling seismograms to obtain the parameters of the earthquake fault motion.

In actual situations, seismometers observe surface waves in addition to the P and S waves, and the subsurface structure is not uniform but inhomogeneous. Recent studies usually supposed a 1D or 3D subsurface structure and synthesized seismograms including surface waves and other waves for the estimation of the CMT solutions. Automated analysis of seismograms recorded globally and regionally is currently being performed. The Global CMT project has reported the CMT solutions of earthquakes that have occurred throughout the world (Fig. 4.5). Those estimated CMT solutions provide fundamental information for earthquake science.

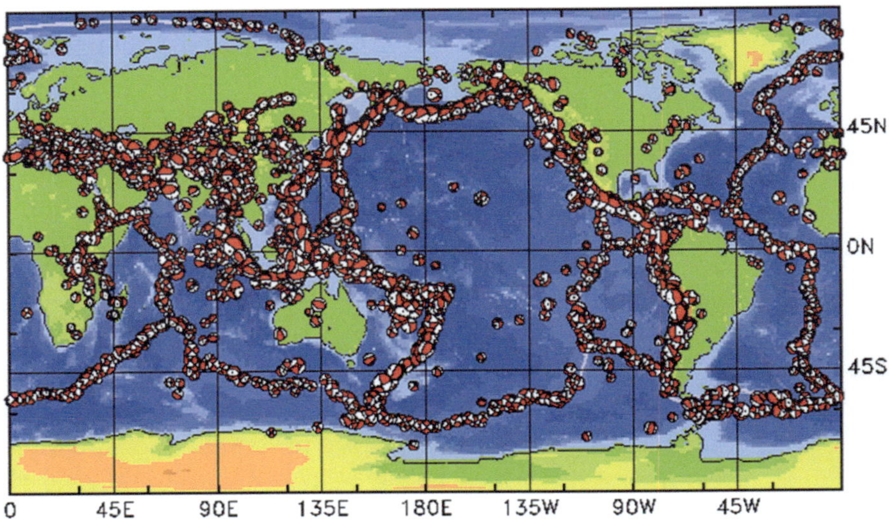

Fig. 4.5 The centroid moment tensor (CMT) solutions (1976–2005) determined by the Global CMT project. This was operated as the Harvard CMT project from 1982 to 2006. This gives a catalog of CMT solutions for large earthquakes ($M > 5$) around the world. (Cited from www.globalcmt.org, accessed 17 Aug 2017)

The CMT solution can be used for rapid tsunami evaluation. It provides information on the fault orientation and its moment $M_0 = \mu DS$ where D is the total slip and S is the area of the fault plane. Using this information, we roughly calculate the permanent sea-bottom displacement as a tsunami source. However, note that the CMT solution cannot determine a unique fault plane but rather two fault plane candidates, nor can it give the fault area S and dislocation D independently. A scaling law of earthquake faults may be useful as a reasonable method to estimate the fault area. We will explain the scaling law in the following sections.

4.2 Finite Fault Model and Scaling Law

When a huge earthquake occurs, we analyze seismograms recorded at some seismic stations to infer the earthquake's source process. The seismic moment can be stably estimated by automated seismogram analysis by assuming a point source model. This is because a point source model (explained in Sect. 4.1) works nicely for synthesizing long-period far-field P, S, and surface waves of which wavelength is longer than the earthquake fault length.

On the other hand, it is usually not so straightforward to estimate the earthquake fault size ($S = LW$, where L and W are the length and width of the fault, respectively) by an automated analysis. Equation (4.56) indicates that the moment tensor per unit volume $M_{pq}^{n+1/2}\delta(\mathbf{x} - \boldsymbol{\xi})$ functions as a stress change. By carefully analyzing the seismograms, such as by analyzing the rupture directivity effects and pulse shape carefully, we can estimate the fault size S and the slip D separately.

In order to reliably predict tsunami height near coasts, we need to set an appropriate earthquake fault model. In particular, for an automated tsunami prediction system, we need to make reasonable assumptions about earthquake fault size. A scaling law applicable to earthquake fault size is useful for assuming the fault size in the automated tsunami prediction system.

A scaling law simply predicts that a large earthquake has a large fault area and a small earthquake has a small fault area. Kanamori and Anderson (1975) made plots of the fault area S and the seismic moment M_0 by compiling the observations of numerous earthquakes with magnitudes ranging from 5.9 to 9.3 (Fig. 4.6). The plot clearly indicates that the fault area S increases as the seismic moment M_0 increases.

What is the physical mechanism behind this observation? Kanamori and Anderson (1975) suggested that the observed relation between fault area and magnitude is reproduced if the stress change on the fault before and after the earthquake is independent of the seismic moment M_0. The stress change is usually referred to as the stress drop.

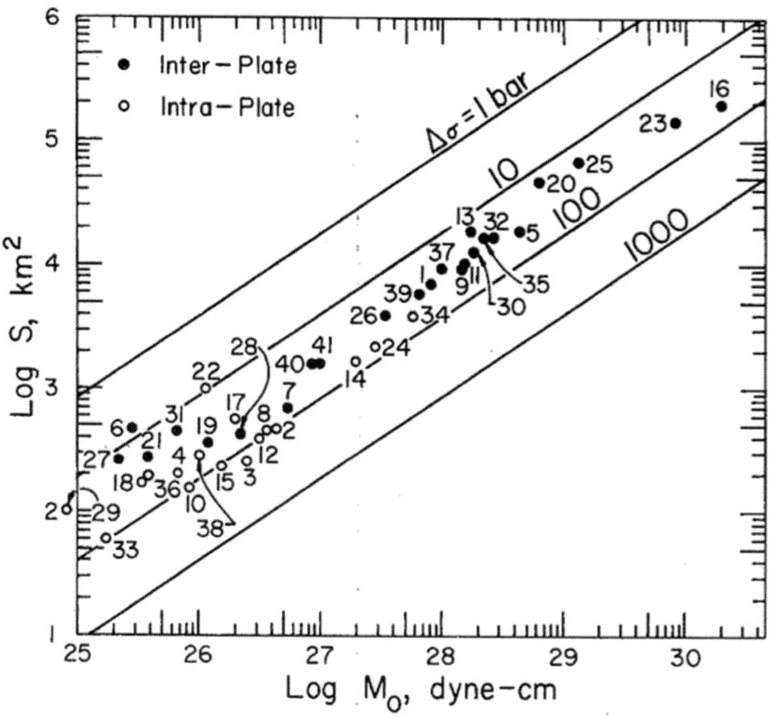

Fig. 4.6 Relation between fault area S and seismic moment M_0. The lines show the theoretical relation based on a constant stress drop $\Delta\sigma$ on a circular fault. (Kanamori and Anderson 1975, copyright by the Seismological Society of America)

4.2.1 Stress Drop

The seismic moment M_0 is one of the most important parameters with which to characterize the earthquake fault:

$$M_0 = \mu DS, \tag{4.17}$$

where D is the average slip on the fault and S is the fault area. The stress drop $\Delta\sigma$ is defined as the average difference between the shear stresses on a fault before an earthquake σ_0 and after the earthquake $\sigma(t=\infty)$ as

$$\Delta\sigma = \frac{1}{S}\iint_{\Sigma}[\sigma_0 - \sigma(t=\infty)]d\Sigma. \tag{4.34}$$

Since the stress drop is defined only by the status before and after the earthquake faulting and it is not affected by how fast the dislocation occurs, it is a static parameter. The stress drop is in general given in the following form:

$$\Delta\sigma = C_0\mu\frac{D}{a}, \tag{4.35}$$

where C_0 is a constant depending on the fault shape, such as a rectangular or circular shape, and a is a characteristic length of the fault plane such as the length of a rectangular fault or the diameter of a circular fault.

As the simplest example, a circular fault buried in an infinite homogeneous medium is considered. At the beginning, we consider an infinite homogeneous medium where uniform shear stress exists in the whole space. Taking the Cartesian coordinates shown in Fig. 4.7, we consider the displacement field given by

$$u_x(x,y,z) = \frac{\sigma_0}{\mu}z, u_y = 0, u_z = 0. \tag{4.36}$$

This displacement field causes uniform shear stress in the whole space (Figure 4.7b):

$$\tau_{xz} = \sigma_0. \tag{4.37}$$

Then, in the uniform stress field, we assume that an earthquake fault occurs at $z = 0$ and $x^2 + y^2 \leq a$ (Fig. 4.7c). After the earthquake occurs, the stress on the fault decreases. We assume that the stress becomes zero on the circular fault after the earthquake. This is given by a boundary condition of the stress as

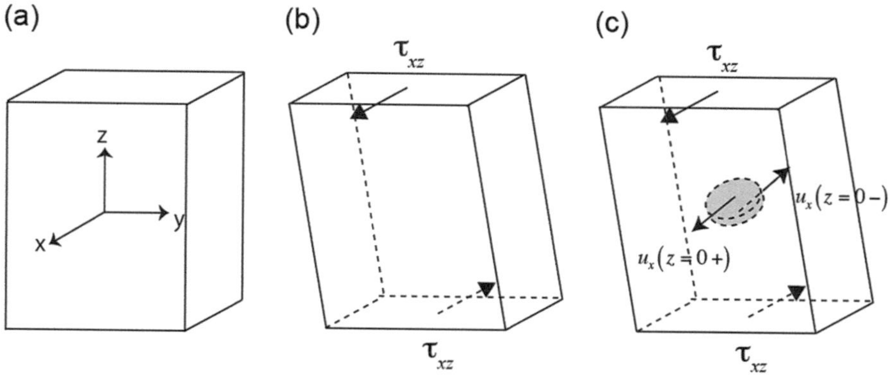

Fig. 4.7 (**a**) A homogeneous medium is considered. The displacement is zero in the whole region. (**b**) The medium is deformed. When the displacement field is given by $u_x(x, y, z) = \Delta\sigma z/\mu$, a homogeneous stress field $\tau_{xz} = \Delta\sigma$ occurs in the whole space. (c) A circular fault with radius a (under a stress-free condition) is made at $z = 0$ in the medium. The displacement and stress fields are perturbed by the fault. The displacement is discontinuous at the up and down sides of the fault, $u_x(x, y, 0+) \neq u_x(x, y, 0-)$

$$\tau_{xz}(x, y, 0) = 0 \quad \text{for } z = 0, x^2 + y^2 \leq a. \tag{4.38}$$

The circular fault is also referred to as a crack. Now, on the fault, the stress changes from σ_0 to 0 in association with the earthquake occurrence. Then, the stress drop $\Delta\sigma$ is $\Delta\sigma = \sigma_0$ according to the definition of the stress drop (Eq. 4.34). Still, the stress field far from the fault is not affected by this earthquake. Hence, the displacement far from the earthquake remains the same as that before the earthquake occurred as

$$u_x(x, y, z) = \frac{\sigma_0}{\mu} z, \quad \text{for } z = \pm\infty. \tag{4.39}$$

When we consider an equilibrium state (no temporal change) after the earthquake occurs, the equation of motion is given by the equilibrium equation

$$0 = \frac{\partial \tau_{xx}}{\partial x} + \frac{\partial \tau_{xy}}{\partial y} + \frac{\partial \tau_{xz}}{\partial z}. \tag{4.40}$$

The stress tensor needs to satisfy Eq. (4.40) in the whole space except on the fault.

We now seek a solution for the displacement field excited by the circular fault. In other words, we solve for the displacement field $u_x(x, y, z)$ that satisfies Eqs. (4.38), (4.39), and (4.40). We might expect there to be an analytical solution for the displacement field $u_x(x, y, z)$. Actually, in a 2-D problem (i.e., $\tau_{xz}(x, y, 0) = 0$ for $-a \leq x \leq a, -\infty \leq y \leq \infty$ instead of Eq. (4.38)), the analytical solution is given by using the elliptic functions (e.g., Hasegawa et al. 2015). However, to my knowledge, no simple analytical solution to the 3-D problem has been found.

If we do not attempt to represent the displacement field in the whole space but only the displacement discontinuity on the circular fault, there is an analytical solution. Eshelby (1957) derived the analytical solution by developing a smart technique. He showed the displacement discontinuity on an elliptical fault when uniform stress is applied. This corresponds to the dislocation caused by the stress change of $\sigma_0 - \sigma(t = \infty) = \sigma_0$. The dislocation on a circular fault with a uniform stress drop $\Delta\sigma$ is given by

$$\begin{aligned} \Delta u_x(x, y) &= u_x(x, y, 0+) - u_x(x, y, 0-) \\ &= \frac{24}{7\pi}\frac{\Delta\sigma}{\mu} a\sqrt{1 - \frac{x^2 + y^2}{a^2}} \quad \text{for } x^2 + y^2 < a^2, \end{aligned} \tag{4.41}$$

(see Eq. (5.7) in Eshelby (1957)). We should note that the dislocation Δu_x that gives a uniform stress drop is not uniform, but the dislocation is spatially varying. Figure 4.8a shows the dislocation distribution on a circular fault given by Eq. (4.41). The dislocation is at its maximum at the center and decreases to zero at the edge of the circle. The maximum dislocation becomes larger as the size of the fault increases (radius a).

It is not difficult to numerically estimate the displacement field once we know the dislocation source representation of Eq. (4.41). The displacement field $u_x(x, y, z)$ is calculated using the code of Okada (1992), where we represent the uniform dislocation as the sum of small subfaults with uniform dislocation. Figures 4.8 b and c show the displacement $u_x(0, 0, z)$ before and after the earthquake, respectively. Before the earthquake, $u_x(x, y, z)$ is given by $u_x(x, y, z) = (\sigma_0/\mu)z$, indicating a constant gradient with respect to the z-axis. After the earthquake (or after the stress drop occurs), the displacement discontinuity occurs at $z = 0$, and the gradient becomes small, becoming zero near the fault ($z\sim - 100 \pm 0$ km in Fig. 4.6c). This means that the strain (stress) is released by the dislocation given by Eq. (4.41).

When the dislocation on the fault is given by Eq. (4.41), the average dislocation on the circular fault is calculated as

$$\begin{aligned} D &= \frac{1}{\pi a^2}\iint_{x^2+y^2 \le a^2} \Delta u_x(x, y) dxdy \\ &= \frac{1}{\pi a^2}\int_0^a dr \int_{-\pi}^{\pi} rd\theta \frac{24}{7\pi}\frac{\Delta\sigma}{\mu} a\sqrt{1 - \frac{r^2}{a^2}} \\ &= \frac{16}{7\pi}\frac{\Delta\sigma}{\mu} a. \end{aligned} \tag{4.42}$$

This indicates that the stress drop $\Delta\sigma$ can be described by using the average slip D and the fault size $2a$ as

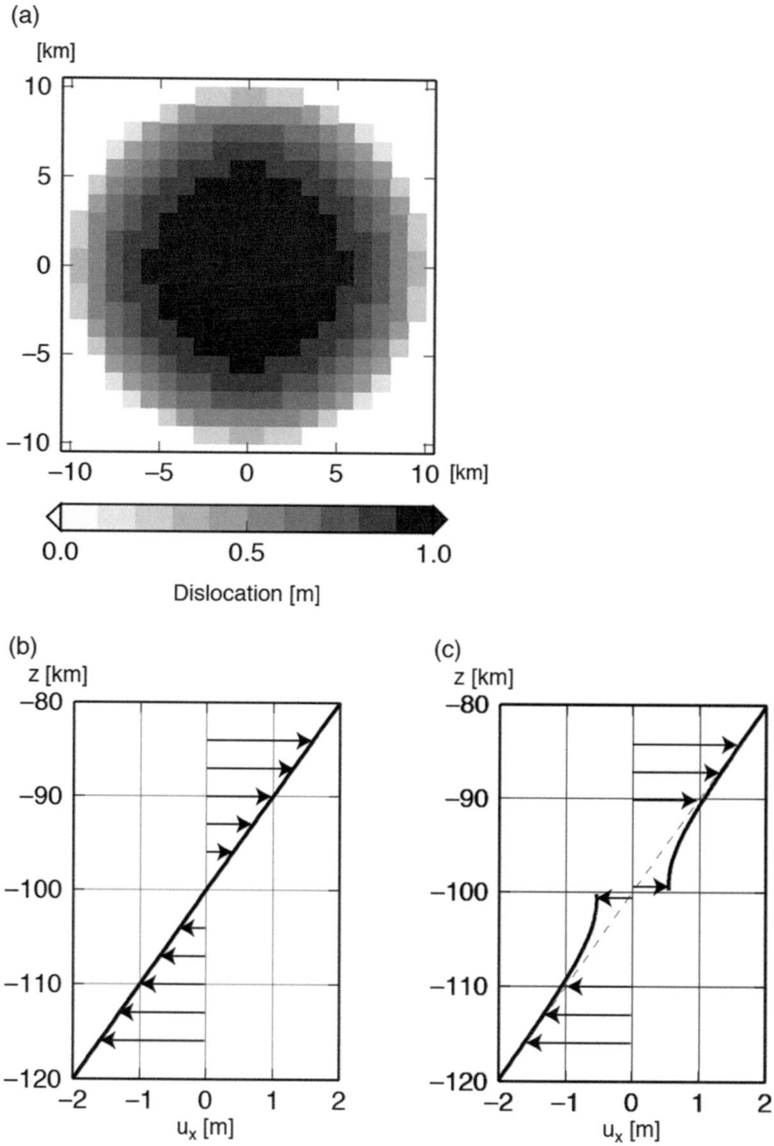

Fig. 4.8 (**a**) The distribution of dislocation given by Eq. (4.41) with $\Delta\sigma = 5$ MPa, $\mu = 50$ GPa, $a = 10$ km. (**b**) The displacement $u_x(0, 0, z)$ before an earthquake occurs. The displacement field is set as $u_x(x, y, z) = (\Delta\sigma/\mu)z$. This gives a uniform shear stress $\tau_{xz} = \Delta\sigma$ in the whole space. (c) The displacement after an earthquake. The earthquake is represented by the dislocation distribution shown in (a) set at $(x, y, z) = (0\text{ km}, 0\text{ km}, -100\text{ km})$. The gradient of the displacement is almost zero at $z = -100 \pm \epsilon$ km, suggesting that the shear strain τ_{xz} is almost zero

$$\Delta\sigma = \frac{7\pi}{8}\mu\frac{D}{2a}. \tag{4.43}$$

This is consistent with the relation between the stress drop and the average slip given by $\Delta\sigma = C_0\mu D/a$ (Eq. 4.35).

4.2.2 *Scaling Law*

When a constant stress change $-\Delta\sigma$ occurs in a circular region with a radius a, Eq. (4.42) gives the average slip D on a circular crack as

$$D = \frac{16}{7\pi}\frac{a\Delta\sigma}{\mu} = \frac{16}{7\pi^{3/2}}\frac{\Delta\sigma}{\mu}\sqrt{S}, \tag{4.44}$$

where S is the crack area. Substituting Eq. (4.44) into $M_0 = \mu DS$ (Eq. 4.17) gives

$$M_0 = \frac{16}{7\pi^{3/2}}\Delta\sigma S^{3/2} \approx 0.41\Delta\sigma S^{3/2}. \tag{4.45}$$

An earthquake model of a constant stress change $-\Delta\sigma$ predicts that $M_0 \propto S^{3/2}$. This reproduces the observed $S - M_0$ relation in Fig. 4.6 nicely when assuming that the stress drop $\Delta\sigma$ ranges from 10 bar (1MPa) to 100 bar (10 MPa) for a wide range of seismic moments M_0 from 10^{18} to 10^{23} N m (10^{25}–10^{30} dyn cm) (M_W is from 5.9 to 9.3). Figure 4.6 suggests that the stress drop $\Delta\sigma$ caused by earthquakes does not systematically vary irrespective of the earthquake fault size.

If we assume that $\Delta\sigma$ is constant and the fault length L is proportional to the width W as $L \propto W$, Eq. (4.44) gives $D \propto L$. Thus, the assumption of a constant stress drop and of similarity in the fault geometry gives the following relations:

$$\begin{aligned} S = LW \propto M_0^{2/3}, \\ D \propto S^{1/2} \propto M_0^{1/3}, \end{aligned} \tag{4.46}$$

and the scaling laws are given by

$$\begin{aligned} S = LW = C_1 M_0^{2/3} \\ D = C_2 M_0^{1/3} \end{aligned} \tag{4.47}$$

where C_1 and C_2 are the constants.

Scaling laws have been investigated in many studies. For example, Murotani et al. (2013) derived the scaling laws by compiling the earthquake fault parameters estimated from tsunami and geodetic data including the huge 2004 Sumatra and

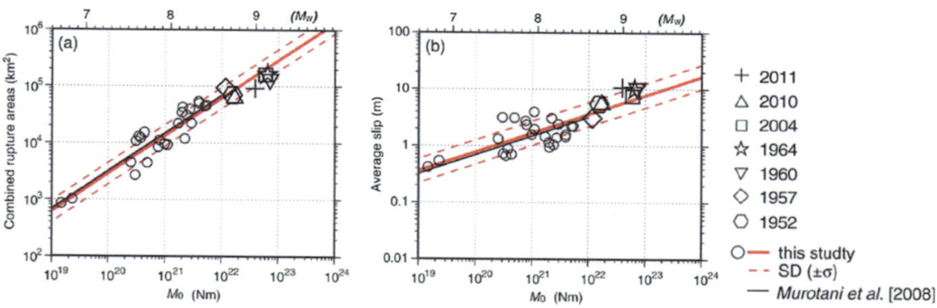

Fig. 4.9 (**a**) Relation between fault area S and seismic moment M_0. (**b**) Relation between average slip D and seismic moment M_0. Red lines indicate the regression lines, and dashed lines show the standard deviation on a log-log scale. (Murotani et al. 2013, copyright by the American Geophysical Union)

2011 Tohoku earthquakes. The relation between the fault area S and M_0 and that between D and M_0 are plotted in Fig. 4.9a and b, respectively.

Figure 4.9 indicated that $M_0 \propto S^{3/2}$ and $D \propto M_0^{1/3}$. The regression analysis gives the coefficients C_1 and C_2 as

$$S\left[\mathrm{km}^2\right] = LW = 1.34 \times 10^{-10} M_0^{2/3}\,[\mathrm{Nm}] \tag{4.48}$$

and

$$D[\mathrm{m}] = 1.66 \times 10^{-7} M_0^{1/3}\,[\mathrm{Nm}]. \tag{4.49}$$

The moment magnitude M_W is defined by using the seismic moment as

$$\log M_0\,[\mathrm{N \cdot m}] = 1.5 M_W + 9.1. \tag{4.18}$$

If the moment magnitude is M_W 9 ($M_0 = 4.0 \times 10^{22}$ Nm), the scaling law (Eq. (4.48)) finds that the fault area $S = 15.7 \times 10^4$ [km^2]. If we assume $L = 2W$, the length of the fault and width are $L = 560$ km and $W = 280$ km. Equation (4.49) gives the slip $D = 6$ m. The fault width of $W = 280$ km may be too large compared to the actual M_W ~9 earthquakes such as the 2004 Sumatra earthquake and the 2011 Tohoku-Oki earthquake. We should note that the scaling law can give only a rough description of the average fault geometries.

Let us consider that the moment magnitude M_W changes to $M_W + \Delta M_W$. The definition of the moment magnitude (Eq. 4.18) gives the seismic moment M_0' corresponding to the moment magnitude $M_W + \Delta M_W$ as

$$\log M_0'[\mathrm{N \cdot m}] = 1.5(M_\mathrm{W} + \Delta M_\mathrm{W}) + 9.1 = \log M_0 + 1.5\Delta M_\mathrm{W},$$
$$M_0' = 10^{1.5\Delta M_\mathrm{W}} M_0. \tag{4.50}$$

When the moment magnitude decreases by 1 ($\Delta M_\mathrm{W} = -1$), the seismic moment decreases by a factor of $10^{-1.5}$~1/32. Since the seismic moment is $M_0 \propto DLW$, the length L, width W, and slip D decrease by a factor of $(10^{-1.5})^{1/3} = 10^{-0.5}$~1/3. The earthquake fault geometry of the M_W 8 is then given by the length $L = 180$ km, the width $W = 90$ km, and the slip $D = 2$ m.

4.2.3 Using the Scaling Law for Tsunami Simulation

The scaling law of the earthquake fault is very useful for rapid tsunami prediction. We can estimate the CMT solution by analyzing seismic waveforms. Also, the seismic wave, which propagates much faster than the tsunami, can deliver the earthquake source information earlier than the tsunami. Hence, we estimate the magnitude of the earthquake by analyzing the observed seismograms promptly after the earthquake and can predict the tsunami height at the coast before the tsunami arrival.

As shown in Sect. 4.1, for far-field P or S waves in seismograms, the seismic source is effectively modeled as a point shear dislocation source. We can obtain the seismic moment and the moment tensor by far-field seismogram analysis. Actually, automated analysis of the global and regional seismograms can estimate the moment tensor for large earthquakes (e.g., Global CMT (http://www.globalcmt.org), USGS CMT (https://hdds.usgs.gov), F-net (Japan) (http://www.fnet.bosai.go.jp/top.php?LANG=en)).

The most important parameters in tsunami simulations are the seismic moment and the location of the centroid moment. The next important parameter is the fault size. To estimate the fault size from the seismic moment, the scaling law is quite useful. By using the CMT solution and the scaling law of the earthquake fault, we roughly set the fault geometry in an automated tsunami prediction algorithm. Some tsunami prediction systems are operating by employing this idea and using the regional and global seismic networks (e.g., Gusman and Tanioka 2014). Tsunami forecast systems using CMT solutions are being operated by the Pacific Tsunami Warning Center (PTWC) (e.g., Wang et al. 2012) and the French Polynesia Tsunami Warning Center (CPPT: Centre Polynésien de Prévention des Tsunamis) (e.g., Clément and Reymond 2015; Jamelot and Reymond 2015). Figure 4.10 shows an example of automated tsunami prediction (Inazu et al. 2016).

The CMT solution cannot determine the fault plane uniquely but can give two candidates for the fault plane because of the symmetry of the moment tensor representations. We need to select an appropriate one from the two candidates. We should note that the scaling law gives only the average fault geometry. Hence, the fault parameters predicted from the scaling law and the actual earthquakes differ

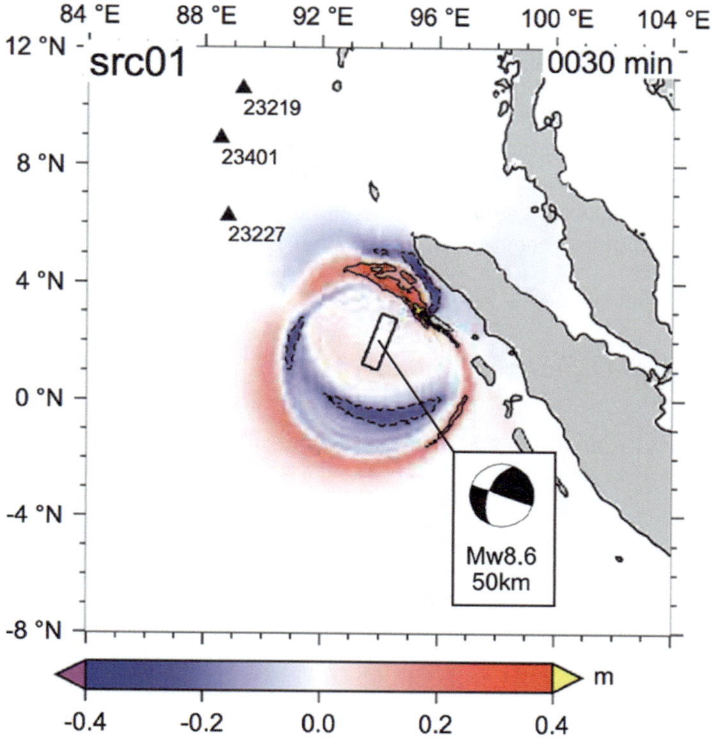

Fig. 4.10 Tsunami simulation using earthquake information. (Inazu et al. 2016, copyright by Springer)

somewhat. Also, mechanical fault models (constant stress change over the fault) with a stress-free surface boundary condition indicate that uniform slip models underestimate the tsunami height (e.g., Geist 1998; Geist and Dmowska 1999). For example, the scaling law predicted an average slip of 6 m for an earthquake of M_W 9, whereas significant slip heterogeneity (the maximum slip was larger than 40 m) was observed at the 2011 Tohoku-Oki earthquake (e.g., Fujii et al. 2011).

4.3 Seismic Wave Simulation

This section illustrates numerical simulations for the excitation of seismic waves and permanent displacement due to a finite earthquake fault. We are able to intuitively understand the seismic wave propagation through animations of the simulation results. The finite difference method is introduced as a numerical simulation method. This method shows the general versatility of numerical methods and is one of the most widely used methods in seismology and tsunami studies. The numerical simulations show that the seismic wavefield drastically changes depending on the time scale of the earthquake rupture; seismic and ocean acoustic waves are excited more when the dislocation occurs more rapidly. On the other hand, the permanent displacement does not depend on the slip rate but depends on the final amount of slip. The balance of the seismic wave excitation and the permanent displacement are important for tsunami monitoring. Strong seismic wave excitation can create noise that interferes with tsunami observation. In contrast, if seismic wave is less excited as in a tsunami earthquake, it would be difficult (but not impossible) to estimate the M_W by automated rapid seismogram analyses.

4.3.1 Finite Difference Method

The equations of motion in a 3-D elastic medium are

$$\begin{aligned}
\rho\frac{\partial v_x}{\partial t} &= \frac{\partial \tau_{xx}}{\partial x} + \frac{\partial \tau_{xy}}{\partial y} + \frac{\partial \tau_{xz}}{\partial z} + f_x, \\
\rho\frac{\partial v_y}{\partial t} &= \frac{\partial \tau_{yx}}{\partial x} + \frac{\partial \tau_{yy}}{\partial y} + \frac{\partial \tau_{yz}}{\partial z} + f_y, \\
\rho\frac{\partial v_z}{\partial t} &= \frac{\partial \tau_{zx}}{\partial x} + \frac{\partial \tau_{zy}}{\partial y} + \frac{\partial \tau_{zz}}{\partial z} + f_z,
\end{aligned} \tag{4.51}$$

and the constitutive laws are

$$\begin{aligned}
\frac{\partial \tau_{xx}}{\partial t} &= (\lambda + 2\mu)\frac{\partial v_x}{\partial x} + \lambda\frac{\partial v_y}{\partial y} + \lambda\frac{\partial v_z}{\partial z}, \\
\frac{\partial \tau_{yy}}{\partial t} &= \lambda\frac{\partial v_x}{\partial x} + (\lambda + 2\mu)\frac{\partial v_y}{\partial y} + \lambda\frac{\partial v_z}{\partial z}, \\
\frac{\partial \tau_{zz}}{\partial t} &= \lambda\frac{\partial v_x}{\partial x} + \lambda\frac{\partial v_y}{\partial y} + (\lambda + 2\mu)\frac{\partial v_z}{\partial z}, \\
\frac{\partial \tau_{yz}}{\partial t} &= \mu\left(\frac{\partial v_y}{\partial z} + \frac{\partial v_z}{\partial y}\right), \\
\frac{\partial \tau_{zx}}{\partial t} &= \mu\left(\frac{\partial v_z}{\partial x} + \frac{\partial v_x}{\partial z}\right), \\
\frac{\partial \tau_{xy}}{\partial t} &= \mu\left(\frac{\partial v_x}{\partial y} + \frac{\partial v_y}{\partial x}\right).
\end{aligned} \tag{4.52}$$

The Lamé parameters of λ and μ are represented as $\lambda = \rho\left(V_P^2 - 2V_S^2\right)$ and $\mu = \rho V_S^2$ when the density ρ and the P and S wave velocities V_P and V_S are given.

We use a staggered grid setting as shown in Fig. 4.11 to represent the differentiation as a finite difference form. The z-axis goes vertically downward. The x- and y-axes run to the north and east, respectively. This coordinate system is often used in seismology (e.g., Aki and Richards 2002). The grid point (i, j, k) is located at $(x, y, z) = (i\Delta x, j\Delta y, k\Delta z)$. The medium parameters of ρ, λ, μ are defined at the grid point

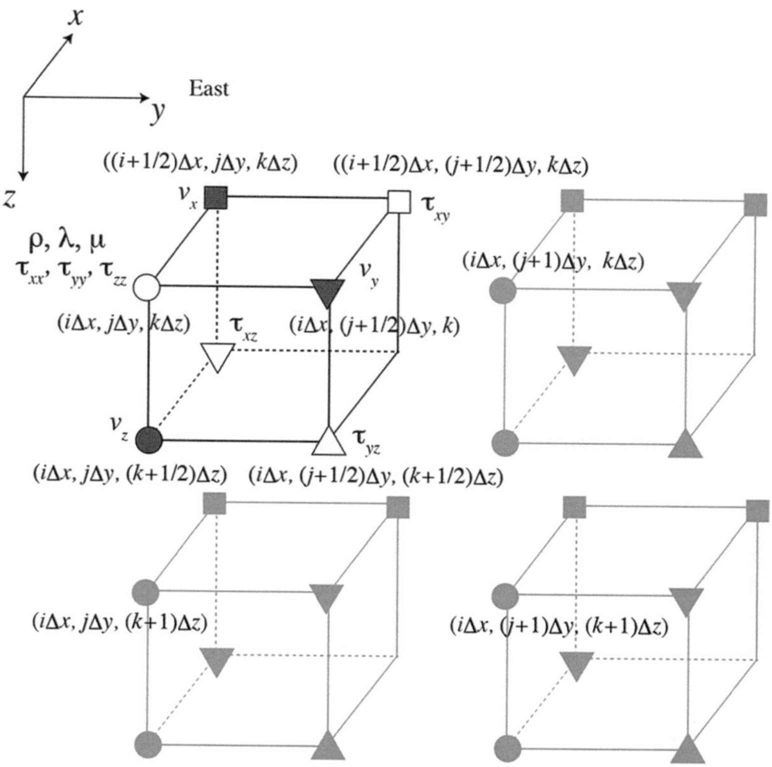

Fig. 4.11 Staggered grid setting for the finite difference approximation

(i, j, k). Dependent variables of v_p and τ_{pq} are defined at the corresponding grid location as follows:

$$\begin{aligned}
v^n_{x,ijk} &= v_x((i+1/2)\Delta x, j\Delta y, k\Delta z, n\Delta t),\\
v^n_{y,ijk} &= v_y(i\Delta x, (j+1/2)\Delta y, k\Delta z, n\Delta t),\\
v^n_{z,ijk} &= v_z(i\Delta x, j\Delta y, (k+1/2)\Delta z, n\Delta t),\\
\tau^{n+1/2}_{xx,ijk} &= \tau_{xx}(i\Delta x, j\Delta y, k\Delta z, (n+1/2)\Delta t),\\
\tau^{n+1/2}_{yy,ijk} &= \tau_{yy}(i\Delta x, j\Delta y, k\Delta z, (n+1/2)\Delta t),\\
\tau^{n+1/2}_{zz,ijk} &= \tau_{zz}(i\Delta x, j\Delta y, k\Delta z, (n+1/2)\Delta t),\\
\tau^{n+1/2}_{xy,ijk} &= \tau_{xy}((i+1/2)\Delta x, (j+1/2)\Delta y, k\Delta z, (n+1/2)\Delta t),\\
\tau^{n+1/2}_{yz,ijk} &= \tau_{yz}(i\Delta x, (j+1/2)\Delta y, (k+1/2)\Delta z, (n+1/2)\Delta t),\\
\tau^{n+1/2}_{zx,ijk} &= \tau_{zx}((i+1/2)\Delta x, j\Delta y, (k+1/2)\Delta z, (n+1/2)\Delta t).
\end{aligned} \tag{4.53}$$

For example, the finite difference approximation with respect to the time derivative of v_x in Eq. (4.51) is given by

$$v_{x,ijk}^{n+1} = v_{x,ijk}^{n} + \frac{\Delta t}{\rho}\left(\frac{\partial \tau_{xx}^{n+1/2}}{\partial x} + \frac{\partial \tau_{xy}^{n+1/2}}{\partial y} + \frac{\partial \tau_{xz}^{n+1/2}}{\partial z} + f_x\right). \tag{4.54}$$

This is the second-order approximation in time. Remembering that a point source located at $\boldsymbol{\xi}$ with the moment tensor M_{pq} is represented as the body force equivalent (Eq. 4.20) as

$$f_p(\mathbf{x}, t) = -\frac{\partial}{\partial x_q}\left[M_{pq}(t)\delta(\mathbf{x} - \boldsymbol{\xi})\right], \tag{4.20}$$

defining $M_{pq}^{n+1/2} = M_{pq}((n + 1/2)\Delta t)$ and substituting $f_x(\mathbf{x}, t)$

$$f_x(\mathbf{x}, t) = -\frac{\partial}{\partial x}[M_{xx}(t)\delta(\mathbf{x} - \boldsymbol{\xi})] - \frac{\partial}{\partial y}\left[M_{xy}(t)\delta(\mathbf{x} - \boldsymbol{\xi})\right] - \frac{\partial}{\partial z}[M_{xz}(t)\delta(\mathbf{x} - \boldsymbol{\xi})],$$

into Eq. (4.54), we obtain

$$\begin{aligned} v_{x,ijk}^{n+1} &= v_{x,ijk}^{n} + \frac{\Delta t}{\rho}\left\{\frac{\partial}{\partial x}\left[\tau_{xx}^{n+1/2} - M_{xx}^{n+1/2}\delta(\mathbf{x} - \boldsymbol{\xi})\right] + \frac{\partial}{\partial y}\left[\tau_{xy}^{n+1/2} - M_{xy}^{n+1/2}\delta(\mathbf{x} - \boldsymbol{\xi})\right]\right. \\ &\qquad \left. + \frac{\partial}{\partial z}\left[\tau_{xz}^{n+1/2} - M_{xz}^{n+1/2}\delta(\mathbf{x} - \boldsymbol{\xi})\right]\right\} \\ &= v_{x,ijk}^{n} + \frac{\Delta t}{\rho}\left\{\frac{\partial \tau'^{n+1/2}_{xx}}{\partial x} + \frac{\partial \tau'^{n+1/2}_{xy}}{\partial y} + \frac{\partial \tau'^{n+1/2}_{xz}}{\partial z}\right\}, \end{aligned} \tag{4.55}$$

where the stress tensor $\tau'^{n+1/2}_{pq}$ is defined as

$$\tau'^{n+1/2}_{pq} = \tau_{pq}^{n+1/2} - M_{pq}^{n+1/2}\delta(\mathbf{x} - \boldsymbol{\xi}). \tag{4.56}$$

Equation (4.56) indicates that the moment tensor $M_{pq}^{n+1/2}$ per unit volume functions as a stress change. Because the minimum unit size of the finite difference representation using the grid points is given by a cubic cell with the volume of $\Delta V = \Delta x \Delta y \Delta z$, we may consider $M_{pq}^{n+1/2}\delta(\mathbf{x} - \boldsymbol{\xi})$ as $M_{pq}^{n+1/2}/\Delta V$ located at $\boldsymbol{\xi}$ as

$$\tau'^{n+1/2}_{pq} = \tau_{pq}^{n+1/2} - M_{pq}^{n+1/2}/\Delta V, \tag{4.57}$$

in the finite difference simulation. Equation (4.57) indicates that the earthquake source represented by a moment tensor has the form of a stress change. Note that $M_{pq}^{n+1/2}/\Delta V$ is not the stress drop explained in Sect. 4.2.1 *Stress Drop*. This is often referred to as a "stress glut" (Aki and Richards 2002).

The spatial differentiation in Eq. (4.55) is then represented in the finite difference form as

$$v_{x,ijk}^{n+1} = v_{x,ijk}^{n} + \frac{\Delta t}{\rho}\left\{\frac{\tau'^{n+1/2}_{xx,i+1jk} - \tau'^{n+1/2}_{xx,ijk}}{\Delta x} + \frac{\tau'^{n+1/2}_{xy,ijk} - \tau'^{n+1/2}_{xy,ij-1k}}{\Delta y} + \frac{\tau'^{n+1/2}_{xz,ijk} - \tau'^{n+1/2}_{xz,ijk-1}}{\Delta z}\right\} \tag{4.58}$$

where the second-order finite difference approximation is used for the spatial differentiation. Similarly, we obtain the finite difference form with respect to $v_{y,ijk}^{n+1}$ and $v_{z,ijk}^{n+1}$ as

$$v_{y,ijk}^{n+1} = v_{y,ijk}^{n} + \frac{\Delta t}{\rho}\left\{\frac{\tau'^{n+1/2}_{yx,ijk} - \tau'^{n+1/2}_{yx,i-1jk}}{\Delta x} + \frac{\tau'^{n+1/2}_{yy,ij+1k} - \tau'^{n+1/2}_{yy,ijk}}{\Delta y} + \frac{\tau'^{n+1/2}_{yz,ijk} - \tau'^{n+1/2}_{yz,ijk-1}}{\Delta z}\right\}, \tag{4.59}$$

and

$$v_{z,ijk}^{n+1} = v_{z,ijk}^{n} + \frac{\Delta t}{\rho}\left\{\frac{\tau'^{n+1/2}_{zx,ijk} - \tau'^{n+1/2}_{zx,i-1jk}}{\Delta x} + \frac{\tau'^{n+1/2}_{zy,ijk} - \tau'^{n+1/2}_{zy,ij-1k}}{\Delta y} + \frac{\tau'^{n+1/2}_{zz,ijk+1} - \tau'^{n+1/2}_{zz,ijk}}{\Delta z}\right\}. \tag{4.60}$$

Equation (4.52) is represented in the finite difference form as

$$\tau_{xx,ijk}^{n+1/2} = \tau_{xx,ijk}^{n-1/2} + \Delta t\left[(\lambda+2\mu)\frac{v_{x,ijk}^{n} - v_{x,i-1jk}^{n}}{\Delta x} + \lambda\frac{v_{y,ijk}^{n} - v_{y,ij-1k}^{n}}{\Delta y} + \lambda\frac{v_{z,ijk}^{n} - v_{z,ijk-1}^{n}}{\Delta z}\right], \tag{4.61}$$

$$\tau_{yy,ijk}^{n+1/2} = \tau_{yy,ijk}^{n-1/2} + \Delta t\left[\lambda\frac{v_{x,ijk}^{n} - v_{x,i-1jk}^{n}}{\Delta x} + (\lambda+2\mu)\frac{v_{y,ijk}^{n} - v_{y,ij-1k}^{n}}{\Delta y} + \lambda\frac{v_{z,ijk}^{n} - v_{z,ijk-1}^{n}}{\Delta z}\right], \tag{4.62}$$

$$\tau_{zz,ijk}^{n+1/2} = \tau_{zz,ijk}^{n-1/2} + \Delta t\left[\lambda\frac{v_{x,ijk}^{n} - v_{x,i-1jk}^{n}}{\Delta x} + \lambda\frac{v_{y,ijk}^{n} - v_{y,ij-1k}^{n}}{\Delta y} + (\lambda+2\mu)\frac{v_{z,ijk}^{n} - v_{z,ijk-1}^{n}}{\Delta z}\right] \tag{4.63}$$

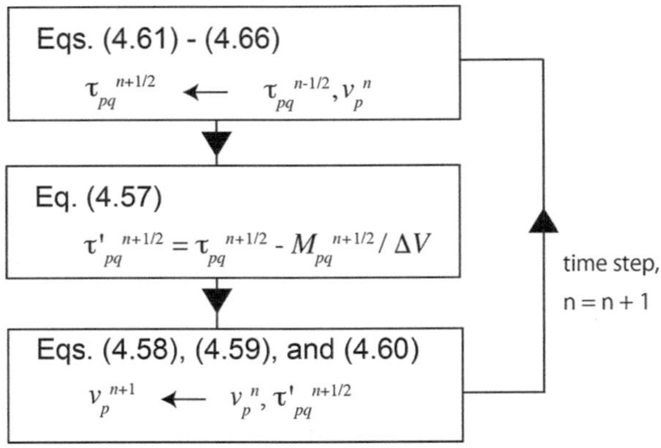

Fig. 4.12 A flow chart for the finite difference numerical simulation

$$\tau_{yz,\,ijk}^{n+1/2} = \tau_{yz,\,ijk}^{n-1/2} + \Delta t \mu \left[\frac{v_{y,\,ijk+1}^{n} - v_{y,\,ijk}^{n}}{\Delta z} + \frac{v_{z,\,ij+1k}^{n} - v_{z,\,ijk}^{n}}{\Delta y} \right], \tag{4.64}$$

$$\tau_{xz,\,ijk}^{n+1/2} = \tau_{xz,\,ijk}^{n-1/2} + \Delta t \mu \left[\frac{v_{x,\,ijk+1}^{n} - v_{x,\,ijk}^{n}}{\Delta z} + \frac{v_{z,\,i+1jk}^{n} - v_{z,\,ijk}^{n}}{\Delta x} \right], \tag{4.65}$$

$$\tau_{xy,\,ijk}^{n+1/2} = \tau_{xy,\,ijk}^{n-1/2} + \Delta t \mu \left[\frac{v_{x,\,ij+1k}^{n} - v_{x,\,ijk}^{n}}{\Delta y} + \frac{v_{y,\,i+1jk}^{n} - v_{y,\,ijk}^{n}}{\Delta x} \right]. \tag{4.66}$$

By using these finite difference equations, we calculate the velocity field v_p and the stress field (τ_{pq}) with the increment of a factional time step as shown in the flow chart in Fig. 4.12.

Usually, the velocity field and stress field at the time $t = 0$ are set at zero, and then the stress tensor τ_{pq} at the time $t = \Delta t/2$ becomes zero based on Eqs. (4.61, 4.62, 4.63, 4.64, 4.65, and 4.66). The stress tensor changes to τ'_{pq} with the moment tensor M_{pq} as the earthquake source according to Eq. (4.57). We then obtain the velocity field v_p at the time $t = \Delta t$, ($n = 1$) by using Eqs. (4.58), (4.59), and (4.60). By repeatedly conducting these arithmetic calculations for the time increment of n, the stress and velocity fields of τ_{pq} and v_p can be numerically calculated by increasing the time step.

4.3.1.1 Effective Medium Parameters

The density ρ and Lamé parameters of λ and μ are defined at the (i, j, k)th grid located at $(x, y, z) = (i\Delta x, j\Delta y, k\Delta z)$ (Fig. 4.11). Because the stresses τ_{xx}, τ_{yy}, and τ_{zz} are also defined at the same grid, Lamé parameters λ and μ of the (i, j, k)th grid located at $(x, y, z) = (i\Delta x, j\Delta y, k\Delta z)$ are used for the calculation of τ_{xx}, τ_{yy}, and τ_{zz}. However, when calculating the stress components of τ_{yz}, τ_{xz}, and τ_{xy}, which are not defined at $(x, y, z) = (i\Delta x, j\Delta y, k\Delta z)$, we should not use the rigidity μ at $(i\Delta x, j\Delta y, k\Delta z)$. For example, we need to use the rigidity $\mu_{i\,j\,+\,1/2\,k\,+\,1/2}$ located at $(i\Delta x, (j+1/2)\Delta y, (k+1/2)\Delta z)$ for the case of the stress τ_{yz} (Eq. 4.64):

$$\tau_{yz,ijk}^{n+1/2} = \tau_{yz,ijk}^{n-1/2} + \Delta t \mu_{i\ j+1/2\ k+1/2}\left[\frac{v_{y,ijk+1}^{n} - v_{y,ijk}^{n}}{\Delta z} + \frac{v_{z,ij+1k}^{n} - v_{z,ijk}^{n}}{\Delta y}\right].$$

Similarly, we need to use the rigidity $\mu_{i\,+\,1/2\,j\,k\,+\,1/2}$ for the stress τ_{xz} and the rigidity $\mu_{i\,+\,1/2\,j\,+\,1/2k}$ for the stress τ_{xy}. This also holds for the density ρ when calculating the particle velocities v_x, v_y, and v_z (Eqs. 4.58, 4.59, and 4.60).

The effective media parameters are used when calculating τ_{yz}, τ_{xz}, τ_{xy}, v_x, v_y, and v_z (see Graves 1996; Moczo et al. 2014). To determine the effective media parameters, we must interpolate the medium parameters such as λ, μ, and ρ. Considering the continuity of the traction and the displacement at the medium discontinuities, the effective medium parameters are given by the harmonic average or the arithmetic average as follows. The theoretical considerations were discussed in Moczo (2002). We only describe the results here. Rigidity is given by the harmonic average as

$$\begin{aligned}
\mu_{i,j+1/2,k+1/2} &= \left[\frac{1}{4}\left(\frac{1}{\mu_{ijk}} + \frac{1}{\mu_{ij+1k}} + \frac{1}{\mu_{ijk+1}} + \frac{1}{\mu_{ij+1k+1}}\right)\right]^{-1}, \\
\mu_{i+1/2,j,k+1/2} &= \left[\frac{1}{4}\left(\frac{1}{\mu_{ijk}} + \frac{1}{\mu_{i+1jk}} + \frac{1}{\mu_{ijk+1}} + \frac{1}{\mu_{i+1jk+1}}\right)\right]^{-1}, \\
\mu_{i+1/2,j+1/2,k} &= \left[\frac{1}{4}\left(\frac{1}{\mu_{ijk}} + \frac{1}{\mu_{i+1jk}} + \frac{1}{\mu_{ij+1k}} + \frac{1}{\mu_{i+1j+1k}}\right)\right]^{-1}.
\end{aligned} \tag{4.67}$$

The density is given by the arithmetic average:

$$\begin{aligned}
\rho_{i+1/2\ j\ k} &= \frac{1}{2}\left(\rho_{i+1\ jk} + \rho_{ijk}\right), \\
\rho_{i\ j+1/2\ k} &= \frac{1}{2}\left(\rho_{ij+1k} + \rho_{ijk}\right), \\
\rho_{i\ j\ k+1/2} &= \frac{1}{2}\left(\rho_{ijk+1} + \rho_{ijk}\right).
\end{aligned} \tag{4.68}$$

4.3.1.2 Moment Tensor Represented by Earthquake Fault Parameters

As shown in Fig. 4.13, a fault plane is defined by strike ϕ_S (the azimuth of the fault measured clockwise from the north) and dip δ (the angle between the horizontal surface and the fault plane). We consider only shear dislocation here. The slip vector is defined by the motion of the hanging wall relative to the foot wall. The slip direction is represented by rake λ (the angle between the slip vector and the strike). The moment tensor of this point shear dislocation is given by (Aki and Richards 2002, Box 4.4)

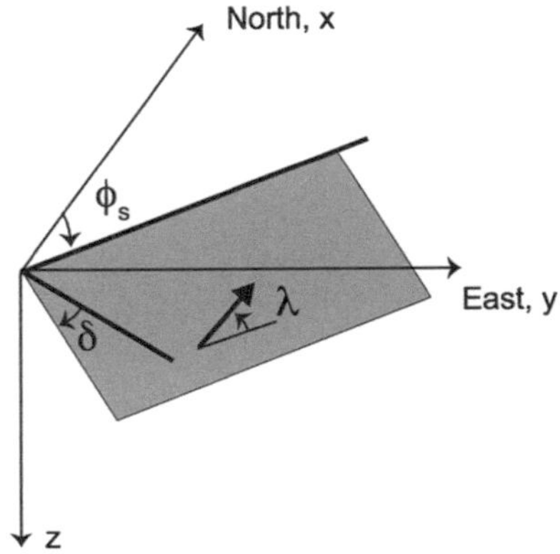

Fig. 4.13 Fault in Cartesian coordinates

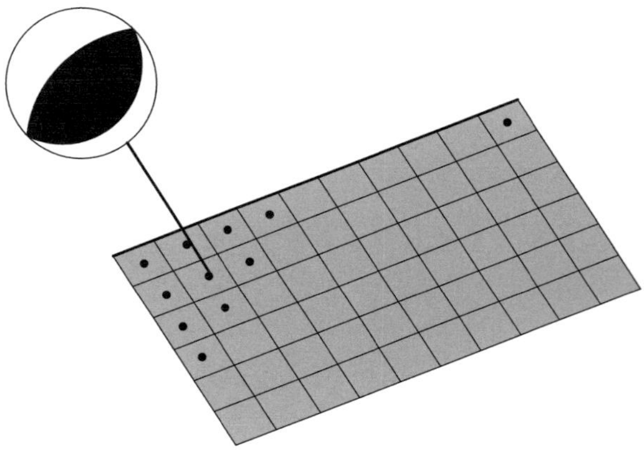

Fig. 4.14 A finite fault represented by numerous point sources

$$\begin{aligned}
M_{xx} &= -M_0(\sin\delta\cos\lambda\sin 2\phi_S + \sin 2\delta\sin\lambda\sin^2\phi_S),\\
M_{xy} &= +M_0\left(\sin\delta\cos\lambda\sin 2\phi_S + \frac{1}{2}\sin 2\delta\sin\lambda\sin 2\phi_S\right),\\
M_{xz} &= -M_0(\mathrm{con}\delta\cos\lambda\cos\phi_S + \cos 2\delta\sin\lambda\sin\phi_S) = M_{zx},\\
M_{yy} &= +M_0(\sin\delta\cos\lambda\sin 2\phi_S - \sin 2\delta\sin\lambda\cos^2\phi_S),\\
M_{yz} &= -M_0(\cos\delta\cos\lambda\sin\phi_S - \cos 2\delta\sin\lambda\cos\phi_S) = M_{zy},\\
M_{zz} &= +M_0\sin 2\delta\sin\lambda.
\end{aligned} \tag{4.69}$$

When considering tsunami excitation, the finite fault size or length and width also matter in addition to the seismic moment, strike, dip, and rake. We may represent the finite size with numerous point shear dislocation sources distributed on the fault plane in order to simulate the permanent sea-bottom displacement and seismic waves. For example, when the total seismic moment is M_0, we may put N point shear dislocation sources on the fault plane as shown in Fig. 4.14. Each shear dislocation source has the moment of M_0/N.

4.3.1.3 Diminishing the Waves Reflected from the Simulation Boundaries

In numerical simulations, we delimit the region for the calculation and set the boundaries of the simulation region. Waves propagating and arriving at the boundaries are reflected at the boundaries. The waves reflected at the boundaries and coming back to the simulation region contaminate the wavefields. These reflected waves are artificial waves that we do not want to reproduce. We intend to minimize the excitation of these unnecessary waves in order to focus the waves that have physical meaning in the simulations. There are various ways of damping the waves reflected at the boundaries of the simulation region such as the one-way propagation method and perfectly matching layer method (e.g., Moczo et al. 2014). We here explain the method of absorbing buffer region that is one of the simplest and most stable methods for diminishing the wave reflection at the simulation boundaries (e.g., Cerjan et al. 1985). The method also has an advantage in that it does not have a high computational cost.

An attenuating medium is supposed where waves gradually attenuate with increasing time as

$$\frac{dy}{dt} = -\alpha y \tag{4.70}$$

where y is the wave amplitude and Eq. (4.70) indicates that $\alpha(>0)$ is the attenuating rate per unit time. The solution of this differential equation is

$$y(t) = y(0)\exp(-\alpha t). \tag{4.71}$$

Also, the finite difference form of this equation is given by

$$y^{n+1} = y^n \exp(-\alpha\Delta t) \approx (1 - \alpha\Delta t)y^n, \tag{4.72}$$

where Δt is the time step of the simulation and $y^n = y(n\Delta t)$. This indicates that the wave amplitude is attenuated by multiplying $(1 - \alpha\Delta t)$ with the amplitude during the time step of Δt. As shown in Fig. 4.15, an attenuating medium is set surrounding the simulation region. The attenuating rate α is set at zero at the start of the attenuation region ($i = N - N_a$) and increases continuously to a larger value more rapidly toward the end of the simulation region ($i = N_a$):

$$\alpha = \begin{cases} 0 & \text{for } i < N - N_a \\ C_0[i - (N - N_a)]^2 & \text{for } i \geq N - N_a. \end{cases} \tag{4.73}$$

4.3.1.4 Further Reading and Open Codes

Seismic wave propagation simulation has been developed rapidly. We explained only the essential parts of the simulation method so that the reader can code it by

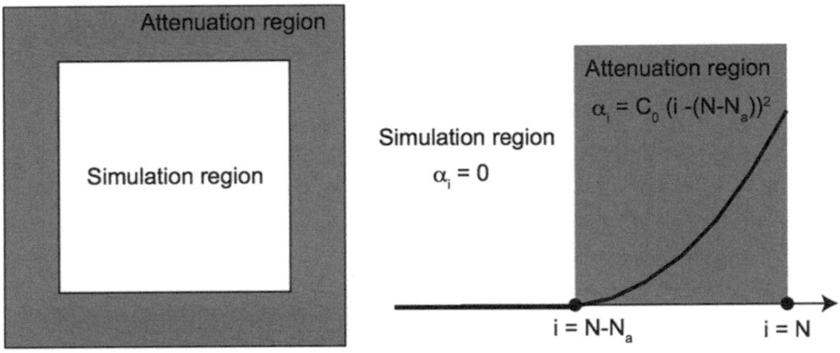

Fig. 4.15 Absorption boundary

themselves. The 2-D simulations are not realistic but are important for theoretical studies. The relation between the 3-D and 2-D wave propagations is explained in Appendix A. For the development of the codes, excellent textbooks (e.g., Fichtner 2010; Moczo et al. 2014) are available for further reading. Some excellent seismic wave simulation codes are provided on the Internet. For example, Maeda et al. (2017) provides OpenSWPC (https://github.com/takuto-maeda/OpenSWPC/releases/).

4.3.2 *Seismic Waves, Ocean Acoustic Waves, and Permanent Displacement*

By using the finite difference method of seismic wave simulation, this section investigates seismic waves excited by a shear dislocation source. We conduct 2-D space (x, z) simulation by assuming that there is no variation along the y-axis (Appendix 4A). An elastic medium is constituted by the sea layer and the uniform crust containing the earthquake fault. The stress is free at the sea surface.

4.3.2.1 Slow Faulting and Rapid Faulting

An earthquake fault with a width of 100 km and a dip of 15 degrees is set. The top of the fault is located 5 km deep below the sea bottom (Fig. 4.16). Although the actual earthquake rupture is characterized by wide-frequency contents from permanent displacement to a high-frequency component (at least ~10 Hz), we employ a simple source time function in order to investigate the mechanism of wave excitation depending on the time scale. We assume that the moment rate $\dot{M}(t)$ is given by the Gaussian function as

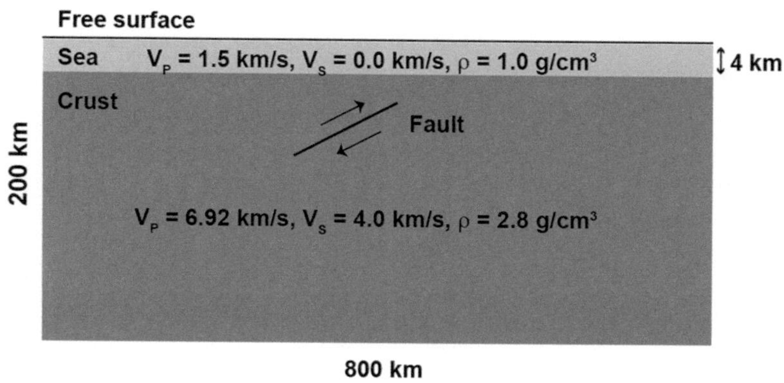

Fig. 4.16 An elastic medium constituted by a water layer and homogeneous crust. The finite size of the earthquake fault is set in the crust

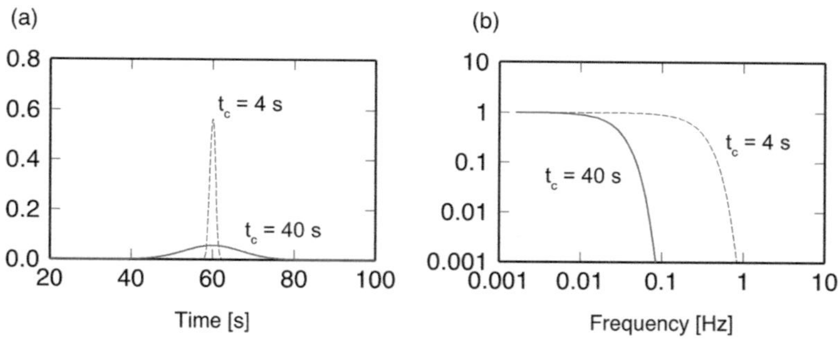

Fig. 4.17 (**a**) Functions of $f(t)$ give the slip rate or moment rate. The parameter t_C that characterizes the duration is set at 4 s (dashed line) and 40 s (solid line) in Eq. (4.74). The parameter t_0 is set at 60 s. (**b**) The spectral amplitudes of the slip rate functions of 4 s (dashed line) and 40 s (solid line). (Saito 2017, copyright by Oxford)

$$\dot{M}(t) = f(t) = \frac{M_0}{\sqrt{\pi}}\frac{4}{t_c}\exp\left[-\frac{(t-t_0)^2}{(t_c/4)^2}\right]. \tag{4.74}$$

The parameter t_c controls the duration of the earthquake slip. The functions with $t_C =$ 4 s and 40 s in Eq. (4.74) are plotted in Fig. 4.17 (a). The spectral amplitude of Eq. (4.74) is

$$\left|\hat{f}(\omega)\right| = M_0 \exp\left[-\frac{1}{4}\left(\frac{t_c}{4}\right)^2\omega^2\right]. \tag{4.75}$$

Figure 4.17 (b) shows the spectral amplitude as a function of the wave frequency. Both the two spectral amplitudes of $t_C = 4$ s and 40 s are the same in the lower-frequency limit since the moment M_0 is the same in the two moment rate functions. The short-duration or rapid faulting of $t_C = 4$ s has a broader spectral content and has much high-frequency components than the slow faulting of $t_C = 40$ s.

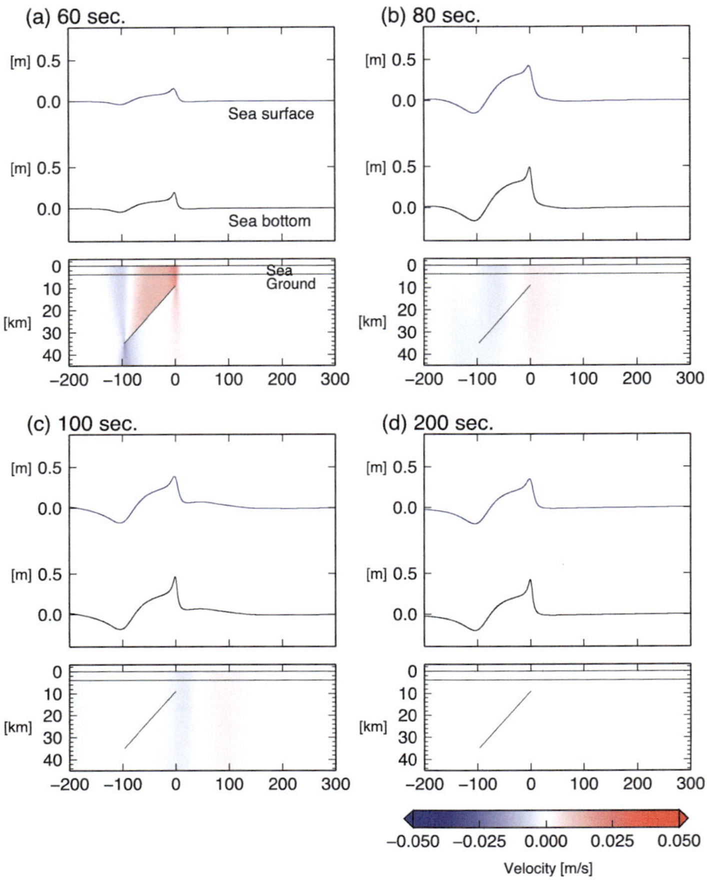

Fig. 4.18 Simulation results of the source duration given by $t_c = 40$ s for various elapsed times, (**a**) 60 s, (**b**) 80 s, (**c**) 100 s, and (**d**) 200 s. Vertical displacement distributions at the sea surface (blue) and sea bottom (black) are plotted in the upper bin. The vertical velocity distribution in the vertical cross section is displayed with a color map in the lower bin. The sea depth is 4 km ($0 < z < 4$ km). The black line in the crust indicates the earthquake fault. (Saito 2017, copyright by Oxford)

First, we conducted a seismic wave simulation when the earthquake faulting is slow or the moment rate function is characterized by $t_c = 40$ s. For simplicity, we neglected the rupture propagation, but the slip starts simultaneously on the fault. In other words, we set the rise time to 40 s, and the rupture velocity was infinite. Figure 4.18 shows the temporal and spatial distributions of the sea-surface height, sea-bottom vertical displacement, and vertical velocity in the sea and crust. At the time of 60 s (Fig. 4.18a), the vertical velocity exists above the earthquake fault, which uplifts the sea bottom and sea surface. The vertical displacement at the sea bottom and sea surface gradually increases with time. At the times of 80 s and 100 s (Fig. 4.18b and c), the slip on the fault has ended, and the velocity distribution caused by the earthquake disappears. If we observe the velocity distribution, we see that a long-wavelength wave propagates horizontally. However, the amplitude of the wave at the sea bottom and sea surface is very small. The wave front is located at 80 km at the time of 80 s and 150 km at the time of 100 s (Fig. 4.18b and c). The propagation velocity is ~3.5 km/s, corresponding to the Rayleigh wave. When the time is 200 s (Fig. 4.18d), the sea-bottom displacement remains above the earthquake fault. This is permanent displacement. Since this simulation neglects gravity, the sea surface does not collapse but remains as a permanent displacement.

Figure 4.19 shows the simulation results when the earthquake faulting occurs quickly or the moment rate function is characterized by $t_c = 4$s. The wave front of the seismic waves is clearly recognized on the sea bottom (black line in the upper panel) at $x = 130$ km at the time of 100 s (Fig. 4.19b) and at $x = 200$ km at the time of 120 s (Fig. 4.19c). This shows that the Rayleigh wave propagates with a speed of ~3.5 km/ s. The velocity distribution in the cross section indicates that considerable amplitude exists both in the sea and the crust for the Rayleigh wave at $x =$ ~100–~200 km in Fig. 4.19c. Unlike the case of the long duration rupture (Fig. 4.18), the amplitude in the sea is much larger than that in the crust for the later waves at 0 km $< x <$ 80 km in Fig. 4.19c and $100 < x < 300$ km in Fig. 4.19d. The later waves are trapped within the sea layer in the case of short-duration earthquake rupture. These waves are the ocean acoustic wave. A simple water-layer model predicts that ocean acoustic waves exist when the wave frequency is higher than $f_{\min} = c_0/(4h_0)$~0.1 Hz (Eq. (3.163) in Sect. 3.3.2 *Ocean Acoustic Waves*). As shown in Fig. 4.17b, the rapid earthquake rupture characterized by $t_c = 4$ s contains significant high-frequency components higher than $f_{\min} = c_0/(4h_0)$~0.1Hz. Hence, this excites considerable ocean acoustic waves, unlike in the case of $t_c = 40$s.

Comparisons of the long-duration and short-duration earthquake ruptures (Figs. 4.18 and 4.19) indicate that the excitation of the ocean acoustic waves strongly depends on the time scale of the earthquake rupture. When the rupture duration is short and the source spectrum contains high-frequency components, the ocean acoustic wave is efficiently excited. On one hand, the permanent sea-bottom displacements (or permanent sea-surface displacement) caused by the long- and short-duration earthquakes are identical. These simulation results are very reasonable. This is because the excitation of the seismic waves is proportional to the moment rate, whereas the permanent displacement is proportional to the moment, as theoretically derived using a simple earthquake source model in Sect. 4.1.2 *Displacement Field from a Point Source*.

4.3.3 Bridging Seismic Wave Simulation to Tsunami Propagation Simulation

In the previous section, we considered the sea-surface height and sea-bottom displacement distributions caused by an earthquake without gravity. In reality, the displaced sea-surface functions as the source of tsunami. Gravity works as the

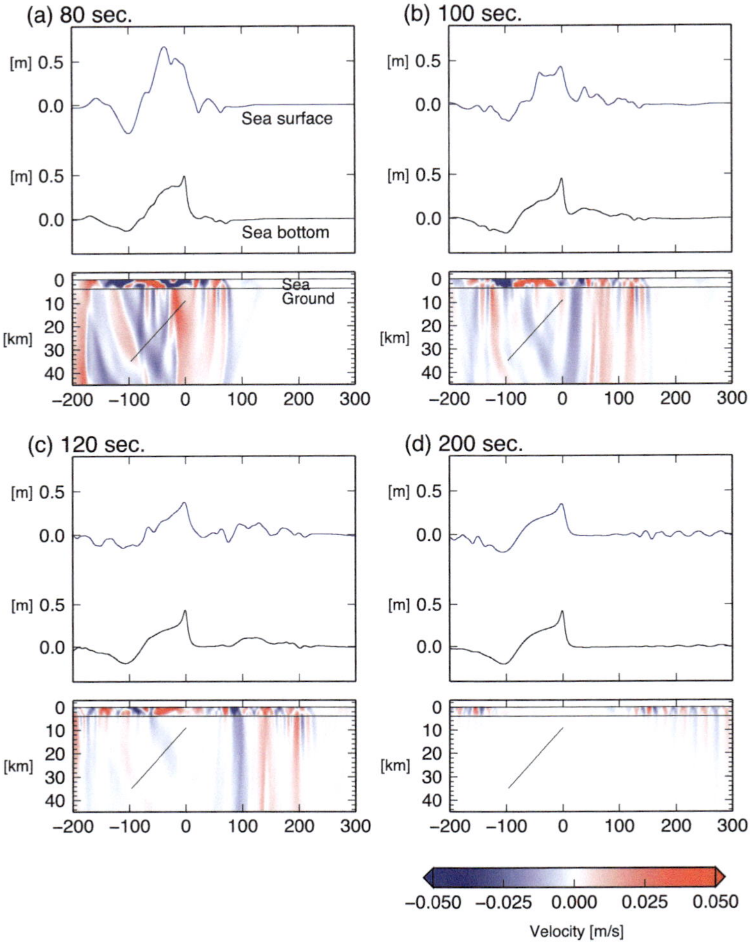

Fig. 4.19 Simulation results of the source duration are given by $t_c = 4$ s for various elapsed times: (**a**) 80 s, (**b**) 100 s, (**c**) 120 s, and (d) 200 s. Vertical displacement distributions at the sea surface (blue) and at the sea bottom (black) are plotted in the upper bin. Vertical velocity distribution in the vertical cross section is displayed with a color map in the lower bin. The sea depth is 4 km ($0 < z < 4$ km). The black line in the crust indicates the earthquake fault. (Saito 2017, copyright by Oxford)

restoring force of tsunami propagation. We here illustrate two methods for bridging the seismic wave simulation to tsunami propagation. One is to simulate tsunami without seismic waves, and the other is to simulate both seismic waves and tsunami.

(i) Only Tsunami Propagation Modeling

This approach is basically the same as the method used in many tsunami studies. Only the permanent displacement at the sea bottom is included but seismic waves are neglected.

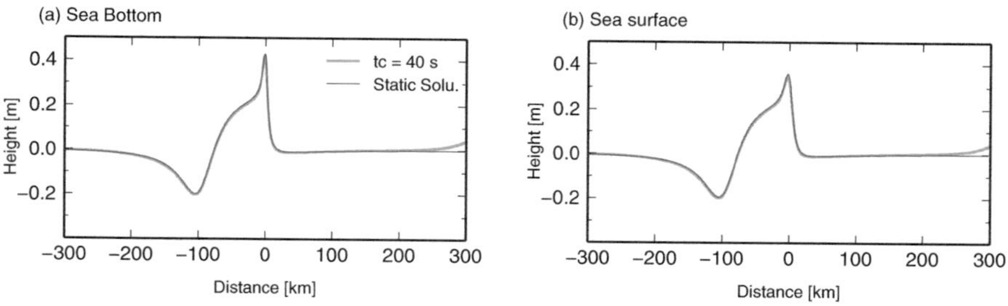

Fig. 4.20 (**a**) The vertical displacement distribution at the sea bottom at the elapsed time of 180 s for the simulation result of source duration tc = 40 s (bold gray line) and the permanent sea-bottom deformation calculated based on the static equation of Okada (1992) (thin red line); (**b**) vertical displacement distribution at the sea surface at the elapsed time of 180 s for the simulation result of source duration tc = 40 s (bold gray line) and the sea-surface deformation calculated based on the static equation and incompressible fluid theory. (Kajiura 1963) (thin red line) (Saito 2017, copyright by Oxford)

The seismic wave propagation simulation gives the permanent vertical displacement at the sea bottom as

$$d(x) = u_z(x, z = z_{\mathrm{bot}}, t = \infty). \tag{4.76}$$

A gray line in Figure 4.20a shows the permanent vertical displacement for the seismic wave simulation shown in Fig. 4.18. Note that the permanent displacements calculated by the slow and rapid rupture (Figs. 4.18 and 4.19) are identical. The maximum displacement reaches at 0.4 m, and the sea bottom ranging −100 km to 0 km is significantly displaced by the earthquake. When the crust is uniform, we can also calculate the permanent vertical displacement $d(x)$ by using the analytical solutions (e.g., Okada 1992). The vertical displacement distribution obtained by the analytical solution is plotted by red line. The excellent agreement between the two lines supports that the seismic wave simulation can correctly calculate the permanent displacement field.

When the sea is incompressible and the sea depth is constant $h = h_0$, the sea-surface displacement $\eta_0(x)$ caused by the sea-bottom displacement $d(x)$ is given by

$$\eta_0(x) = \frac{1}{2\pi}\int_{-\infty}^{\infty} dk \exp[ikx]\frac{\widehat{d}(k)}{\cosh\,(kh_0)}, \tag{4.77}$$

where $\widehat{d}(k)$ is the spatial Fourier transform of $d(x)$ (e.g., Takahashi 1942; Kajiura 1963). Equation (4.77) indicates that the sea layer of $h = h_0$ works as a spatial low-pass filter for the input of the sea-bottom displacement $d(x)$ and the output of the sea-surface displacement $\eta_0(x)$. The filter is often referred to as a Kajiura filter by researchers. We treat this theoretical background in more detail in the next chapter (Chap. 5: *Tsunami Generation*). The red line in Figure 4.20b shows the vertical displacement $\eta_0(x)$ at the sea surface calculated from the sea-bottom displacement $d(x)$ with Eq. (4.77). The maximum displacement at the sea surface is slightly smaller than that at the sea bottom, and the displacement distribution at the sea surface is smoother than that at the sea bottom because of the spatial low-pass filter effects (Kajiura filter). The gray line in Figure 4.20b shows the permanent vertical displacement $u_z(x, z = z_{sur}, t = \infty)$ ($z = z_{sur}$ is the location of the sea surface) numerically calculated by the seismic wave simulation in Fig. 4.18. The excellent agreement between the two lines indicates that we may use either the analytical method using the Kajiura filter or numerical simulations of seismic wave propagation for calculating the initial tsunami height distribution.

In the tsunami propagation simulation, the sea-surface displacement given by $\eta_0(x)$ (Eq. 4.77) is then used as the initial tsunami height distribution $\eta(x, t=0) = \eta_0(x)$. Alternatively, $u_z(x, z = z_{sur}, t = \infty)$ numerically calculated by the seismic wave simulation can be used for the initial tsunami height distribution. The tsunami propagation is calculated based on the 1-D linear dispersive equations:

$$\frac{\partial \eta(x,t)}{\partial t} + \frac{\partial}{\partial x}(h_0 v(x,t)) = 0, \tag{4.78}$$

and

$$\frac{\partial v(x,t)}{\partial t} + g_0 \frac{\partial \eta}{\partial x} = \frac{1}{3} h_0 \frac{\partial}{\partial t} \frac{\partial^2}{\partial x^2}(h_0 v(x,t)), \tag{4.79}$$

where $\eta(x, t)$ is the sea-surface height distribution, h_0 is the sea depth, and $v(x, t)$ is the horizontal velocity averaged over the sea depth. The details of the derivation of Eqs. (4.78) and (4.79) and the numerical calculation scheme are explained in Chap. 6: *Propagation Simulation*. Figure 4.21 shows the sea-surface height distribution calculated by the tsunami simulation. The sea-surface height $\eta(x, t)$ propagates horizontally as tsunami with the velocity of ~0.2 km/s (the tsunami propagates about 180 km during 900 s). At the time $t = 600$ and 900 s, we recognize a dispersion; the short-wavelength tsunamis follow long-wavelength tsunami.

(ii) Both Seismic Wave and Tsunami Propagation Modeling

This approach intends to include the effects of seismic waves and ocean acoustic waves in addition to tsunami for simulating sea-surface displacement. This method assumes a compressible sea for calculating seismic waves and ocean acoustic waves but assumes an incompressible sea for calculating tsunami.

The vertical velocity at the sea surface $v_z(x, z_{sur}, t)$ is calculated by the seismic wave simulation (Fig. 4.19). Then, we conduct a tsunami propagation simulation using $v_z(x, z_{sur}, t)$ as tsunami source. During the time Δt from $t_{i-1} = (i-1)\Delta t$ to $t_i = i\Delta t$, the height change $v_z(x, z_{sur}, t_i)\Delta t$ is added to the sea-surface height as

$$\eta(x, t_i) = \eta^*(x, t_i) + v_z(x, z_{sur}, t_i)\Delta t, \tag{4.80}$$

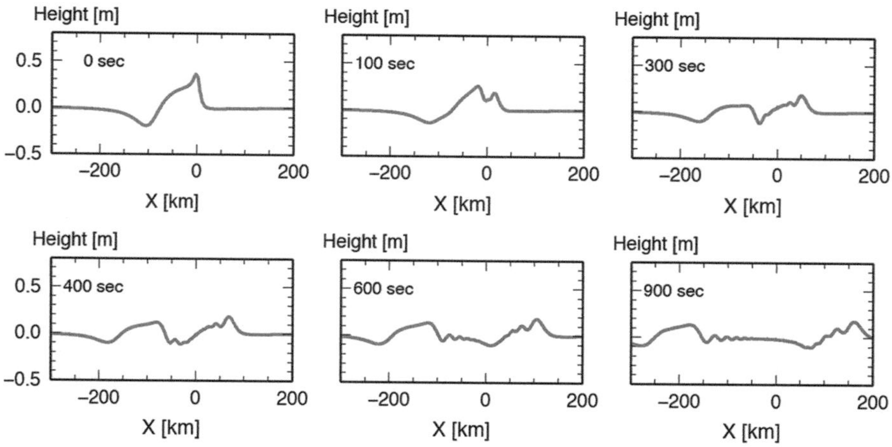

Fig. 4.21 Tsunami propagation from the initial height distribution $\eta_0(x)$ (Eq. 4.77). The propagation in an incompressible sea is calculated with Eqs. (4.78) and (4.79)

where $\eta^*(x, t_i)$ is the tsunami height distribution numerically simulated from the tsunami height and velocities at the previous time $t = t_{i-1}$. The height change $v_z(x, z_{sur}, t_i)\Delta t$ at each time step functions as the source.

Figure 4.22 shows the sea-surface height distribution calculated by this method. Since we set the origin time as the zero in the moment rate function shown in Fig. 4.17, no displacement appears at time $t = 0$ s. At time $t = 64$ s (4 s after the moment rate function takes a peak value), a large sea-surface displacement appears. This displacement is larger than the maximum height in the tsunami simulation in Fig. 4.21. Also, we recognize short-wavelength waves in the times of $t = 70 - 150$ s. These short-wavelength waves are seismic waves or ocean acoustic waves. After the seismic and ocean acoustic waves moved away ($t > \sim 400$ s), the sea-surface height is quite similar to those in the tsunami simulation results in Fig. 4.21.

Figure 4.23 compares the simulation results for these two approaches: (i) only tsunami propagation modeling and (ii) both seismic wave and tsunami propagation modeling. Because the moment rate function is included in the method (ii), the time difference (60 s) between the two methods is corrected in Fig. 4.23. The comparison indicates that at the time of 70 s (Fig. 4.23a), there is a significant discrepancy between the two methods. The sea-surface height by the method (ii) (red line) contains higher frequency and shorter wavelength components. This difference comes from the ocean acoustic waves and seismic waves. At the elapsed time of 160 s (Fig. 4.23b), a difference still exists between the two methods. The difference gradually decreases as the time elapses. At the time of 320 s (Fig. 4.23c), after the seismic and ocean acoustic waves move away from the simulation region, the two simulation results show excellent agreement.

This comparison also suggests that seismic waves and tsunami are almost independent of each other. In discussing this, we should carefully use the terms "sea-surface height change," "tsunami," and "seismic wave." "Sea-surface height

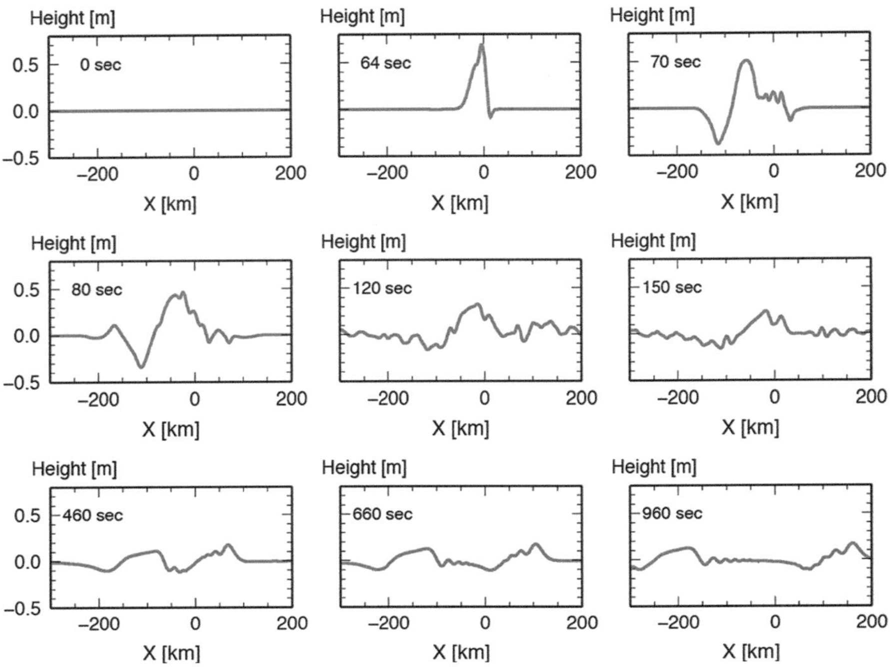

Fig. 4.22 Temporal and spatial change of the sea-surface displacement. These are calculated based on the tsunami propagation in an incompressible sea Eqs. (4.78) and (4.79) with a source given by Eq. (4.80) that includes seismic waves and ocean acoustic waves in compressible sea

change" is considered to be the vertical displacement at the sea surface. "Tsunami" is the wave caused by gravity as the restoring force. "Seismic wave" is the wave caused by elasticity of the crust and compressibility of the seawater as the restoring force. Note that, in general, the "seismic wave" is caused by elasticity, but gravity also matters, in particular for long-period seismic waves. Also, elasticity matters somewhat for tsunamis. However, the dominant driving force is gravity for tsunami and is elasticity for seismic waves. The sea-surface height change is not only caused by tsunami. Also, seismic waves ($< \sim 10$ s) can considerably contribute to the sea-surface height change, as shown in Fig. 4.23a. However, we can say that seismic waves do not significantly affect tsunami because the tsunami plotted with red and gray lines is almost identical in Fig. 4.23c. This means that we may consider only tsunami without considering seismic waves when we assess tsunami inundation near coasts or calculate tsunami far from the earthquake fault. However, when we analyze the records observed near or inside the earthquake source (e.g., sea-surface displacement and sea-bottom pressure change), we need to consider both seismic waves and tsunami because the records contain both of them.

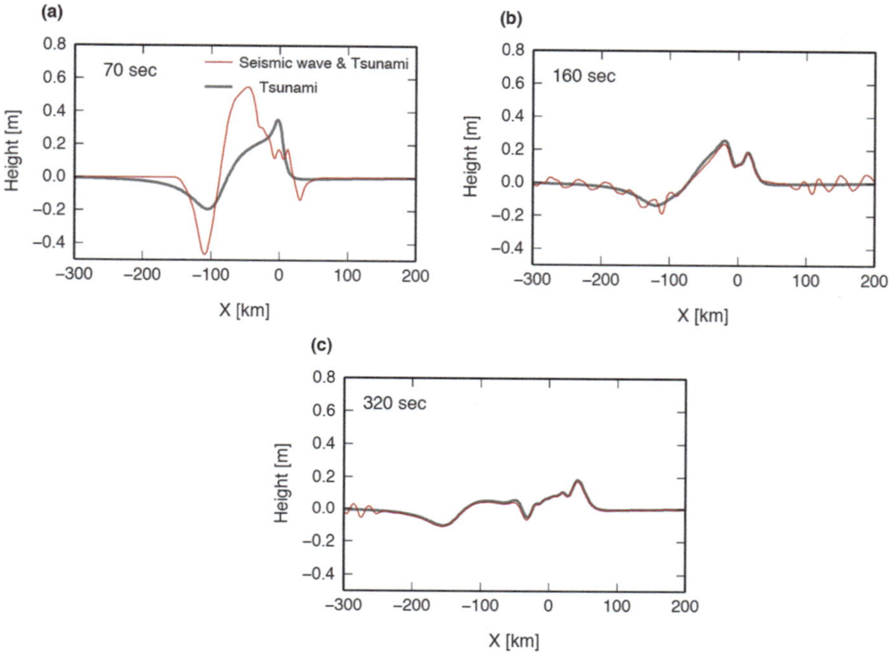

Fig. 4.23 Sea-surface height distributions caused by the earthquake dislocation characterized by the source duration $t_c = 4$ s in Eq. (4.74). Both seismic waves and tsunamis are included with Eq. (4.80) (red line), and only tsunami is included with Eq. (4.77) (gray line). Sea-surface heights at the elapsed times of (**a**) 70 s, (**b**) 160 s, and (**c**) 320 s

Appendix A: Seismic Wave Propagation in 2-D Space: P-SV Problem

Seismic wave propagation in 2-D space is unrealistic. We should not directly compare the seismograms simulated in 2-D space with observed seismograms because the observed seismograms are in 3-D space. Nevertheless, the wave propagation in 2-D space is very useful for studies of the physical mechanism of seismic wave propagation. Since the 2-D and 3-D propagations have different dimensions, we need to carefully define each parameter and its units. This appendix illustrates a 2-D seismic wave propagation problem by deriving the propagation equations and source parameters from those of the 3-D simulation.

At the beginning, the 2-D space (x, z) is defined based on the 3-D space (x, y, z) as shown in Fig. 4.24. An earthquake fault is supposed in 3-D space with the strike and rake defined as $\phi_s = \pi/2$ and $\lambda = \pi/2$, respectively (see also Fig. 4.14). Also, we assume that the fault length L is much longer than the fault width W. Then, the fault is located in the 3-D space (x, y, z) as shown in Fig. 4.24. In the (x, z) space, the fault is represented as a line segment characterized by the length (width) W and the dip δ. In the (x, y) space, the fault ranges from $y = -L/2$ to $y = L/2$ where $L \gg W$. The moment time function is given by

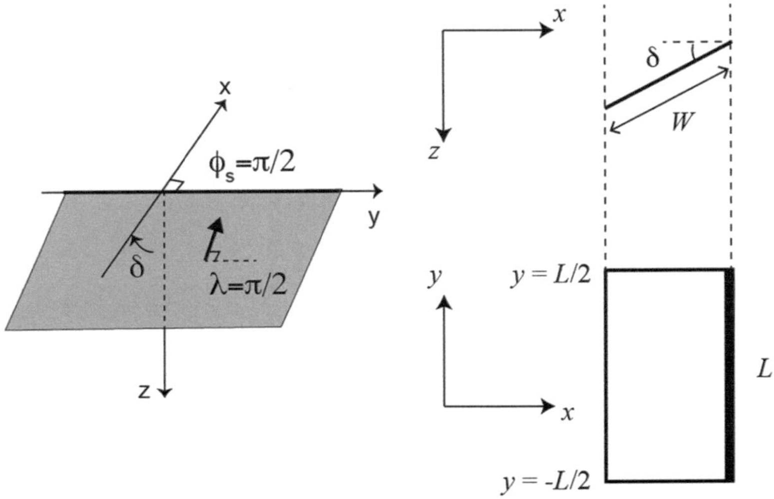

Fig. 4.24 Defining the 2-D (x, z) space from the 3-D space (x, y, z). We set the strike and rake to be $\phi_s = \pi/2$ and $\lambda = \pi/2$, respectively, in Fig. 4.14. Also, the fault length L is much longer than the fault width W. We define the 2-D space (x, z) as the plane $y = 0$ in the 3-D space (x, y, z)

$$M(t) = \mu LWD(t) \tag{A.1}$$

where $D(t)$ is the slip on the fault. Defining the 2-D space (x, z) at the plane $y = 0$ in the 3-D space (x, y, z), we consider the wave propagation in the 2-D space (x, z).

Due to the symmetry of the problem (now we suppose $\lambda = \pi/2$), the wavefield in the 2-D space (x, z) is represented by (v_x, v_z) and $v_y = 0$, and the spatial derivative with respect to the y-axis is zero $\partial/\partial y = 0$. Then, the equation of motion (Eq. 4.51) is reduced to

$$\begin{aligned} \rho\frac{\partial v_x}{\partial t} &= \frac{\partial \tau_{xx}}{\partial x} + \frac{\partial \tau_{xz}}{\partial z} + f_x, \\ \rho\frac{\partial v_z}{\partial t} &= \frac{\partial \tau_{zx}}{\partial x} + \frac{\partial \tau_{zz}}{\partial z} + f_z. \end{aligned} \tag{A.2}$$

The stress components τ_{xx}, $\tau_{xz} = \tau_{zx}$, and τ_{zz} are given by the constitutive law Eq. (4.52) as

$$\begin{aligned} \frac{\partial \tau_{xx}}{\partial t} &= (\lambda + 2\mu)\frac{\partial v_x}{\partial x} + \lambda\frac{\partial v_z}{\partial z}, \\ \frac{\partial \tau_{zz}}{\partial t} &= \lambda\frac{\partial v_x}{\partial x} + (\lambda + 2\mu)\frac{\partial v_z}{\partial z}, \\ \frac{\partial \tau_{zx}}{\partial t} &= \mu\left(\frac{\partial v_z}{\partial x} + \frac{\partial v_x}{\partial z}\right). \end{aligned} \tag{A.3}$$

Note that in our definition of the 2-D space, the dimensions of all the parameters in (4A.2) and (4A.3) are common with those in the 3-D space. For example, the unit of

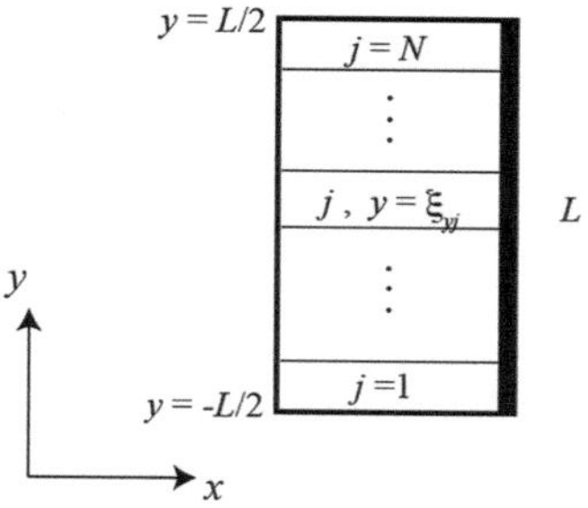

Fig. 4.25 Dividing a fault into N small subfaults along the y-axis. The centroid of the jth subfault is located at (ξ_x, ξ_{yj}, ξ_z)

density ρ is [kg/m^3], and that of the body force (f_x and f_z) is [N/m^3] also in the 2-D problem.

For the 3-D space (x, y, z), the body force $f_p(\mathbf{x}, t)$ corresponding to the moment tensor $M_{pq}(t)$ is given by (4.20):

$$f_p(\mathbf{x}, t) = -\frac{\partial}{\partial x_q}\left[M_{pq}(t)\delta(\mathbf{x} - \boldsymbol{\xi})\right]. \tag{4.20}$$

As shown in Fig. 4.25, the fault is divided into N subfaults along the fault length (along the y-axis). For the jth subfault, the centroid is located at (ξ_x, ξ_{yj}, ξ_z). The y coordinates ξ_{yj} change according to the index j. Each subfault is $\Delta L = L/N$ in length and W in width. The equivalent body force for the jth subfault $f_{p,j}(\mathbf{x}, t)$ is given by

$$f_{p,j}(\mathbf{x}, t) = \frac{1}{N} f_p\left(\mathbf{x}, t; \xi_{yj}\right), \tag{A.4}$$

where

$$\begin{aligned} f_p\left(\mathbf{x}, t; \xi_{yj}\right) = & -\frac{\partial}{\partial x}\left[M_{px}(t)\delta(x - \xi_x)\delta\left(y - \xi_{yj}\right)\delta(z - \xi_z)\right] \\ & -\frac{\partial}{\partial y}\left[M_{py}(t)\delta(x - \xi_x)\delta\left(y - \xi_{yj}\right)\delta(z - \xi_z)\right] \\ & -\frac{\partial}{\partial z}\left[M_{pz}(t)\delta(x - \xi_x)\delta\left(y - \xi_{yj}\right)\delta(z - \xi_z)\right]. \end{aligned} \tag{A.5}$$

Summing up the contributions from each subfault, we obtain

$$\begin{aligned} f_p(\mathbf{x}, t) &= \sum_{j=1}^{N} f_{p,j}(\mathbf{x}, t) = \sum_{j=1}^{N} \frac{1}{N} f_p\left(\mathbf{x}, t; \xi_{yj}\right) \\ &= \sum_{j=1}^{N} \frac{M(t)}{N} \frac{f_p\left(\mathbf{x}, t; \xi_{yj}\right)}{M(t)} \\ &= \mu D W \sum_{j=1}^{N} \frac{L}{N} \frac{f_p\left(\mathbf{x}, t; \xi_{yj}\right)}{M(t)} \\ &= \mu D W \sum_{j=1}^{N} \Delta L \frac{f_p\left(\mathbf{x}, t; \xi_{yj}\right)}{M(t)}. \end{aligned} \tag{A.6}$$

Limiting $N \to \infty$ gives

$$
\begin{aligned}
f_p(\mathbf{x}, t) &= \mu DW \int_{-L/2}^{L/2} d\xi_y \frac{f_p(\mathbf{x}, t; \xi_y)}{M(t)} \\
&= \mu DW \int_{-L/2}^{L/2} d\xi_y \left\{ -\frac{\partial}{\partial x}\left[\frac{M_{px}(t)}{M(t)} \delta(x-\xi_x)\delta(y-\xi_y)\delta(z-\xi_z)\right] \right. \\
&\quad -\frac{\partial}{\partial y}\left[\frac{M_{py}(t)}{M(t)} \delta(x-\xi_x)\delta(y-\xi_y)\delta(z-\xi_z)\right] \\
&\quad \left. -\frac{\partial}{\partial z}\left[\frac{M_{pz}(t)}{M(t)} \delta(x-\xi_x)\delta(y-\xi_y)\delta(z-\xi_z)\right] \right\}.
\end{aligned}
\tag{A.7}
$$

We calculate

$$
\begin{aligned}
& f_p(\mathbf{x}, t) \\
&= \mu DW \left\{ -\frac{\partial}{\partial x}\left[\frac{M_{px}(t)}{M(t)} \delta(x-\xi_x)\delta(z-\xi_z)\right] \right\} \int_{-L/2}^{L/2} d\xi_y \delta(y-\xi_y) \\
&\quad + \mu DW \left\{ -\left[\frac{M_{py}(t)}{M(t)} \delta(x-\xi_x)\delta(z-\xi_z)\right] \right\} \int_{-L/2}^{L/2} d\xi_y \frac{\partial}{\partial y} \delta(y-\xi_y) \\
&\quad + \mu DW \left\{ -\frac{\partial}{\partial z}\left[\frac{M_{pz}(t)}{M(t)} \delta(x-\xi_x)\delta(z-\xi_z)\right] \right\} \int_{-L/2}^{L/2} d\xi_y \delta(y-\xi_y)
\end{aligned}
$$

Finally, setting $L \to \infty$, we obtain

$$
\begin{aligned}
f_p(\mathbf{x}, t) = &-\frac{\partial}{\partial x}\left[\mu DW \frac{M_{px}(t)}{M(t)} \delta(x-\xi_x)\delta(z-\xi_z)\right] \\
&-\frac{\partial}{\partial z}\left[\mu DW \frac{M_{pz}(t)}{M(t)} \delta(x-\xi_x)\delta(z-\xi_z)\right].
\end{aligned}
\tag{A.8}
$$

Since, in our problem, the subscript p takes the value $p = x$ and z, substituting $\lambda = \pi/2$ and $\phi_s = \pi/2$ into Eq. (4.62), we obtain

$$
\begin{aligned}
\frac{M_{xx}}{M(t)} &= -\sin 2\delta, \\
\frac{M_{xz}}{M(t)} &= -\cos 2\delta, \\
\frac{M_{zz}}{M(t)} &= \sin 2\delta.
\end{aligned}
\tag{A.9}
$$

Substituting Eqs. (A.9) into (4A.8) and comparing the resultant equation with Eq. (4.20), the moment tensor in the 2-D space (x, z) should be defined as

$$\begin{aligned} M_{xx}^{2D} &= -\mu DW \sin 2\delta, \\ M_{xz}^{2D} &= M_{zx}^{2D} = -\mu DW \cos 2\delta, \\ M_{zz}^{2D} &= \mu DW \sin 2\delta, \end{aligned} \tag{A.10}$$

when the fault width is W and the dip is δ (as shown in Fig. 4.24). We should note that the dimension of the 2-D moment tensor defined in Eq. (A.10) is different from that in 3-D space. In the 2-D space, the moment is defined as $M^{2D} = \mu DW$. The units of M^{2D} are given by [N], whereas the moment in 3-D space is [N m].

In summary, for the 2-D space (x, z), the earthquake fault is described by the fault width W and the dip δ. The moment tensor [N] is given by Eq. (A.10). The equivalent body force [Nm^{-3}] in the 2-D space is given by

$$f_p(\mathbf{x}, t) = -\frac{\partial}{\partial x_q}\left[M_{pq}^{2D}(\boldsymbol{\xi}, t)\delta(\mathbf{x} - \boldsymbol{\xi})\right], \tag{A.11}$$

which is in the same form as Eq. (4.20) in 3-D space, but the delta function is defined in 2-D space, $\delta(\mathbf{x} - \boldsymbol{\xi}) = \delta(x - \xi_x)\delta(z - \xi_z)$. The equation of motion and the constitutive law are given by Eqs. (A.2) and (A.3), respectively.

References

Aki K, Richards P (2002) Quantitative seismology. University Science Books, Mill Valley

Aki K (1966) Generation and propagation of G waves from the Niigata earthquake of June 16, 1964. Part 2. Estimation of earthquake moment, released energy, and stress-strain drop from the G wave spectrum. Bull Earthquake Res Inst 44:73–88

Burridge R, Knopoff L (1964) Body force equivalents for seismic dislocations. Bull Seismol Soc Am 54(6A):1875–1888

Cerjan C, Kosloff D, Kosloff R, Reshef M (1985) A nonreflecting boundary condition for discrete acoustic and elastic wave equations. Geophysics 50(4):705–708

Clément J, Reymond D (2015) New tsunami forecast tools for the French Polynesia tsunami warning system part I: moment tensor, slowness and seismic source inversion. Pure Appl Geophys 172:791–804. https://doi.org/10.1007/s00024-014-0888-6

Eshelby JD (1957) The determination of the elastic field of an ellipsoidal inclusion, and related problems. In Proceedings of the royal society of London A: mathematical, physical and engineering sciences 241 1226 376-396). The Royal Society London

Fichtner A (2010) Full seismic waveform modelling and inversion. Springer Science & Business Media, New York

Fujii Y, Satake K, Sakai SI, Shinohara M, Kanazawa T (2011) Tsunami source of the 2011 off the Pacific coast of Tohoku earthquake. Earth Planets and Space 63(7):55. https://doi.org/10.5047/eps.2011.06.010

Geist EL (1998) Local tsunamis and earthquake source parameters. Adv Geophys 39:117–209. https://doi.org/10.1016/S0065-2687(08)60276-9

Geist EL, Dmowska R (1999) Local tsunamis and distributed slip at the source. Pure Appl Geophys 154:485–512. https://doi.org/10.1007/978-3-0348-8679-6_6

Graves RW (1996) Simulating seismic wave propagation in 3D elastic media using staggered-grid finite differences. Bull Seismol Soc Am 86(4):1091–1106

Gusman AR, Tanioka Y (2014) W phase inversion and tsunami inundation modeling for tsunami early warning: case study for the 2011 Tohoku event. Pure Appl Geophys 171(7):1409–1422. https://doi.org/10.1007/s00024-013-0680-z

Hasegawa A, Sato H, Nishimura T (2015) Introduction to modern earth science series, vol. 6, Seismology, Kyoritsu Shuppan (in Japanese)

Inazu D, Pulido N, Fukuyama E, Saito T, Senda J, Kumagai H (2016) Near-field tsunami forecast system based on near real-time seismic moment tensor estimation in the regions of Indonesia, the Philippines, and Chile. Earth, Planets and Space 68(1):73. https://doi.org/10.1186/s40623-016-0445-x

Jamelot A, Reymond D (2015) New tsunami forecast tools for the French Polynesia tsunami warning system Part II: numerical modelling and tsunami height estimation. Pure Appl Geophys 172:805–819. https://doi.org/10.1007/s00024-014-0997-2

Kajiura K (1963) The leading wave of a tsunami. Bull Earthquake Res Inst 41:535–571

Kanamori H, Anderson DL (1975) Theoretical basis of some empirical relations in seismology. Bull Seismol Soc Am 65(5):1073–1095

Maeda T, Takemura S, Furumura T (2017) OpenSWPC: an open-source integrated parallel simulation code for modeling seismic wave propagation in 3D heterogeneous viscoelastic media. Earth, Planets and Space 69(1):102. https://doi.org/10.1186/s40623-017-0687-2

Maruyama T (1963) On the force equivalents of dynamical elastic dislocations with reference to the earthquake mechanism. Bull Earthquake Res Inst 41:467–486

Murotani S, Satake K, Fujii Y (2013) Scaling relations of seismic moment, rupture area, average slip, and asperity size for M~9 subduction-zone earthquakes. Geophys Res Lett 40(19):5070–5074. https://doi.org/10.1002/grl.50976

Moczo P (2002) 3D heterogeneous staggered-grid finite-difference modeling of seismic motion with volume harmonic and arithmetic averaging of elastic moduli and densities. Bull Seismol Soc Am 92(8):3042–3066

Moczo P, Kristek J, Gális M (2014) The finite-difference modelling of earthquake motions: waves and ruptures. Cambridge University Press, Cambridge

Okada Y (1992) Internal deformation due to shear and tensile faults in a half-space. Bull Seismol Soc Am 82(2):1018–1040

Saito T (2017) Tsunami generation: validity and limitations of conventional theories. Geophys J Int 210:1888–1900. https://doi.org/10.1093/gji/ggx275

Takahashi R (1942) On seismic sea waves caused by deformations of the sea bottom. Bull Earthquake Res Inst 20:357–400

Wang D et al (2012) Real-time forecasting of the April 11, 2012 Sumatra tsunami. Geophys Res Lett 39(19):L19601. https://doi.org/10.1029/2012GL053081

Chapter 5
Tsunami Generation

Abstract This chapter theoretically investigates tsunami generation. When an earthquake occurs in an offshore region, seismic waves, ocean acoustic waves, and tsunami are excited. Although the compressibility and elasticity of the sea layer are important for the propagation of ocean acoustic waves and high-frequency seismic waves, we may assume that the sea layer is incompressible for tsunami. This chapter is based on incompressible fluid dynamics. The theory gives the analytical solutions for tsunami generation and propagation, by which we would be able to understand the mechanism behind these phenomena in addition to describing the motion. Section 5.1 explains the difference between ocean acoustic waves and tsunami. In Sect. 5.2, a linear potential theory is formulated for the tsunami generation process in a water with uniform depth. Analytical solutions for the sea-surface displacement, velocity, and pressure field in the seawater are derived. In Sect. 5.3, we examine the analytical solutions for tsunami generation and propagation. The mathematical equations can directly provide us with a clear perspective on the tsunami mechanism. In Sect. 5.4, we bridge the gap between the analytical solutions derived under a constant sea-depth assumption and tsunami simulations with realistic bathymetry. The theoretical background of the initial conditions in the numerical simulations is explained.

Keywords Linear potential theory · Incompressible fluid · Initial tsunami height distribution · Dynamic pressure change · Static pressure change

5.1 Ocean Acoustic Waves and Tsunami: Different Driving Forces

Earthquakes excite seismic waves and tsunami. Seawater is characterized by zero rigidity. In other words, shear stress is zero in the sea, even if shear strain is considerably large. As a result, the *S* wave that is excited by shear stress cannot propagate through the sea layer. Only the *P* wave can propagate through the sea layer. The ocean acoustic wave is nothing but the *P* wave trapped within the sea layer

T. Saito, *Tsunami Generation and Propagation*, Springer Geophysics,
https://doi.org/10.1007/978-4-431-56850-6_5

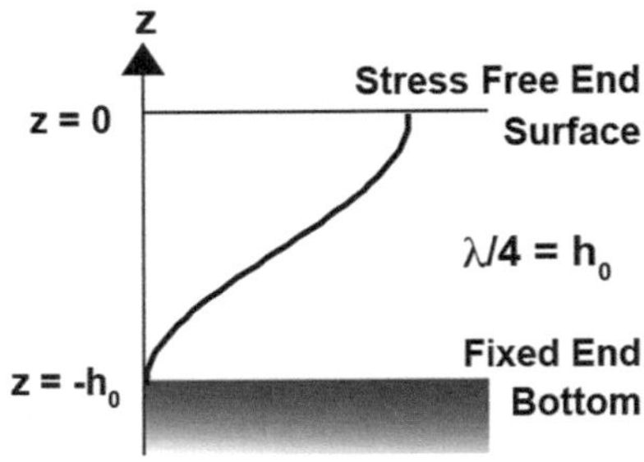

Fig. 5.1 A simple model for a fundamental mode in a region bounded by a free end and a fixed end

from the viewpoint of elastic wave theory. If the period is longer, the *P* wave wavelength becomes longer, which results in the energy extending significantly into the crust. The wave may be considered as a seismic *P* wave rather than an ocean acoustic wave, because the restoring force of the wave is mainly caused by the elasticity of the crust rather than the water layer.

Let us roughly estimate the longest period of the ocean acoustic waves. When the phase velocity of the ocean acoustic wave is c_0 and the period is T_0, the wavelength is given by $\lambda = c_0 T_0$. Assuming a boundary condition at the sea surface as a stress-free boundary and one at the sea bottom as a rigid boundary (fixed end), the fundamental mode of the ocean acoustic wave is given by $\lambda/4 = h_0$ where h_0 is the sea depth (Fig. 5.1). This gives the period of the ocean acoustic wave as $T_0 = 4h_0/c_0$. This is the period of the fundamental mode or the mode of the longest period in this system. When considering the average depth of the Pacific Ocean to be $h_0 = 4000$ m and the ocean acoustic wave velocity to be $c_0 = 1.5$ km/s, then the period is $T_0 \sim 10$ s. This indicates that the period of the ocean acoustic wave should be shorter than $T_0 < \sim 10$ s so that the *P* wave is trapped within the sea layer. When the sea depth becomes smaller, the characteristic period becomes smaller. If the sea depth is $h_0 = 1000$ m, the characteristic time is $T_0 \sim 2.5$s. The derivation here is a very simple one. Section 3.3.2 *Ocean Acoustic Waves* describes the theory of ocean acoustic waves and their critical time based on a more general approach.

We then consider tsunami. Near the earthquake fault, the fault slip causes permanent displacement. The gravity collapses the permanent displacement of the sea surface as shown in Fig. 5.2. The sea layer cannot keep the displacement distribution (deformation) at the sea surface since the water layer does not hold shear stress. As a result, the water volume uplifted by the earthquake flows downward. This downward flow of large volume of water causes long-wavelength horizontal flow in the sea layer from the bottom to the surface. This volume fluctuation propagates a long distance as a tsunami. Apparently, the main force causing the tsunami is gravity. The tsunami propagation velocity is roughly given by $\sqrt{g_0 h_0}$ when the wavelength is much longer than the sea depth (Chap. 3). If we consider the tsunami wavelength to be $\lambda = 100$ km and $h_0 = 4000$ m, the dominant period is $T_0 = \lambda/\sqrt{g_0 h_0}\tilde{\ }500\,\mathrm{s}$.

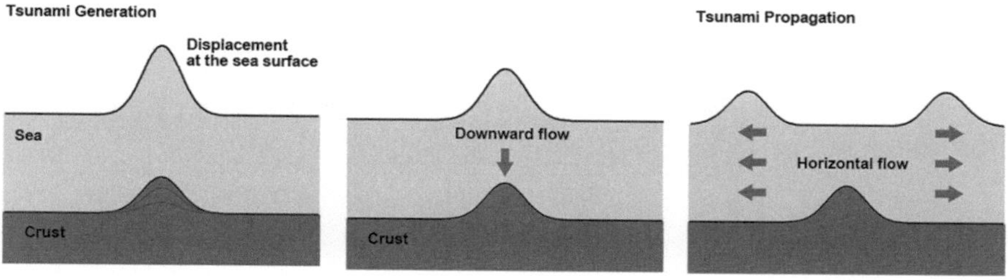

Fig. 5.2 A schematic illustration of tsunami generation and propagation

As stated above, earthquakes excite both ocean acoustic waves and tsunami. Note that both ocean acoustic waves and tsunami produce the sea-surface displacement. However, the excitation mechanisms are totally different. The difference originates from the restoring force. The seismic wave (ocean acoustic wave) requires the seawater to exhibit elasticity. Tsunami propagation requires gravity as a restoring force. As a consequence, the time scale (the dominant period) of the tsunami is often longer than ~10 min, which is much longer than that of the ocean acoustic wave < ~10 s. This substantial difference in the time scales allows us to consider the tsunami and ocean acoustic waves separately in many practical cases. Section 4.3.2*: Seismic Waves, Ocean Acoustic Waves, and Permanent Displacement* described the seismic wave propagation in elastic media. This chapter considers tsunami.

5.2 Linear Potential Theory

The tsunami generation process is described by a linear incompressible fluid theory in which the elasticity or compressibility of the seawater is neglected. Considering appropriate boundary conditions at the sea surface using the gravitational acceleration g_0 and giving the sea bottom displacement as a source of the tsunami, the tsunami generation theory describes the water motion in the sea layer. A solution represented in the wavenumber-frequency domain was found in Takahashi (1942) for the case of a constant sea depth. Kajiura (1963) derived an analytical solution for the vertical displacement at the sea surface caused by instantaneous sea-bottom deformation (i.e., the sea-bottom displacement represented by a step function with respect to time). The analytical solution is often referred to as Kajiura's equation, which is frequently used in the setting of the initial tsunami height distribution for tsunami propagation simulations (e.g., Tanioka and Seno 2001). The solutions are applicable just after the sea-bottom deformation ends but cannot describe the wavefield in the sea layer during the generation (e.g., Kervella et al. 2007). Saito (2013) derived more general solutions, which can describe the water particle motion through the sea layer during the tsunami generation. In the following sections, the theory is illustrated.

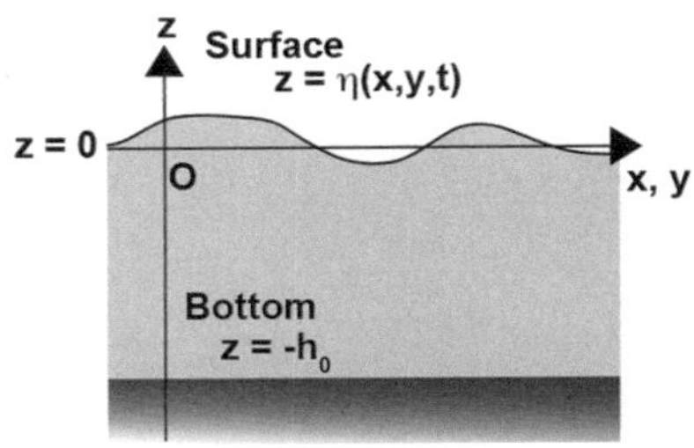

Fig. 5.3 Coordinates used in the formulation

5.2.1 Governing Equations

Cartesian coordinates (Fig. 5.3) are used for the formulation. The z-axis runs vertically upward, and the x- and y-axes lie in a horizontal plane. The sea surface at rest is located at $z = 0$, and the flat sea bottom is at $z = -h_0$. The sea-surface height is given by $z = \eta\ (x, y, t)$.

The particle velocity in the fluid is given by the vector $\mathbf{v}\ (\mathbf{x}, t) = v_x\mathbf{e_x} + v_y\mathbf{e_y} + v_z\mathbf{e_z}$ where $\mathbf{x} = x\mathbf{e_x} + y\mathbf{e_y} + z\mathbf{e_z}$ and $\mathbf{e_x}$, $\mathbf{e_y}$, and $\mathbf{e_z}$ are the basis unit vectors in the x-, y-, and z-axes, respectively. The sea-surface height $\eta(x, y, t)$ is assumed to be small enough compared with the water depth, i.e., $|\eta| \ll h_0$. An incompressible medium does not allow dilatation as

$$\nabla \cdot \mathbf{v} = 0. \tag{5.1}$$

When an irrotational flow is assumed (rot $\mathbf{v}(\mathbf{x}, t) = 0$), the velocity vector is represented using the velocity potential $\phi(\mathbf{x}, t)$ as

$$\mathbf{v}(\mathbf{x}, t) = \nabla\phi(\mathbf{x}, t). \tag{5.2}$$

Substitution of Eq. (5.2) into Eq. (5.1) gives

$$\Delta\phi(\mathbf{x}, t) = 0. \tag{5.3}$$

In an incompressible medium, the velocity potential $\phi(\mathbf{x}, t)$ needs to satisfy the Laplace equation.

We then consider the equation of motion in the sea layer. The linear equation of motion is given by

$$\frac{\partial \mathbf{v}(\mathbf{x}, t)}{\partial t} = -\frac{1}{\rho_0}\nabla p(\mathbf{x}, t) + \mathbf{g}_0, \tag{5.4}$$

where ρ_0 is the water density that is assumed to be constant, $p(\mathbf{x}, t)$ is the pressure, and $\mathbf{g}_0$ is the body force due to gravity. This equation is also derived if we assume a constant density through the medium and a shear-wave velocity or rigidity of zero in the equation for the elastic medium (Eqs. (3.14) and (3.29)).

Consider a case when the pressure field is independent of time, that is, when the water is static or in an equilibrium state. Equation (5.4) becomes

$$0 = -\frac{1}{\rho_0}\nabla p_0(\mathbf{x}) + \mathbf{g}_0. \tag{5.5}$$

The pressure $p_0(\mathbf{x})$ is referred to as hydrostatic pressure. The pressure $p(\mathbf{x}, t)$ in Eq. (5.4) is represented by the sum of the hydrostatic pressure $p_0(\mathbf{x})$ and the fluctuation $p_e(\mathbf{x}, t)$ caused by the fluid motion as

$$p(\mathbf{x}, t) = p_0(\mathbf{x}) + p_e(\mathbf{x}, t). \tag{5.6}$$

Substituting Eq. (5.6) into (5.4) and using Eq. (5.5) gives the following equation

$$\begin{aligned}\frac{\partial \mathbf{v}(\mathbf{x}, t)}{\partial t} &= -\frac{1}{\rho_0}\nabla p_0(\mathbf{x}) - \frac{1}{\rho_0}\nabla p_e(\mathbf{x}, t) + \mathbf{g}_0 \\ &= -\frac{1}{\rho_0}\nabla p_e(\mathbf{x}, t).\end{aligned}$$

Then, using Eq. (5.2), we obtain

$$\nabla\frac{\partial \phi(\mathbf{x}, t)}{\partial t} = -\frac{1}{\rho_0}\nabla p_e(\mathbf{x}, t).$$

Since that $p_e(\mathbf{x}, t)$ should be zero when the system is in an equilibrium state, the pressure due to the motion $p_e(\mathbf{x}, t)$ is represented by the velocity potential $\phi(\mathbf{x}, t)$ as

$$p_e(\mathbf{x}, t) = -\rho_0\frac{\partial \phi(\mathbf{x}, t)}{\partial t}. \tag{5.7}$$

Let us consider the boundary condition of the velocity potential $\phi(\mathbf{x}, t)$ at the sea surface $z = 0$. The vertical displacement at the surface $\eta(x, y, t)$ is given by an integration of the vertical velocity at the surface as

$$\eta(x, y, t) = \int_{-\infty}^{t} v_z(x, y, z = 0, t)dt.$$

By using the velocity potential (5.2), the surface $\eta(x, y, t)$ is represented as

$$\eta(x, y, t) = \int_{-\infty}^{t} \left.\frac{\partial \phi(\mathbf{x}, t)}{\partial z}\right|_{z=0} dt. \tag{5.8}$$

Differentiating Eq. (5.8) with respect to time gives

$$\frac{\partial \eta(x, y, t)}{\partial t} = \left.\frac{\partial \phi(\mathbf{x}, t)}{\partial z}\right|_{z=0}. \tag{5.9}$$

This represents the relation between the motion and the shape of the surface. It is referred to as the kinematic boundary condition.

We then consider the pressure at the surface. When the sea-surface height is $\eta(x, y, t)$, the pressure at $z = 0$ is given by $p_e(z = 0) = \rho_0 g_0 \eta(x, y, t)$ (Eq. (3.34)). Considering that the pressure is represented by the velocity potential in Eq. (5.7), the velocity potential satisfies

$$-\left.\frac{\partial \phi(\mathbf{x}, t)}{\partial t}\right|_{z=0} = g_0 \eta(x, y, t), \tag{5.10}$$

at $z = 0$. Since this condition is related to the motion and the force (pressure), this is referred to as the dynamic boundary condition. By using kinematic and dynamic boundary conditions (Eqs. (5.9) and (5.10)) and deleting the surface height $\eta(x, y, t)$, we obtain

$$\left.\frac{\partial^2 \phi(\mathbf{x}, t)}{\partial t^2}\right|_{z=0} + g_0 \left.\frac{\partial \phi(\mathbf{x}, t)}{\partial z}\right|_{z=0} = 0, \tag{5.11}$$

as the boundary condition of the velocity potential $\phi(\mathbf{x}, t)$ at the surface. The boundary condition of Eq. (5.11) at the sea surface is common in fluid dynamics (e.g., Kambe 2007; Pedlosky 2013).

Then, we consider the boundary condition at the sea bottom ($z = -h_0$). In many fluid dynamic problems, assuming the continuity of the vertical displacement on a rigid sea bottom, the vertical velocity is set as zero at the sea bottom. This is a boundary condition for the propagation problem. On the other hand, in the tsunami generation problem, we describe the motion at the sea bottom as the tsunami source (e.g., Takahashi 1942). Supposing the permanent vertical displacement at the sea bottom is given by $d(x, y)$, the vertical velocity at the sea bottom is represented as $d(x, y)\chi(t)$, where a function $\chi(t)$ has a dimension of inverse of the time. The function $\chi(t)$ represents the time evolution of the displacement at the bottom. Since it represents the rate of sea-bottom displacement, we refer to $\chi(t)$ as the rate function. The boundary condition of the velocity potential at the bottom is given by

$$v_z(x, y, z = -h_0, t) = \left.\frac{\partial \phi(\mathbf{x}, t)}{\partial z}\right|_{z=-h_0} = d(x, y)\chi(t). \tag{5.12}$$

The rate function $\chi(t)$ needs to satisfy

$$\int_{-\infty}^{\infty} \chi(t)dt = 1. \tag{5.13}$$

For example, when the sea-bottom deformation occurs with an infinitely short duration time at $t = 0$, the function $\chi(t)$ is given by the delta function:

$$\chi(t) = \delta(t). \tag{5.14}$$

On one hand, the sea-bottom deformation occurs during the finite duration t_c, which is closely related to the rise time in the earthquake fault motion. The function $\chi(t)$ is, for example, given by

$$\chi(t) = \frac{1}{\sqrt{\pi}} \frac{4}{t_c} \exp\left[-\left(\frac{t - t_0}{t_c/4} \right)^2 \right], \tag{5.15}$$

where t_c determines the duration and t_0 controls the time when the deformation rate reaches its maximum. Figure 5.4a shows the function $\chi(t)$ when $t_c = 40$ s and $t_0 = 30$ s.

We have listed all the ingredients of the tsunami generation formulation. The wave motion $\mathbf{v}(\mathbf{x}, t)$ and the pressure field $p_e(\mathbf{x}, t)$ can be described by Eqs. (5.2) and (5.7), respectively, using the velocity potential $\phi(\mathbf{x}, t)$. The velocity potential $\phi(\mathbf{x}, t)$ satisfies the Laplace equation (Eq. (5.3)) through the water medium. The boundary conditions at the surface and the bottom are given by Eqs. (5.11) and (5.12), respectively. To find the solution of the velocity potential that satisfies Eqs. (5.3), (5.11), and (5.12) is a mathematical problem in the study of tsunami generation and propagation.

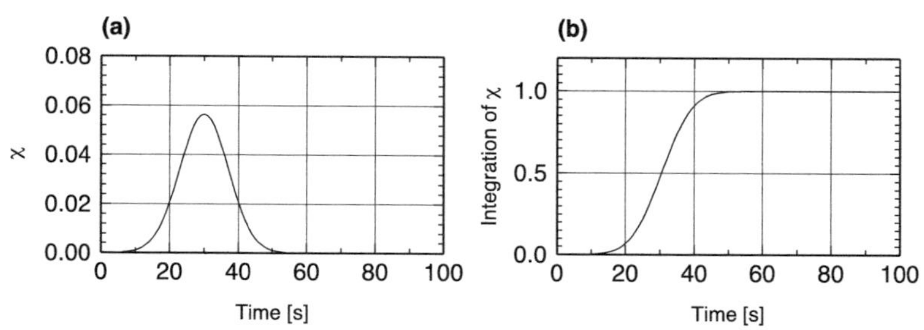

Fig. 5.4 (**a**) A rate function $\chi(t)$ and (**b**) an integration of $\chi(t)$ with respect to time t. The parameters are set at $t_c = 40$ s and $t_0 = 30$ s in Eq. (5.15)

5.2.2 Solving Laplace Equation with Boundary Conditions

By taking the 2-D Fourier transform in space-wavenumber domain and the Fourier transform in time-angular frequency domain, we transform $\phi(x, y, z, t)$ to $\widehat{\phi}(k_x, k_y, z, \omega)$ as,

$$\widehat{\phi}(k_x, k_y, z, \omega) = \int\int_{-\infty}^{\infty} dxdy \int_{-\infty}^{\infty} dt\, e^{-i(k_x x + k_y y)} e^{i\omega t} \phi(x, y, z, t). \tag{5.16}$$

The corresponding inverse Fourier transform is

$$\phi(x, y, z, t) = \frac{1}{(2\pi)^2} \int\int_{-\infty}^{\infty} dk_x dk_y \frac{1}{2\pi} \int_{-\infty}^{\infty} d\omega\, e^{i(k_x x + k_y y)} e^{-i\omega t} \widehat{\phi}(k_x, k_y, z, \omega). \tag{5.17}$$

By using this definition of the Fourier transform, we rewrite the Laplace equation (Eq. (5.3)) as

$$\frac{d^2}{dz^2} \widehat{\phi}(k_x, k_y, z, \omega) = k^2 \widehat{\phi}(k_x, k_y, z, \omega), \tag{5.18}$$

where $k^2 = {k_x}^2 + {k_y}^2$. Since, in Eq. (5.18), the values of k_x, k_y, and ω are considered to be constant values rather than independent variables, we used a representation of d/dz instead of $\partial/\partial z$. A general solution of this ordinary differential equation of (5.18) is given by

$$\widehat{\phi}(k_x, k_y, z, \omega) = A \cosh(kz) + B \sinh(kz). \tag{5.19}$$

The coefficients A and B should be chosen so as to satisfy the boundary conditions.

The boundary condition at the surface (Eq. (5.11)) is given by

$$\left(\frac{d}{dz} - \frac{\omega^2}{g_0} \right) \widehat{\phi}(k_x, k_y, z, \omega) \bigg|_{z=0} = 0, \tag{5.20}$$

and that at the bottom (Eq. (5.12)) is given by

$$\frac{d}{dz} \widehat{\phi}(k_x, k_y, z, \omega) \bigg|_{z=-h_0} = \widehat{d}(k_x, k_y) \widehat{\chi}(\omega), \tag{5.21}$$

in the wavenumber-angular frequency domain when the Fourier transform (Eq. (5.16)) is used, where $\widehat{d}(k_x, k_y)$ is the 2-D spatial Fourier transform of $d(x, y)$ and $\widehat{\chi}(\omega)$ is the Fourier transform of $\chi(t)$.

Substituting Eq. (5.19) into Eq. (5.20) and Eq. (5.21) gives

$$-\frac{\omega^2}{g_0}A + kB = 0,$$

and

$$-kA\sinh kh_0 + kB\cosh kh_0 = \widehat{d}\left(k_x, k_y\right)\widehat{\chi}(\omega).$$

Then, the coefficients A and B are given by

$$A = \frac{1}{(k\omega^2/g_0)\cosh kh_0 - k^2\sinh kh_0} k\widehat{d}\left(k_x, k_y\right)\widehat{\chi}(\omega),$$

and

$$B = \frac{1}{(k\omega^2/g_0)\cosh kh_0 - k^2\sinh kh_0} \frac{\omega^2}{g_0}\widehat{d}\left(k_x, k_y\right)\widehat{\chi}(\omega).$$

By substituting the estimated A and B into Eq. (5.19), we obtain

$$\widehat{\phi}\left(k_x, k_y, z, \omega\right) = \frac{1}{k}\frac{\omega^2\sinh(kz) + g_0 k\cosh(kz)}{\omega^2\cosh(kh_0) - g_0 k\sinh(kh_0)}\widehat{d}\left(k_x, k_y\right)\widehat{\chi}(\omega). \tag{5.22}$$

We then obtain a solution in time and space by the inverse Fourier transform as follows:

$$\begin{aligned}\phi(x, y, z, t) = \frac{1}{2\pi}\int_{-\infty}^{\infty} e^{-i\omega t}\widehat{\chi}(\omega)d\omega \frac{1}{(2\pi)^2}\int_{-\infty}^{\infty}\int_{-\infty}^{\infty} dk_x dk_y\, e^{i\left(k_x x + k_y y\right)} \\ \frac{1}{k}\frac{\omega^2\sinh(kz) + g_0 k\cosh(kz)}{\omega^2\cosh(kh_0) - g_0 k\sinh(kh_0)}\widehat{d}\left(k_x, k_y\right).\end{aligned} \tag{5.23}$$

This is a formal expression of the velocity potential using the inverse Fourier transform. This formal expression was shown in Takahashi (1942) using cylindrical coordinates. This form is still not suitable for examining the mechanism of the tsunami generation process. If the integration of the angular frequency ω were performed, we would be able to interpret the analytical solution in order to understand the mechanism of tsunami generation and propagation. The main difficulty with respect to the integration of the angular frequency ω is that the residue theorem is not applicable when $\chi(t)$ is an arbitrary function or is given by the delta function $\chi(t) = \delta(t)$.

In order to conduct the integration with respect to the angular frequency using the residue theorem, we first consider a special case in which the rate function $\chi(t)$ is represented as

$$\chi(t) = \begin{cases} \frac{1}{T} & \text{for } 0 < t < T \\ 0 & \text{for others} \end{cases} \\ = \frac{1}{T}[H(t) - H(t-T)], \tag{5.24}$$

where a function $H(t)$ is defined as

$$H(t) = \begin{cases} 0 & \text{for } t \leq 0 \\ 1 & \text{for } t > 0. \end{cases} \tag{5.25}$$

The Fourier transform of the rate function is given by

$$\hat{\chi}(\omega) = \int_{-\infty}^{\infty} \chi(t) e^{i\omega t} dt = \frac{1}{i\omega T}\left(e^{i\omega T} - 1\right). \tag{5.26}$$

Substitution of Eq. (5.26) into Eq. (5.23) gives Eq. (5.27) and Eq. (5.28)

$$\begin{aligned} &\phi(x,y,z,t) \\ &= \frac{1}{(2\pi)^2} \int_{-\infty}^{\infty} \int_{-\infty}^{\infty} dk_x dk_y \, e^{i(k_x x + k_y y)} \frac{1}{k} \frac{\hat{d}(k_x,k_y)}{\cosh(kh_0)} \\ &\qquad \frac{1}{2\pi} \frac{i}{T} \int_{-\infty}^{\infty} d\omega \, e^{-i\omega t} \frac{1 - e^{i\omega T}}{\omega} \frac{\omega^2 \sinh(kz) + g_0 k \cosh(kz)}{\omega^2 - g_0 k \tanh(kh_0)} \\ &= \frac{1}{(2\pi)^2} \int_{-\infty}^{\infty} \int_{-\infty}^{\infty} dk_x dk_y \, e^{i(k_x x + k_y y)} \frac{1}{k} \frac{\hat{d}(k_x,k_y)}{\cosh(kh_0)} \\ &\quad \frac{1}{2\pi} \frac{i}{T} \left[\int_{-\infty}^{\infty} d\omega \frac{e^{-i\omega t}}{\omega} \frac{\omega^2 \sinh(kz) + g_0 k \cosh(kz)}{\omega^2 - g_0 k \tanh(kh_0)} \right. \\ &\qquad \left. - \int_{-\infty}^{\infty} d\omega \frac{e^{-i\omega(t-T)}}{\omega} \frac{\omega^2 \sinh(kz) + g_0 k \cosh(kz)}{\omega^2 - g_0 k \tanh(kh_0)} \right]. \end{aligned} \tag{5.27}$$

We perform an integration with respect to ω by using the residue theorem in the complex ω plane by recognizing that the poles are located at $\omega = 0, \pm\sqrt{g_0 k \tanh(kh_0)}$ (Fig. 5.5). The detailed procedure is illustrated in Appendix 5A. The result of the integrations is

$$\begin{aligned} &\phi(x,y,z,t) \\ &= -\frac{1}{(2\pi)^2} \int_{-\infty}^{\infty} \int_{-\infty}^{\infty} dk_x dk_y \, e^{i(k_x x + k_y y)} \frac{\hat{d}(k_x,k_y)}{k} \\ &\quad \left[\frac{\cosh(kz)}{\sinh(kh_0)} \frac{H(t-T) - H(t)}{T} \right. \\ &\qquad \left. - \left(\frac{\cosh(kz)}{\sinh(kh_0)} + \frac{\sinh(kz)}{\cosh(kh_0)} \right) \frac{H(t)\cos(\omega_0 t) - H(t-T)\cos(\omega_0 (t-T))}{T} \right], \end{aligned} \tag{5.28}$$

where the angular frequency ω_0 is given by

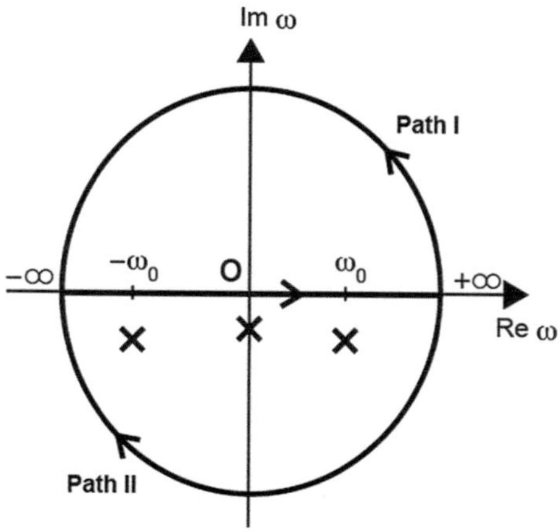

Fig. 5.5 Poles in the complex ω plane for tsunami generation and propagation

$$\omega_0 = \sqrt{g_0 k \tanh(kh_0)}. \tag{5.29}$$

Equation (5.29) is the dispersion relation of ocean waves propagating using gravity as the restoring force (see Eq. (3.47)).

Now, defining a function $\psi(x, y, z, t)$ as

$$\begin{aligned}\psi(x, y, z, t) = &-\frac{1}{(2\pi)^2}\int_{-\infty}^{\infty}\int_{-\infty}^{\infty} dk_x dk_y\, e^{i(k_x x + k_y y)}\frac{\widehat{d}(k_x, k_y)}{k}\\ &\left[\frac{\cosh(kz)}{\sinh(kh_0)} - \left(\frac{\cosh(kz)}{\sinh(kh_0)} + \frac{\sinh(kz)}{\cosh(kh_0)}\right)\cos(\omega_0 t)\right]H(t),\end{aligned} \tag{5.30}$$

we rewrite Eq. (5.28) as

$$\phi(x, y, z, t) = \frac{\psi(x, y, z, t) - \psi(x, y, z, t - T)}{T}. \tag{5.31}$$

Equations (5.30) and (5.31) represent the velocity potential for the sea-bottom deformation when the rate function of the sea-bottom deformation is given by a special form of Eq. (5.24).

5.2.2.1 Impulse Response

We then consider a general solution of the velocity potential for any rate function. When T approaches zero in Eq. (5.24), the rate function $\chi(t)$ approaches the delta function as

$$\chi(t) = \lim_{T\to 0} \frac{H(t) - H(t-T)}{T} = \delta(t). \tag{5.32}$$

Therefore, we obtain the velocity potential for the delta function type rate function (or impulsive rate function) by making T approach zero in Eq. (5.31). The corresponding velocity potential function $\phi_{\text{impulse}}(t)$ is given by

$$\phi_{\text{Impulse}}(\mathbf{x}, t) = \lim_{T\to 0} \frac{\psi(\mathbf{x}, t) - \psi(\mathbf{x}, t-T)}{T} = \frac{\partial \psi(\mathbf{x}, t)}{\partial t}. \tag{5.33}$$

Substituting Eq. (5.30) into Eq. (5.33), we obtain

$$\begin{aligned}
&\phi_{\text{Impulse}}(\mathbf{x}, t) \\
&\quad = \frac{\partial \psi(\mathbf{x}, t)}{\partial t} \\
&\quad = -\frac{1}{(2\pi)^2} \int_{-\infty}^{\infty} \int_{-\infty}^{\infty} dk_x dk_y \, e^{i(k_x x + k_y y)} \frac{\widehat{d}(k_x, k_y)}{k} \left[\frac{\cosh(kz)}{\sinh(kh_0)} \delta(t) \right. \\
&\qquad + \left(\frac{\cosh(kz)}{\sinh(kh_0)} + \frac{\sinh(kz)}{\cosh(kh_0)} \right) \omega_0 \sin(\omega_0 t) H(t) \\
&\qquad \left. - \left(\frac{\cosh(kz)}{\sinh(kh_0)} + \frac{\sinh(kz)}{\cosh(kh_0)} \right) \cos(\omega_0 t) \delta(t) \right] \\
&\quad = -\frac{1}{(2\pi)^2} \int_{-\infty}^{\infty} \int_{-\infty}^{\infty} dk_x dk_y \, e^{i(k_x x + k_y y)} \frac{\widehat{d}(k_x, k_y)}{k} \left[\frac{\cosh(kz)}{\sinh(kh_0)} \delta(t) \right. \\
&\qquad + \left(\frac{\cosh(kz)}{\sinh(kh_0)} + \frac{\sinh(kz)}{\cosh(kh_0)} \right) \omega_0 \sin(\omega_0 t) H(t) \\
&\qquad \left. - \left(\frac{\cosh(kz)}{\sinh(kh_0)} + \frac{\sinh(kz)}{\cosh(kh_0)} \right) \delta(t) \right],
\end{aligned}$$

where we set $t = 0$ in the term including $\delta(t)$. We calculate using a trigonometric sum identity as

$$\begin{aligned}
&\phi_{\text{Impulse}}(\mathbf{x}, t) \\
&= \frac{1}{(2\pi)^2} \int_{-\infty}^{\infty} \int_{-\infty}^{\infty} dk_x dk_y \, e^{i(k_x x + k_y y)} \frac{\widehat{d}(k_x, k_y)}{\cosh(kh_0)} \\
&\quad \left\{ -\frac{\omega_0}{k} \left[\cosh(kz) \frac{\cosh(kh_0)}{\sinh(kh_0)} + \sinh(kz) \right] \sin(\omega_0 t) H(t) + \frac{1}{k} \sinh(kz) \delta(t) \right\} \\
&= \frac{1}{(2\pi)^2} \int_{-\infty}^{\infty} \int_{-\infty}^{\infty} dk_x dk_y \, e^{i(k_x x + k_y y)} \frac{\widehat{d}(k_x, k_y)}{\cosh(kh_0)} \\
&\quad \left\{ -\frac{\omega_0}{k} \frac{\cosh[k(z + h_0)]}{\sinh kh_0} \sin(\omega_0 t) H(t) + \frac{1}{k} \sinh(kz) \delta(t) \right\}.
\end{aligned} \tag{5.34}$$

Equation (5.34) is the solution of the velocity potential with respect to the delta function or the impulse response. The solution with respect to any rate function $\chi(t)$ is given by the convolution:

$$\phi(\mathbf{x}, t) = \int_{-\infty}^{\infty} \phi_{\text{Impulse}}(\mathbf{x}, t-\tau)\chi(\tau)d\tau. \tag{5.35}$$

5.2.2.2 Velocity Field

Since we have obtained an analytical representation of the velocity potential (5.34), the velocity field is given by the gradient of the velocity potential (Eq. (5.2)). Using the horizontal gradient ∇_H defined as

$$\nabla_H = \frac{\partial}{\partial x}\mathbf{e_x} + \frac{\partial}{\partial y}\mathbf{e_y}, \tag{5.36}$$

the horizontal velocity vector $\mathbf{v}_H(\mathbf{x}, t)$ is represented as the following equation

$$\begin{aligned} \mathbf{v}_H(\mathbf{x}, t) &= \nabla_H \phi_{\text{Impulse}}(\mathbf{x}, t) \\ &= \frac{1}{(2\pi)^2}\int_{-\infty}^{\infty}\int_{-\infty}^{\infty} dk_x dk_y\, e^{i(k_x x + k_y y)}\frac{\widehat{d}(k_x, k_y)}{\cosh(kh_0)} \\ &\left\{-i\omega_0\frac{\mathbf{k}_H}{k}f_H(k, z, h_0)\sin(\omega_0 t)H(t) + \frac{i\mathbf{k}_H}{k}\sinh(kz)\delta(t)\right\}, \end{aligned} \tag{5.37}$$

where $\mathbf{k}_H$ is defined as $\mathbf{k}_H \equiv k_x\mathbf{e_x} + k_y\mathbf{e_y}$ and the function $f_H(k, z, h_0)$ is defined as

$$f_H(k, z, h_0) = \frac{\cosh[k(z+h_0)]}{\sinh(kh_0)}. \tag{5.38}$$

The vertical velocity $v_z(\mathbf{x}, t)$ is represented as

$$\begin{aligned} v_z(\mathbf{x}, t) &= \frac{\partial \phi_{\text{Impulse}}(\mathbf{x}, t)}{\partial z} \\ &= \frac{1}{(2\pi)^2}\int_{-\infty}^{\infty}\int_{-\infty}^{\infty} dk_x dk_y\, e^{i(k_x x + k_y y)}\frac{\widehat{d}(k_x, k_y)}{\cosh(kh_0)} \\ &\{-\omega_0 f_z(k, z, h_0)\sin(\omega_0 t)H(t) + \cosh(kz)\delta(t)\}, \end{aligned} \tag{5.39}$$

where the function $f_z(k, z, h_0)$ is defined as

$$f_z(k,z,h_0) = \frac{\sinh[k(z+h_0)]}{\sinh(kh_0)}. \tag{5.40}$$

The functions $f_H(k,z,h_0)$ and $f_z(k,z,h_0)$ indicate the velocity-amplitude distributions for the horizontal and vertical components, respectively, at a depth z in a sea of a constant depth h_0. These velocity distributions are peculiar to the propagation process. The same functions appear in Eqs. (3.53) and (3.55) for the propagation problem in Chap. 3. The distribution functions will be explained later (Fig. 5.13).

The vertical displacement at the surface, or the tsunami height $\eta(x,y,t)$, is given by the velocity potential (see Eq. (5.10)) as

$$\begin{aligned}\eta(x,y,t) &= -\frac{1}{g_0}\frac{\partial\phi_{\text{Impulse}}(\mathbf{x},t)}{\partial t}\bigg|_{z=0}\\ &= \frac{1}{(2\pi)^2}\int_{-\infty}^{\infty}\int_{-\infty}^{\infty} dk_x dk_y\, e^{i(k_x x+k_y y)}\frac{\widehat{d}(k_x,k_y)}{\cosh(kh_0)}\cos(\omega_0 t)H(t).\end{aligned} \tag{5.41}$$

The pressure change due to the fluid motion $p_e(\mathbf{x},t)$ is represented by the velocity potential (Eq. (5.7)) as

$$\begin{aligned}&p_e(\mathbf{x},t)\\ &= -\rho_0\frac{\partial\phi_{\text{Impulse}}(\mathbf{x},t)}{\partial t}\\ &= \frac{1}{(2\pi)^2}\int_{-\infty}^{\infty}\int_{-\infty}^{\infty} dk_x dk_y\, e^{i(k_x x+k_y y)}\frac{\rho_0\widehat{d}(k_x,k_y)}{\cosh(kh_0)}\\ &\quad\left\{\frac{\omega_0^2}{k}\frac{\cosh[k(z+h_0)]}{\sinh kh_0}\cos(\omega_0 t)H(t)\right.\\ &\qquad\left.+\frac{\omega_0}{k}\frac{\cosh[k(z+h_0)]}{\sinh kh_0}\sin(\omega_0 t)\delta(t)-\frac{1}{k}\sinh(kz)\cdot\frac{d\delta(t)}{dt}\right\}.\end{aligned}$$

Using the dispersion relation $\omega_0^2 = g_0 k\tanh kh_0$ (Eq. (5.29)), we rewrite Eq. (5.42),

$$\begin{aligned}p_e(\mathbf{x},t) = &\frac{1}{(2\pi)^2}\int_{-\infty}^{\infty}\int_{-\infty}^{\infty} dk_x dk_y \exp[i(k_x x+k_y y)]\\ &\times\frac{\rho_0\widehat{d}(k_x,k_y)}{\cosh(kh_0)}\left\{g_0\frac{\cosh[k(z+h_0)]}{\cosh kh_0}\cos(\omega_0 t)H(t)-\frac{1}{k}\sinh(kz)\cdot\frac{d\delta(t)}{dt}\right\}.\end{aligned} \tag{5.42}$$

In particular, the pressure change at the sea bottom $z = -h_0$ is given by Eq. (5.43)

$$\begin{aligned}p_e(\mathbf{x},t)|_{z=-h_0} = &\frac{1}{(2\pi)^2}\int_{-\infty}^{\infty}\int_{-\infty}^{\infty} dk_x dk_y \exp[i(k_x x+k_y y)]\\ &\times\frac{\rho_0\widehat{d}(k_x,k_y)}{\cosh(kh_0)}\left\{g_0\frac{\cos(\omega_0 t)}{\cosh(kh_0)}H(t)+\frac{1}{k}\sinh(kh_0)\cdot\frac{d\delta(t)}{dt}\right\}\end{aligned} \tag{5.43}$$

The analytical solutions of (5.37), (5.39), (5.41), and (5.42) describe the fluid and surface motion for the tsunami generation and propagation. It is noteworthy that only the tsunami height (Eq. (5.41)) is given by a single term that represents propagating waves characterized by the dispersion relation of Eq. (5.29). The particle velocity (Eqs. (5.37) and (5.39)) and the pressure (Eqs. (5.42) and (5.43)) are represented by the two terms. One of the terms represents propagating waves characterized by the dispersion relation of Eq. (5.29). The propagation term includes gravity acceleration g_0. The other term represents the contribution directly from the source. It is independent of gravity and does not represent the propagating waves. This term is necessary to reproduce the velocity and pressure field during the tsunami generation. The analytical solutions clearly indicate the generation process and the propagation process via the two different terms. Also, it is clear that gravity (g_0) plays an important role in the propagation, whereas the gravity does not appear in the generation. We will discuss the resultant equations in more detail in 5.3.2. *Analytical Solutions.*

5.3 Generation

5.3.1 *Visualization*

When computers were not sufficiently powerful, we relied mainly on analytical solutions to obtain images of wavefields. However, advances have made it possible for computers to visualize the wavefield, which greatly helps to our understanding tsunamis. By conducting an integration over the wavenumber (k_x, k_y) in Eqs. (5.37), (5.39), and (5.41) (a FFT subroutine is used for the integration), we can describe the spatial and temporal distribution of the velocity field and sea-surface height. As a simple example, sea-bottom motion is assumed to be given by the vertical velocity at $z = -h_0$ as

$$v_z(\mathbf{x}, t)\big|_{z=-h_0} = d_0 e^{-x^2/(L/2)^2} \frac{H(t) - H(t-T)}{T}, \tag{5.44}$$

where the permanent displacement at the sea bottom is given by $d(x, y) = d_0 \exp\left[-x^2/(L/2)^2\right]$ and the velocity does not change during the time T. The spatial Fourier transform (defined by Eq. (5.16)) of $d(x, y)$ is

$$\hat{d}\left(k_x, k_y\right) = \frac{\sqrt{\pi} d_0 L}{2} e^{-(Lk_x)^2/16} \cdot 2\pi\delta\left(k_y\right),$$

where $\int_{-\infty}^{\infty} dy \exp(-iky) = 2\pi\delta\left(k_y\right)$ is used. Substituting $\hat{d}\left(k_x, k_y\right)$ into Eqs. (5.37), (5.39), and (5.41) and convoluting the corresponding rate function, we describe the wavefield of tsunami generation. Discrete integration is conducted with respect to the wavenumber k_x. We visualize two cases, large (long-wavelength) and small (short-wavelength) source area cases, in the following.

To represent a large source area generating long-wavelength tsunamis, we set the parameters in Eq. (5.44) as $d_0 = 1$ m, $L = 60$ km, and $T = 31$ s. This source region ($L = 60$ km) is sufficiently larger than the sea depth h_0 (=4 km). Figure 5.6 shows the height η (upper panel) and the velocity distribution in the sea (lower panel) for each elapsed time. As the sea bottom moves according to the boundary conditions in Eq. (5.44) ($T < 31$ s), it produces a vertical flow through the sea layer from the bottom to the surface over the source region (Fig. 5.6a, b, and c). The vertical flow uplifts the sea surface. When the sea-bottom motion ends ($T = 31$ s), the vertical velocity induced by the sea bottom disappears. The uplifted sea surface then begins to collapse, generating a descending flow (Fig. 5.6d). The descending water displaces a massive amount of water sideways from the source region, and the water moves with a dominant horizontal flow (Fig. 5.6e, f). The horizontal propagation of the displacement is the tsunami propagation. The horizontal velocity is always positive (negative) where the location x is positive (negative). These snapshots intuitively illustrate that gravity plays an important role in tsunami propagation. If gravity did not exist (or $g_0 = 0$), the uplifted sea surface would not collapse and the tsunami would not propagate.

We then consider a small source area generating short-wavelength tsunamis, setting the parameter $L = 20$ km, but the other parameters remain the same. The results are shown in Fig. 5.7. As the sea bottom moves ($T < 31$ s), the vertical velocity from the sea bottom produces a vertical flow of water over the source region (Fig. 5.7a, b, and c). Note that the maximum height at the surface (0.8 m) is smaller than in the case with a larger source ($L = 60$ km, Fig. 5.6). This indicates that the vertical velocity is less efficiently excited in a smaller source region. This is because the horizontal flow is more excited (see the flow vector around the sea bottom in Fig. 5.7a, b, and c). After the sea-bottom motion ends ($T > 31$ s), the collapse of the uplifted sea surface accelerates, which causes a descending flow (Fig. 5.7d). The descending water displaces the water sideways, and the water moves with a horizontal flow. Unlike the large source case where the horizontal flow dominates (Fig. 5.6), the vertical component of the velocity is also significant in Fig. 5.7e, f. The water height at the surface indicates a dispersive character: a short-wavelength tsunami propagates more slowly than a long-wavelength one. The horizontal component of the velocity distribution oscillates between positive and negative values. Prograde rotation of the velocity vector occurs near the surface. The velocity distribution accompanied by the short-wavelength tsunami is localized near the surface (Fig. 5.7f). This is because a short-wavelength tsunami cannot displace a large volume of water unlike a long-wavelength tsunami. The comparison of Figs. 5.6 and 5.7 shows that the tsunami generation and propagation characteristics strongly depend on the source size/depth ratio (60/4 and 20/4 in Figs. 5.6 and 5.7, respectively).

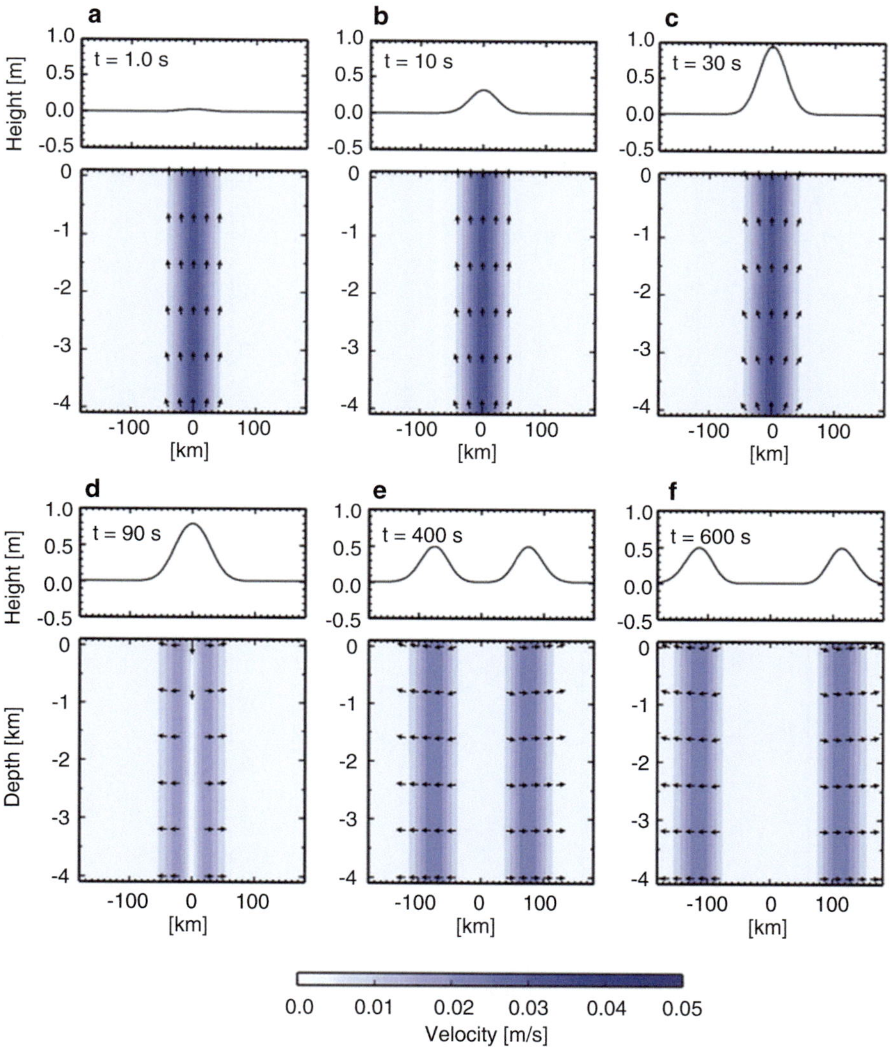

Fig. 5.6 Surface height and velocity distribution for a large source ($L = 60$ km). The permanent sea-bottom deformation is given by $d(x, y) = d_0 \exp[-x^2/(L/2)^2]$, where $d_0 = 1$ m, and the source duration is $T = 31$ s. The water height at the surface is shown in the upper panel, and the velocity distribution from the sea bottom to the sea surface is shown in the lower panel, for each elapsed time t from the start of sea-bottom deformation, (**a**) $t = 1$ s, (**b**) 10 s, (**c**) 30 s, (**d**) 90 s, (**e**) 400 s, and (**f**) 600 s. The vectors indicate the direction of the flow velocity in the fluid in the lower panel (Saito 2013, copyright by Springer)

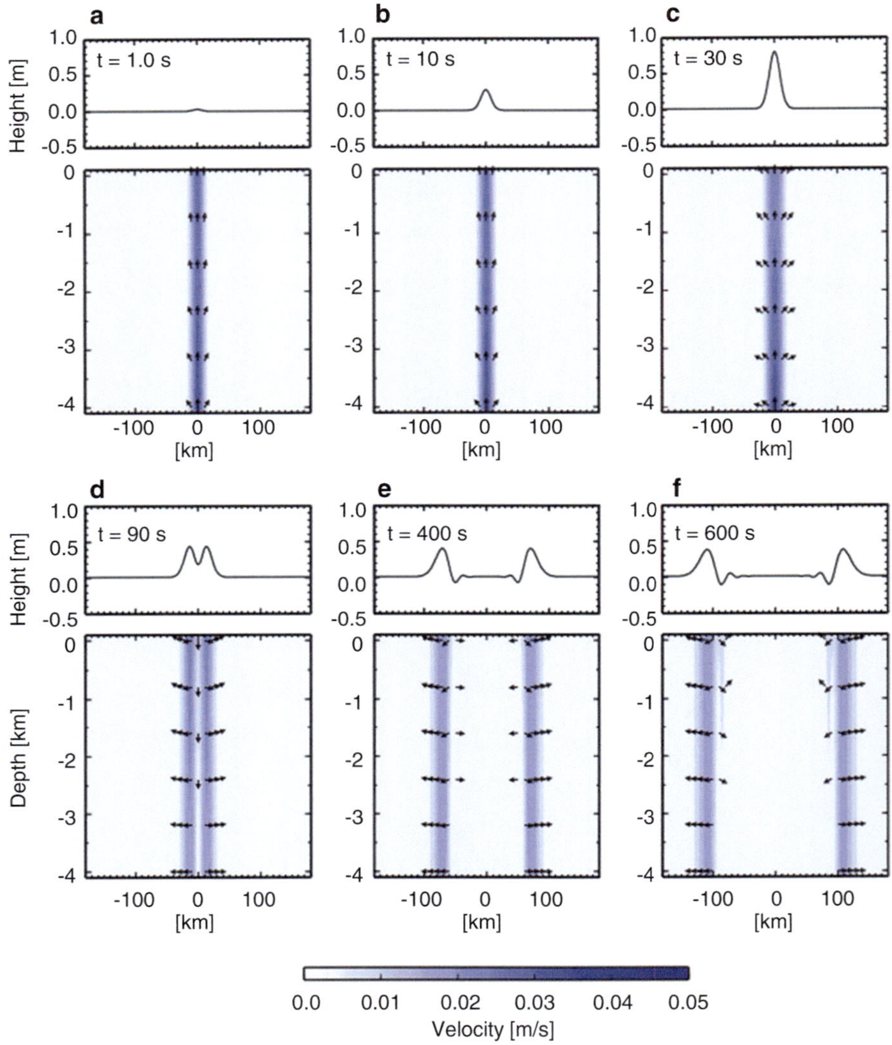

Fig. 5.7 Surface height and velocity distribution for a large source ($L = 20$ km). See Fig. 5.6 for details

5.3.2 Analytical Solutions

Then, we investigate tsunami generation and propagation by examining the analytical solutions (Eqs. (5.37), (5.39), and (5.41)).

5.3.2.1 Generation Process

We have considered tsunami generation to be caused by sea-bottom deformation. This is schematically shown in Fig. 5.8a. The resultant analytical solutions for the velocity potential, the horizontal components of the velocity field, the vertical component of the velocity field, the tsunami height, and the sea-bottom pressure change are listed again below.

$$\phi(t) = \frac{1}{(2\pi)^2}\int_{-\infty}^{\infty}\int_{-\infty}^{\infty} dk_x dk_y\, e^{i(k_x x + k_y y)}\,\frac{\widehat{d}(k_x,k_y)}{\cosh kh_0}\left\{-\frac{\omega_0}{k}\frac{\cosh[k(z+h_0)]}{\sinh kh_0}\sin(\omega_0 t)H(t) + \frac{1}{k}\sinh(kz)\delta(t)\right\}, \tag{5.45}$$

$$\mathbf{v}_H(\mathbf{x},t) = \frac{1}{(2\pi)^2}\int_{-\infty}^{\infty}\int_{-\infty}^{\infty} dk_x dk_y\, e^{i(k_x x + k_y y)}\,\frac{\widehat{d}(k_x,k_y)}{\cosh kh_0}\left\{-i\omega_0\frac{\mathbf{k}_H}{k}\frac{\cosh[k(z+h_0)]}{\sinh kh_0}\sin(\omega_0 t)H(t) + \frac{i\mathbf{k}_H}{k}\sinh(kz)\delta(t)\right\}, \tag{5.46}$$

$$v_z(\mathbf{x},t) = \frac{1}{(2\pi)^2}\int_{-\infty}^{\infty}\int_{-\infty}^{\infty} dk_x dk_y\, e^{i(k_x x + k_y y)}\,\frac{\widehat{d}(k_x,k_y)}{\cosh kh_0}\left\{-\omega_0\frac{\sinh[k(z+h_0)]}{\sinh kh_0}\sin(\omega_0 t)H(t) + \cosh(kz)\delta(t)\right\}, \tag{5.47}$$

$$\eta(x,y,t) = \frac{1}{(2\pi)^2}\int_{-\infty}^{\infty}\int_{-\infty}^{\infty} dk_x dk_y\, e^{i(k_x x + k_y y)}\,\frac{\widehat{d}(k_x,k_y)}{\cosh kh_0}\cos(\omega_0 t)H(t), \tag{5.48}$$

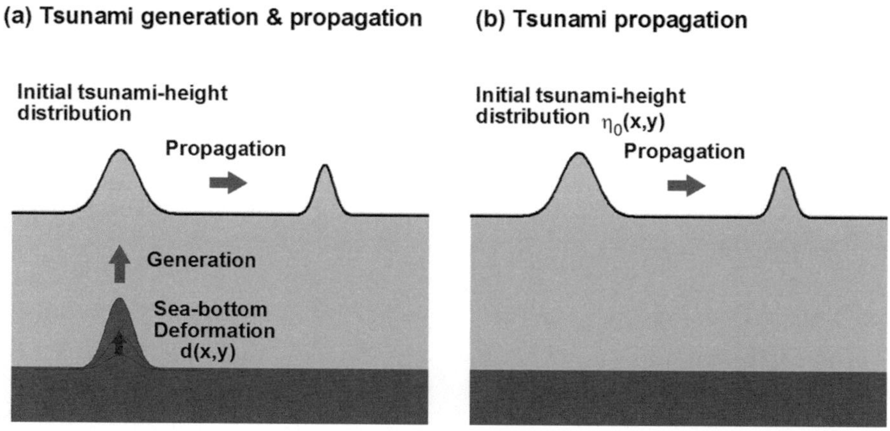

Fig. 5.8 (**a**) Tsunami generation due to sea-bottom deformation and its propagation. (**b**) Tsunami propagation from an initial tsunami height distribution (Saito 2013, copyright by Springer)

$$p_e(\mathbf{x},t)|_{z=-h_0} = \frac{1}{(2\pi)^2}\int_{-\infty}^{\infty}\int_{-\infty}^{\infty} dk_x dk_y\, e^{i(k_x x + k_y y)} \frac{\rho_0 \widehat{d}(k_x,k_y)}{\cosh kh_0} \left\{ g_0 \frac{\cos(\omega_0 t)}{\cosh(kh_0)} H(t) + \frac{1}{k}\sinh(kh_0)\cdot\frac{d\delta(t)}{dt} \right\}. \tag{5.49}$$

In contrast to this situation, if we set the initial tsunami height distribution $\eta_0(x, y)$ as shown in Fig. 5.8b, we simulate only the propagation process without considering the generation process. By comparing these two cases (Fig. 5.8a, b) and investigating the difference, we can examine the generation process. The solutions obtained from the initial tsunami height distribution $\eta_0(x, y)$ are (Appendix 5B)

$$\phi_0(t) = \frac{1}{(2\pi)^2}\int_{-\infty}^{\infty}\int_{-\infty}^{\infty} dk_x dk_y\, e^{i(k_x x + k_y y)} \widehat{\eta}_0(k_x,k_y) \left\{ -\frac{\omega_0}{k}\frac{\cosh[k(z+h_0)]}{\sinh kh_0} \sin(\omega_0 t) H(t) \right\}, \tag{5.50}$$

$$\mathbf{v}_H(\mathbf{x},t) = \frac{1}{(2\pi)^2}\int_{-\infty}^{\infty}\int_{-\infty}^{\infty} dk_x dk_y\, e^{i(k_x x + k_y y)} \widehat{\eta}_0(k_x,k_y) \left\{ -i\omega_0 \frac{\mathbf{k}_H}{k}\frac{\cosh[k(z+h_0)]}{\sinh kh_0} \sin(\omega_0 t) H(t) \right\}, \tag{5.51}$$

$$\mathrm{v}_z(\mathbf{x},t) = \frac{1}{(2\pi)^2}\int_{-\infty}^{\infty}\int_{-\infty}^{\infty} dk_x dk_y\, e^{i(k_x x + k_y y)} \widehat{\eta}_0(k_x,k_y) \left\{ -\omega_0 \frac{\sinh[k(z+h_0)]}{\sinh kh_0} \sin(\omega_0 t) H(t) \right\}, \tag{5.52}$$

$$\eta(\mathbf{x},t) = \frac{1}{(2\pi)^2}\int_{-\infty}^{\infty}\int_{-\infty}^{\infty} dk_x dk_y\, e^{i(k_x x + k_y y)} \widehat{\eta}_0(k_x,k_y) \cos(\omega_0 t) H(t), \tag{5.53}$$

$$p_e(x,y,z=-h_0,t) = \frac{1}{(2\pi)^2}\int_{-\infty}^{\infty}\int_{-\infty}^{\infty} dk_x dk_y\, e^{i(k_x x + k_y y)} \rho_0 g_0 \widehat{\eta}_0(k_x,k_y) \frac{\cos(\omega_0 t)}{\cosh(kh_0)} H(t), \tag{5.54}$$

where $\widehat{\eta}_0(k_x, k_y)$ is the 2-D spatial Fourier transform of the initial tsunami height distribution $\eta_0(x, y)$.

5.3.2.2 Gravity

Comparison of the solutions with and without the generation process (e.g., a comparison of (5.45) and (5.50)) indicates that the generation process is described by the term represented by the delta function in Eqs. (5.45, 5.46, 5.47), and (5.49).

These terms do not contain the acceleration of gravity g_0. The generation process is basically independent of gravity. This is a fundamental feature of the generation process in contrast to the propagation process. This theoretical consequence supports the idea that the permanent sea-surface displacement calculated without considering gravity (e.g., Okada 1985) works as an initial tsunami height distribution. We also discuss the methods in 4.3.2 *Seismic Waves, Ocean Acoustic Waves, and Permanent Displacement*.

5.3.2.3 Factor of 1/ cosh (kh_0)

The factor $\widehat{d}(k_x, k_y)/\cosh(kh_0)$ appears in the generation process, while $\widehat{\eta}_0(k_x, k_y)$ appears in the solutions without the generation process. This suggests that the factor $\widehat{d}(k_x, k_y)/\cosh(kh_0)$ contains the information about the generation process. The initial tsunami height distribution caused by the sea-bottom deformation is not identical to the sea-bottom deformation. The factor 1/ cosh (kh_0) works as a spatial low-wavelength pass filter. This is often called Kajiura filter (Kajiura 1963). Figure 5.9 shows the factor 1/ cosh (kh_0). This filter removes the short-wavelength components ($k > 1/h_0$) as a low-wavelength pass filter. Due to this low-wavelength pass filter, the initial tsunami height distribution (sea-surface displacement distribution) becomes smooth.

Kajiura (1963) conducted the integration over the wavenumber domain and derived the analytical solution in the spatial domain as

$$\begin{aligned}\bar{\eta}(x, y, 0) &= \frac{1}{(2\pi)^2}\int_{-\infty}^{\infty}\int_{-\infty}^{\infty} dk_x dk_y\, e^{i(k_x x + k_y y)}\frac{1}{\cosh(kh_0)} \\ &= \frac{1}{2\pi}\int_0^{\infty} dk\, J_0(kr)\frac{k}{\cosh(kh_0)}.\end{aligned}$$

Using a series of exponential (Eq. (1.232.2) in Gradshteyn and Ryzihk 2000):

$$\frac{1}{\cosh x} = \operatorname{sech} x = 2\sum_{n=0}^{\infty}(-1)^n \exp[-(2n+1)x],$$

we obtain

$$\begin{aligned}\bar{\eta}(x, y, 0) &= \frac{1}{\pi}\sum_{n=0}^{\infty}(-1)^n\int_0^{\infty} dk \exp[-(2n+1)kh_0]k\, J_0(kr) \\ &= \frac{1}{\pi r^2}\sum_{n=0}^{\infty}(-1)^n\int_0^{\infty} d\bar{k} \exp\left[-(2n+1)\frac{h_0}{r}\bar{k}\right]\bar{k}\, J_0(\bar{k}).\end{aligned}$$

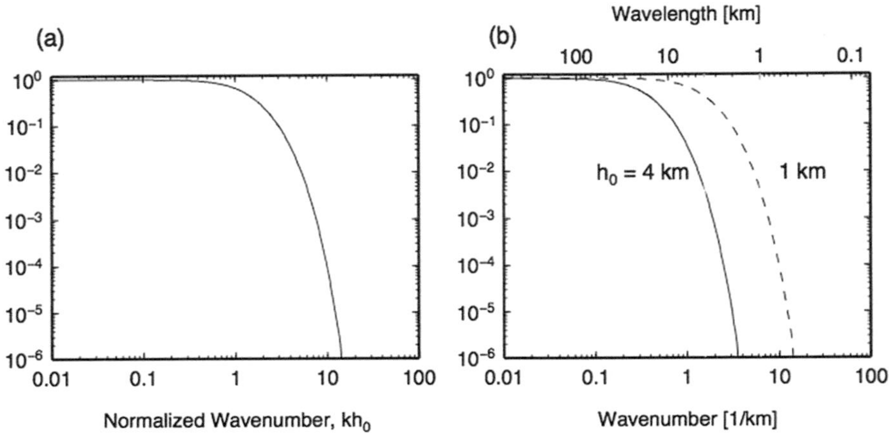

Fig. 5.9 (**a**) A function of $1/\cosh(kh_0)$ with respect to the normalized wavenumber kh_0 where k is the wavenumber of the sea-bottom deformation and h_0 is the sea depth. (**b**) A function of $1/\cosh(kh_0)$ with respect to the wavenumber k for the sea depth of $h_0 = 4$ km (solid line) and 1 km (dashed line). The function of $1/\cosh(kh_0)$ works as a low-wavelength pass filter that removes the high wavenumber components $k \gg 1/h_0$

The integration is conducted using a formula (Eq. (6.623.2) in Gradshteyn and Ryzihk 2000):

$$\int_0^\infty d\bar{k} \exp(-\bar{h}\bar{k})\bar{k} J_0(\bar{k}) = \frac{\bar{h}}{(\bar{h}^2 + 1)^{3/2}}.$$

Then, we obtain

$$\begin{aligned} \bar{\eta}(x, y, 0) &= \frac{1}{\pi r^2}\sum_{n=0}^{\infty}(-1)^n(2n+1)\frac{h_0}{r}\frac{1}{\left[(2n+1)^2(h_0/r)^2+1\right]^{3/2}} \\ &= \frac{1}{\pi h_0^2}\sum_{n=0}^{\infty}\frac{(-1)^n(2n+1)}{\left[(r/h_0)^2+(2n+1)^2\right]^{3/2}}. \end{aligned} \tag{5.55}$$

This is the vertical displacement distribution at the sea surface for the sea-bottom deformation occurring at a point given by $d_z(x, y) = \delta(x)\delta(y)$. Equation (5.55) is often referred to as Kajiura's equation (Kajiura 1963). Figure 5.10 shows Eq. (5.55) with respect to the distance normalized by the sea depth, r/h_0.

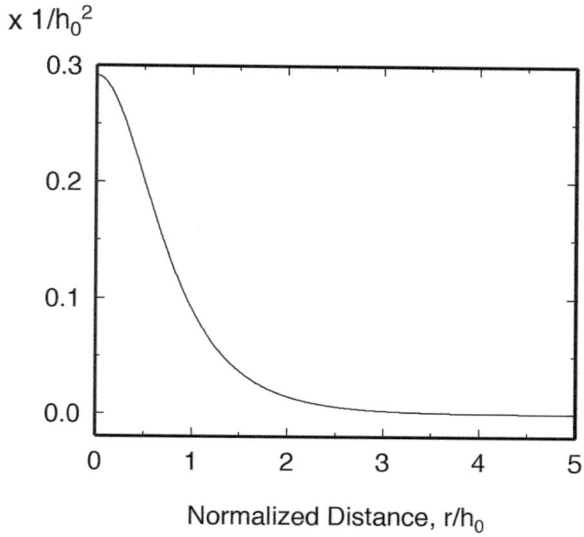

Fig. 5.10 Sea-surface height distribution with respect to the point source vertical displacement at the sea bottom given by $\delta(x)\delta(y)$. The distance r is given by $r=\sqrt{x^2+y^2}$, and the sea depth is h_0

5.3.2.4 Conservation of Displaced Water Volume

When the sea-bottom displacement is given by $d(x, y)$, the total volume brought by the sea-bottom movement into the sea layer is given by the integral over 2-D horizontal space as

$$\begin{aligned}\iint_{-\infty}^{\infty} d(x,y)dxdy &= \iint_{-\infty}^{\infty} dxdy\frac{1}{(2\pi)^2}\iint_{-\infty}^{\infty} dk_xdk_y\,\widehat{d}\left(k_x,k_y\right)e^{i\left(k_xx+k_yy\right)}\\ &=\iint_{-\infty}^{\infty} dk_xdk_y\,\delta(k_x)\delta\left(k_y\right)\widehat{d}\left(k_x,k_y\right)\\ &=\widehat{d}(0,0).\end{aligned} \tag{5.56}$$

On the other hand, the total displaced water volume at the sea surface is given by

$$\begin{aligned}\iint_{-\infty}^{\infty} \eta(x,y,0)dxdy &= \iint_{-\infty}^{\infty} dxdy\frac{1}{(2\pi)^2}\iint_{-\infty}^{\infty} dk_xdk_y\frac{\widehat{d}\left(k_x,k_y\right)}{\cosh(kh_0)}e^{i\left(k_xx+k_yy\right)}\\ &=\iint_{-\infty}^{\infty} dk_xdk_y\,\delta(k_x)\delta\left(k_y\right)\frac{\widehat{d}\left(k_x,k_y\right)}{\cosh(kh_0)}\\ &=\widehat{d}(0,0).\end{aligned} \tag{5.57}$$

Even though the sea-surface height distribution $\eta(x, y, 0)$ has fewer short-wavelength components due to the spatial filtering effect of $1/\cosh(kh_0)$, the total displaced volume at the sea surface is identical to the total volume provided at the sea bottom. This is because the theory assumes the seawater to be an incompressible fluid (no dilatation), which preserves the total volume of the medium. The total displaced water volume is a conserved quantity for tsunami generation and propagation. This volume can be a good measure for representing the magnitude of a tsunami (e.g., Satake and Kanamori 1991). We should also note that during the propagation, there is the energy conservation by the continuity equation of energy in 2-D form (3.2.3: *Energy Density and Energy Flux Density for Incompressible Fluid*).

5.3.2.5 Initial Velocity Distribution

Equations (5.46) and (5.47) indicate that the velocity distribution has a term including the delta function when the generation process is considered. On the other hand, when the generation process is neglected and only the propagation process is considered, the velocity distribution does not have the term including the delta function (Eqs. (5.51) and (5.52)). These delta-function terms originate from the delta function in the velocity potential (Eq. (5.45)), which represents the sea-bottom deformation as the source. Figure 5.11 shows the amplitudes of the horizontal components of the source terms (proportional to $|\sinh(kz)|$, see Eq. (5.46)) and the vertical component (proportional to $|\cosh(kz)|$, see Eq. (5.47)) for the values of λ=20 and 60 km, normalized by the vertical component at the sea bottom (proportional to $|\cosh(kh_0)|$, see Eq. (5.47)). The vertical component (solid line) decreases from the sea bottom to the sea surface. The horizontal component (dashed line) decreases more rapidly than the vertical component and become almost zero at the sea surface. These are the velocity distributions appearing only during the source process time (see also figures a, b, and c in Figs. 5.6 and 5.7). These terms become zero in the propagation process.

The first term including the step function $H(t)$ in Eqs. (5.46) and (5.47) represents the propagation, but the second term of the delta function $\delta(t)$ does not contribute to the propagation. This indicates that the velocity field caused by the sea-bottom deformation results in the initial tsunami height distribution but does not directly contribute to the propagation. This may defy an intuitive sense that we should set the initial horizontal velocity in the tsunami propagation simulation according to the horizontal velocity distribution caused by the sea-bottom deformation. The analytical solutions of the incompressible fluid theory suggest that we need to set the initial tsunami height distribution as

$$\eta(x,y,0) = \frac{1}{(2\pi)^2}\int_{-\infty}^{\infty}\int_{-\infty}^{\infty} dk_x dk_y\, e^{i\left(k_x x + k_y y\right)}\,\frac{\widehat{d}\left(k_x,k_y\right)}{\cosh\left(kh_0\right)},$$

and we need to set the initial velocity distribution to zero:

$$v_z(x,y,z,0) = 0, \text{and } \mathbf{v_H}(x,y,z,0) = 0$$

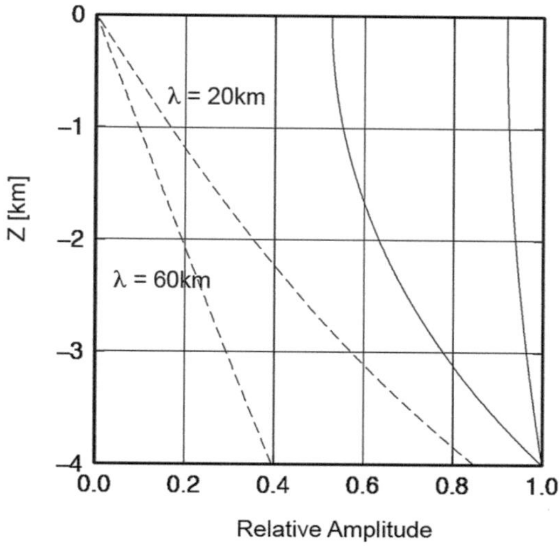

Fig. 5.11 Amplitude of the source term in the horizontal component (dashed lines) and the vertical component (solid lines) normalized by the vertical component at the sea bottom. The cases of $\lambda = 20$ and 60 km (which correspond to $k = 0.3$ and $0.1\ \mathrm{km}^{-1}$ when the sea depth is set at 4 km) are plotted (Saito 2013, copyright by Springer)

in tsunami propagation simulations in order to calculate the tsunami η and velocity v_z and $\mathbf{v_H}$ outside the source duration (see Saito 2013). This procedure is the same as that usually used in past studies involving 2-D tsunami simulations (e.g., Fujii and Satake 2008; Saito and Furumura 2009) where the initial velocity distribution is set at zero in the whole space. Equations (5.46) and (5.47) are fundamental equations indicating that we need to make the initial velocity distribution zero but set only the initial tsunami height distribution for the propagation process. These are derived based on a constant sea-depth model. When the sea bottom is not flat, Tanioka and Satake (1996) developed a method to include the horizontal "displacement" into the initial tsunami height distribution. We explain this in Sect. 5.4*: Bridging Generation to Propagation*. On one hand, Song et al. (2017) took the horizontal "velocity" also into account as the initial condition for tsunami simulations. However, the numerical simulations including earthquake faulting, tsunami generation, and propagation suggest that the contribution of the horizontal velocity is negligible in the initial condition for the tsunami simulations (Lotto et al. 2017).

5.3.2.6 Propagation

A comparison of Eqs. (5.46) and (5.51) indicates that they share the distribution function of the horizontal velocity $f_H(k, z, h_0)$ (Eq. (5.58)). The function, hence, is peculiar to the propagation process. The same is true for the function of the vertical velocity $f_z(k, z, h_0)$ (Eq. (5.59)). The functions are

$$f_H(k, z, h_0) \equiv \frac{\cosh[k(z + h_0)]}{\sinh(kh_0)}, \tag{5.58}$$

and

$$f_z(k, z, h_0) \equiv \frac{\sinh[k(z + h_0)]}{\sinh(kh_0)}. \tag{5.59}$$

These indicate the velocity distributions for the horizontal and vertical components, respectively, at a depth z in a sea layer of a constant depth h_0. The same functions are also derived in Eqs. (3.50) and (3.52) for the tsunami propagation process in Chap. 3: *Propagation of Tsunami and Seismic Waves*. The propagation process is typically classified into short-wavelength waves (surface- or deep-water waves) or long-wavelength waves (shallow-water waves) in hydrodynamics (e.g., Gill 1982; Kambe 2007). Figure 5.12a shows the distribution function of the horizontal velocity $f_H(k, z, h_0)$ for values of $k = 6.3$, 16, 0.1 km^{-1}. These values correspond to the wavelengths of 1, 4, and 60 km for a water depth of 4 km. When the wavelength is $\lambda = 1$ km (short-wavelength waves), the horizontal component of the velocity is distributed only in the shallower regions ($z < 1$ km). As the wavelength increases, the distribution function extends deeper. When the wavelength is $\lambda = 60$ km (long-wavelength waves), much greater than the water depth h_0 of 4 km, the horizontal velocity component shows a large value (2.3~2.5) over the entire depth, suggesting that horizontal flow occurs in the entire seawater layer. Figure 5.12b shows the distribution function of the vertical velocity $f_z(k, z, h_0)$. Similar to the case of the horizontal component, when the wavelength is $\lambda = 1$ km, the vertical component of the velocity is distributed only in the shallower region ($z < 1$ km). Unlike in the case of the horizontal components, when the wavelength is $\lambda = 60$ km, much greater than the water depth, the vertical component of the velocity decreases with increasing depth and becomes zero at the sea bottom.

5.3.3 Pressure Change

We investigate the pressure change at the sea bottom during tsunami generation based on incompressible fluid theory. It should be noted that assuming a compressible sea is more realistic than assuming an incompressible sea (e.g., Nosov 1999). The seawater compressibility mainly affects the excitation of ocean acoustic waves in high-frequency pressure change ($f > c_0/(4h_0)$) where h_0 is the sea depth and c_0 is the ocean acoustic wave phase velocity (see 3.3.2 *Ocean Acoustic Waves*). The compressibility can also affect longer-period wavefield $f \ll c_0/(4h_0)$, but the effect is minor (e.g., Yamamoto 1982). Therefore, the tsunami generation theory that assumes incompressible fluid provides us with a fundamental concept in tsunami

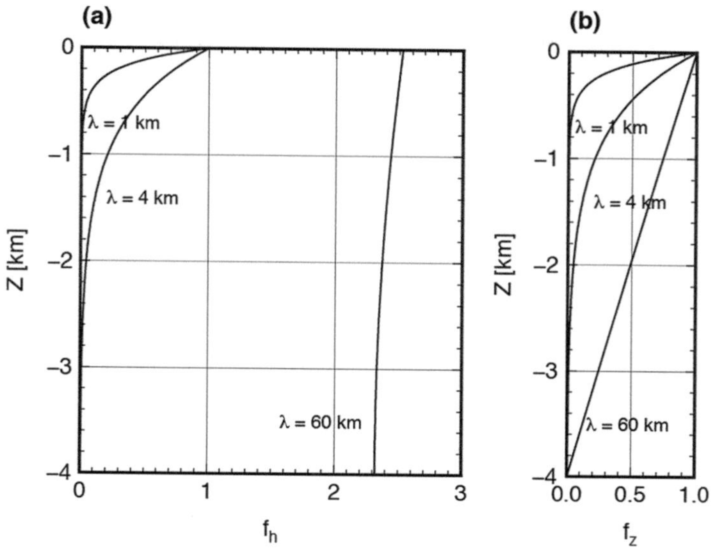

Fig. 5.12 Velocity distribution functions of (**a**) the horizontal component of the velocity (Eq. (5.38)) and (**b**) the vertical component of the velocity (Eq. (5.40)), plotted as functions of the depth, z, with a water depth of $h_0 = 4$ km. The distribution functions for $\lambda = 1,\ \ 4$, and 60 km (corresponding to $k = 6.28,\ \ 1.57$, and $0.1\ \mathrm{km}^{-1}$, respectively) are plotted

generation. Here, we consider the ocean-bottom pressure change based on the incompressible fluid theory. This is a good approximation for longer-period wavefield $f \ll c_0/(4h_0)$.

Considering the load due to the sea-surface height $\eta(x, y, t)$ on the sea bottom, most past tsunami studies employed a simple relation in the analysis of the ocean-bottom pressure change p_e at the sea bottom $z = \ -h$ as

$$p_e(x, y, z = -h, t) \approx \rho_0 g_0 \eta(x, y, t), \tag{5.60}$$

where ρ_0 is the seawater density and η is the sea-surface height change. This relation works reasonably when the ocean-bottom pressure gauges are located far from the source region. However, when the ocean-bottom pressure gauges are located inside the source region or near the earthquake hypocenter, this relation is too simple to represent the relation between the pressure change p_e and the sea-surface height η. We need to consider two additional effects in practice.

5.3.3.1 Permanent Sea-Bottom Deformation

As shown in Fig. 5.13, inside the source region, the vertical displacement permanently remains on the sea bottom. The hydrostatic relation (Eq. (3.31)) gives the pressure change Δp as

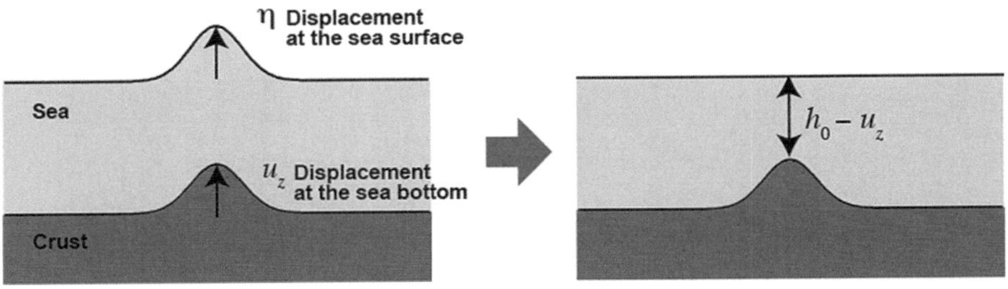

Fig. 5.13 A schematic illustration of the sea-surface displacement η and the sea bottom displacement u_z. (Left) The total sea-depth change caused by the sea-surface and the sea-bottom displacement is $\eta - u_z$. (Right) When the time elapses, the sea-surface displacement η becomes zero. Then the sea depth h_0 changes to $h_0 - u_z$

$$\Delta p(x, y, z) = -\rho_0 g_0 \Delta z. \tag{5.61}$$

When an observation point at the sea bottom uplifts by $u_z(x, y, z = -h, t)$, the station location changes from $(x, y, -h)$ to $(x, y, -h + u_z)$. Then, the pressure at the sea bottom changes by $\Delta p = -\rho_0 g_0 u_z(x, y, z = -h, t)$. Considering this effect in Eq. (5.60), the pressure change is given by

$$p_e(x, y, z = -h, t) = \rho_0 g_0 [\eta(x, y, t) - u_z(x, y, z = -h, t)]. \tag{5.62}$$

Note that the value of $\eta(x, y, t) - u_z(x, y, -h, t)$ is the net sea-depth change at (x, y).

Consider a case in which the sea-bottom deformation occurs instantaneously; in other words, the rate function $\chi(t)$ is given by the delta function as $\chi(t) = \delta(t)$. At the time just after the sea-bottom deformation ends, the tsunami height η is almost the same as the sea-bottom displacement: $\eta(x, y, t \to 0) \approx u_z(x, y, z = -h, t \to 0)$. The slight difference between $\eta(x, y, t \to 0)$ and $u_z(x, y, z = -h, t \to 0)$ is due to a finite sea-depth effect (Fig. 5.9). On the other hand, when we consider that enough time elapses, $t \to \infty$, the sea-surface height or tsunami disappears, $\eta(x, y, t \to \infty) = 0$ (Fig. 5.13). Hence, Eq. (5.62) predicts that the pressure change at the sea bottom behaves as follows:

$$p_e(x, y, z = -h, t) = \begin{cases} 0 & t \to 0 \\ -\rho_0 g_0 u_z(x, y, z = -h, t) & t \to \infty. \end{cases} \tag{5.63}$$

Figure 5.14 shows the observed ocean-bottom pressure records of the 2003 Tokachi-Oki earthquake. In the records of PG1 and PG2, which were located inside the focal area, we recognize that the pressure change behaves as predicted by Eq. (5.63): the pressure change when the earthquake occurred (indicated by vertical dashed lines) was minor, and an offset appeared when ~40 min elapsed.

If we look carefully around the time of the generation at station PG1, we also find high-frequency pressure fluctuation. This pressure fluctuation is caused by another

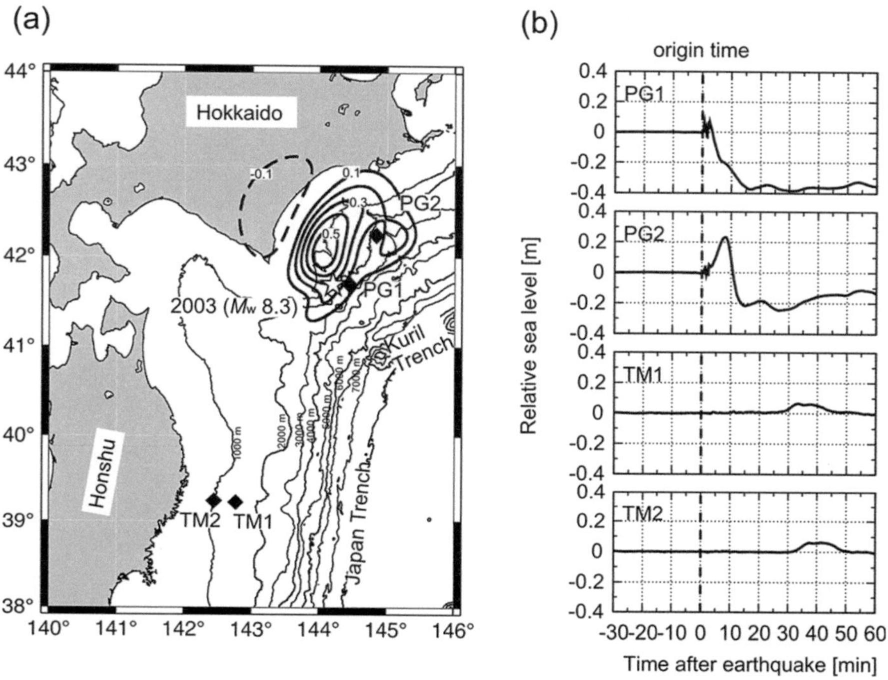

Fig. 5.14 (**a**) Locations of stations of ocean-bottom pressure records PG1, PG2, TM1, and TM2 (black diamonds) and the source region of the 2003 Tokachi-Oki earthquake (black contours). Solid contours show uplift, and dashed contours show subsidence. The contour interval is 0.1 m. The epicenter is indicated by a star. (**b**) Ocean-bottom pressure records. The ocean tide was removed, and a low-pass filter with a cutoff period of 60 s was applied (Tsushima et al. 2012, copyright by the American Geophysical Union)

mechanism. This record shows that the pressure fluctuation is small compared to the pressure changes caused by tsunami and permanent sea-bottom deformation. However, note that the theory described below predicts that this kind of pressure change can be more significant depending on the earthquake rupture.

5.3.3.2 Dynamic Pressure Change

The solution of Eq. (5.49) predicts that the motion of the fluid excites a pressure change. When the vertical velocity at the sea bottom is represented by the Fourier transform as

$$
\begin{aligned}
v_z(x, y, z = -h_0, t) &= d(x, y)\chi(t) \\
&= \frac{1}{(2\pi)^2} \int \int_{-\infty}^{\infty} dk_x dk_y\, e^{i(k_x x + k_y y)} \hat{d}(k_x, k_y)\chi(t),
\end{aligned} \tag{5.64}
$$

the sea-bottom pressure change is then rewritten as

$$
\begin{aligned}
& p_e(\mathbf{x}, t)\big|_{z=-h_0} \\
& = \frac{1}{(2\pi)^2} \int_{-\infty}^{\infty} \int_{-\infty}^{\infty} dk_x dk_y\, e^{i(k_x x + k_y y)} \frac{\rho_0 \hat{d}(k_x, k_y)}{\cosh(kh_0)} \\
& \left\{ \frac{g_0}{\cosh(kh_0)} \int_{-\infty}^{t} \cos(\omega_0(t - \tau))\, \chi(\tau) d\tau + \frac{1}{k} \sinh(kh_0) \cdot \int_{-\infty}^{\infty} \frac{d\delta(t - \tau)}{d(t - \tau)} \chi(\tau) d\tau \right\}.
\end{aligned}
$$

The integration of the second term on the right-hand side is calculated by partial integration, and we obtain

$$
\begin{aligned}
& p_e(\mathbf{x}, t)\big|_{z=-h_0} \\
& = \frac{1}{(2\pi)^2} \int_{-\infty}^{\infty} \int_{-\infty}^{\infty} dk_x dk_y\, e^{i(k_x x + k_y y)} \frac{\rho_0 \hat{d}(k_x, k_y)}{\cosh(kh_0)} \\
& \left\{ \frac{g_0}{\cosh(kh_0)} \int_{-\infty}^{t} \cos(\omega_0(t - \tau))\, \chi(\tau) d\tau + \frac{1}{k} \sinh(kh_0) \cdot \frac{d\chi(t)}{dt} \right\} \\
& = \frac{1}{(2\pi)^2} \int_{-\infty}^{\infty} \int_{-\infty}^{\infty} dk_x dk_y\, e^{i(k_x x + k_y y)} \\
& \left\{ \frac{\rho_0 g_0 \hat{\eta}(k_x, k_y, t)}{\cosh(kh_0)} + \rho_0 h_0 \frac{\tanh kh_0}{kh_0} \hat{a}_z^{bot}(k_x, k_y, t) \right\},
\end{aligned} \tag{5.65}
$$

where $\hat{\eta}(k_x, k_y, t)$ is the 2-D spatial Fourier transform of the tsunami height $\eta(x, y, t)$ as

$$
\hat{\eta}(k_x, k_y, t) = \int \int_{-\infty}^{\infty} dk_x dk_y\, e^{-i(k_x x + k_y y)} \eta(x, y, t), \tag{5.66}
$$

and $\hat{a}_z^{bot}(k_x, k_y, t)$ is the 2-D spatial Fourier transform of the vertical acceleration at the sea bottom $a_z^{bot}(x, y, t)$ (see Eq. (5.64)):

$$
\begin{aligned}
\hat{a}_z^{bot}(k_x, k_y, t) &= \int \int_{-\infty}^{\infty} dk_x dk_y\, e^{-i(k_x x + k_y y)} a_z^{bot}(x, y, t) \\
&= \hat{d}(k_x, k_y) \frac{d\chi(t)}{dt}.
\end{aligned} \tag{5.67}
$$

In particular, when the tsunami wavelength or the wavelength of the sea-bottom deformation is much greater than the sea depth $kh_0 \ll 1$, Eq. (5.65) is approximated to

$$p_e(\mathbf{x},t)\big|_{z=-h_0} \approx \rho_0 g_0 \eta(x,y,t) + \rho_0 h_0 a_z^{bot}(x,y,t) \ \text{ for } kh_0 \ll 1. \tag{5.68}$$

Equations (5.65) and (5.68) indicate that the sea-bottom pressure change is caused not only by the static loading caused by the tsunami height $\eta(x, y, t)$ (the first term of Eq. (5.65) or (5.68)) but also by the contribution of the acceleration at the sea bottom (the second term of Eq. (5.65) or (5.68)). Since the second term is given by the acceleration or motion of the sea bottom, we referred to this pressure change as the dynamic pressure change. Results of numerical simulations also indicated that the pressure change was caused not only by static pressure change but also by dynamic pressure change (e.g., Kakinuma and Akiyama 2007).

Note that the dynamic pressure is independent of gravity but is excited by the motion of the sea and the crust. In the term $\rho_0 h_0 a_z^{bot}(x,y,t)$, the value of $\rho_0 h_0$ is the mass of the water column per unit area (Fig. 5.15). Therefore, $\rho_0 h_0 a_z^{bot}(x,y,t)$ is the action-reaction force generated when the sea bottom is uplifted with the acceleration $a_z^{bot}(x,y,t)$. Basically, this originates from the inertia of the water column motion (Filloux 1982). The factor of tanhkh_0/kh_0 in Eq. (5.65) indicates the effect of the finite size of the sea-bottom displacement wavelength relative k to the sea depth h_0.

Taking the pressure changes due to the permanent sea-bottom displacement (i) into account in Eq. (5.65), we represent the sea-bottom pressure change caused by the sea-bottom deformation $u_z(x, y, -h, t)$ as

$$\begin{aligned} p_e(\mathbf{x},t)\big|_{z=-h_0} &= \frac{1}{(2\pi)^2}\int_{-\infty}^{\infty}\int_{-\infty}^{\infty} dk_x dk_y\, e^{i(k_x x + k_y y)} \\ &\quad \left\{\frac{\rho_0 g_0 \widehat{\eta}(k_x,k_y,t)}{\cosh(kh_0)} + \rho_0 h_0 \frac{\tanh kh_0}{kh_0}\widehat{a}_z^{bot}(k_x,k_y,t)\right\} \\ &\quad - \rho_0 g_0 u_z(x,y,z=-h,t) \\ &\approx \rho_0 g_0\{\eta(x,y,t) - u_z(x,y,-h)\} + \rho_0 h_0 a_z^{bot}(x,y,t) \ \text{ for } kh_0 \ll 1. \end{aligned} \tag{5.69}$$

Figure 5.16 shows an example of the actual observed ocean-bottom pressure record (Kubota et al. 2015). A low-pass filter of 100 s was applied to the ocean-bottom pressure records. This record clearly indicates a large dynamic pressure

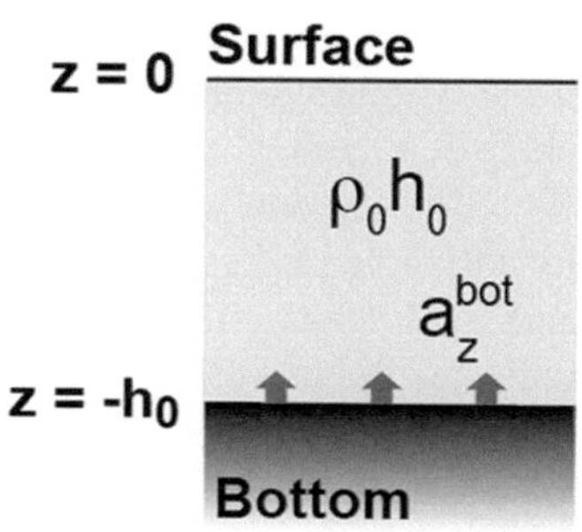

Fig. 5.15 The mass of the water column is $\rho_0 h_0$ per unit area where ρ_0 is the water density and h_0 is the water column height. The vertical acceleration at the bottom is $a_z^{bot}(x,y,t)$

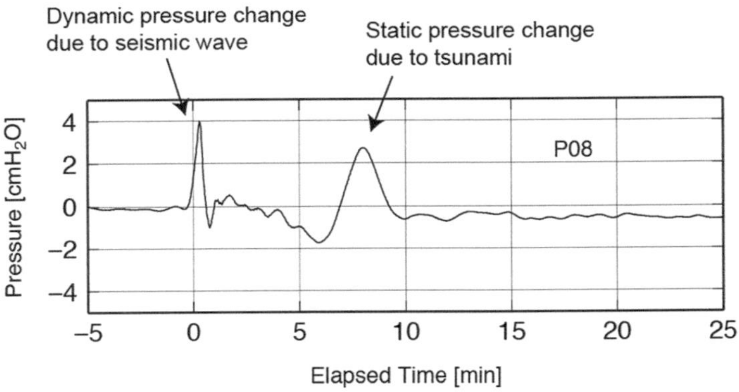

Fig. 5.16 Ocean-bottom pressure record observed at the gauge located ~50 km away from the hypocenter of a moderate-size earthquake (M_W 7.0) that occurred on July 10, 2011, offshore from the coast of northeastern Honshu, Japan. The pressure change of 1 cmH_2O corresponds to the hydrostatic pressure due to a water depth of 1 cm (1 $cmH_2O = 98$ Pa) (Kubota et al. 2015 Copyright by the American Geophysical Union)

change caused by seismic waves in addition to the static pressure change caused by the tsunami. The dynamic pressure change may function as significant noise when analyzing the pressure records within a short elapsed time (a few minutes after the earthquake occurs). On the other hand, because the dynamic pressure change represents the sea-bottom acceleration, we expect that the dynamic component in the ocean-bottom pressure records can be used for the estimation of the kinematic parameters of the earthquake fault motion (e.g., rise time and source time duration). Investigating how to use the dynamic component as seismic signals would be an interesting challenge (e.g., An et al. 2017; Kubota et al. 2017).

5.3.3.3 Comparison with Elastic Dynamic Theory

The incompressible fluid theory has been employed without considering seawater elasticity. It would be important to validate whether the dynamic pressure change appears in the seismic wave simulations. Supposing a 2-D space composed of an elastic water layer and an elastic crust (Fig. 5.17a), the sea-bottom pressure change caused by earthquake fault motion was simulated. The gray line in Fig. 5.17b (under the blue and red lines) shows the temporal variation of the sea-bottom pressure change p_e inside the focal area when the earthquake fault has a fault width $W = 100$ km and a rupture duration of $t_c = 40$ s (the station location is indicated by a triangle in Fig. 5.17a). The dynamic pressure change begins to increase at 40 s and takes a maximum value of 0.008 MPa at the elapsed time of 55 s. The blue line indicates the temporal change of the sea-bottom acceleration $a_z^{bot}(x, y, t)$; here the dimension is converted to pressure according to $\rho_0 h_0 a_z^{bot}(x, y, t)$. The red line

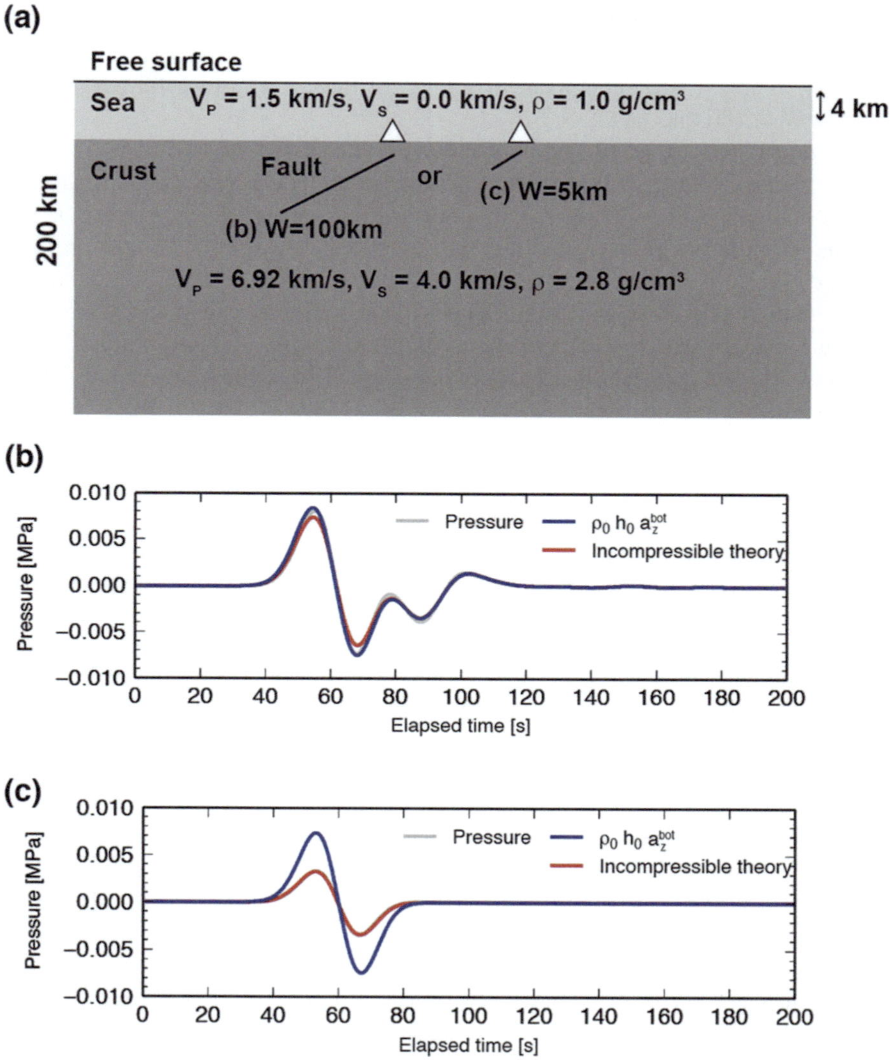

Fig. 5.17 (**a**) A schematic illustration of the simulation settings. Elastic medium consisting of a water layer and homogeneous crust. The finite size of the earthquake fault is set in the crust. A large earthquake (fault width of $W = 100$ km) and small earthquake ($W = 5$ km) are supposed. Locations of hypothesized stations are plotted by triangles. (**b**) Sea-bottom pressure change (gray) on the sea bottom inside the source region for the large earthquake ($W = 100$ km). The pressure change predicted by a simple relation $\rho_0 h_0 a_z^{bot}$ (blue) and Eq. (5.65) assuming the incompressible fluid theory (red) where a_z^{bot} is the acceleration on the sea bottom. (**c**) Same as (b) but for the small earthquake ($W = 5$ km). The red line and gray line (under the red line) are almost identical (Saito 2017 copyright by Oxford Academic)

indicates the temporal change of the sea-bottom acceleration $a_z^{bot}(x, y, t)$ where the dimension is converted to pressure using a spatial low-pass filter of $\rho_0 h_0$ tanh kh_0/kh_0 (see Eq. (5.65)). Excellent agreement among the three lines was confirmed for the case of $W = 100$ km.

A comparison is also shown for the case when the earthquake fault width is short ($W = 5$ km) in Fig. 5.17c. We found that the simple theoretical relation $\rho_0 h_0 a_z^{bot}$ (blue) does not agree with the sea-bottom pressure change (gray, but this is not visible in the figure because the gray line is under the red line), while the pressure change obtained by applying the spatial low-pass filter of $\rho_0 h_0$ tanh kh_0/kh_0 to the sea-bottom acceleration a_z^{bot} (red) shows good agreement with the sea-bottom pressure change (gray). The prediction based on the incompressible fluid theory can be confirmed in the elastic dynamic simulation. This supports the validity of the incompressible fluid theory.

5.3.3.4 Dynamic Pressure Change and Static Pressure Change

When using ocean-bottom pressure gauges deployed inside the focal area, even after removing the ocean acoustic waves with an appropriate low-pass filter, the pressure records still include the contribution of the dynamic pressure change caused by the sea-bottom acceleration. Figure 5.18 shows an example of a synthesized ocean-bottom pressure record inside the focal area. The red line shows the pressure change including both the dynamic and static contributions, whereas the bold gray line shows the pressure change including only the static contribution. The static pressure change (bold gray line) is zero at the elapsed time of zero because the vertical displacements at the sea surface and sea bottom are almost the same. When enough time elapses, the pressure change shows an offset value of -10 mH_2O. This is because the sea bottom is uplifted by 10 m. On the other hand, if we observe the pressure change including the dynamic contribution (red line), it shows a large impulsive signal at the elapsed time of 80 s. This does not reflect the sea-surface height change, but this large impulsive signal is excited by the dynamic pressure change. We should take care not to consider this pressure change to be the sea-surface height change just by using a simple hydrostatic relation $p_e = \rho_0 g_0 \eta$. The dynamic pressure change was actually observed by the ocean-bottom pressure gauge deployed in deep ocean (see Fig. 5.16). If this dynamic pressure change is mistakenly considered to be the tsunami height in real-time tsunami monitoring, it may cause incorrect tsunami prediction at the coast. It would be challenging to develop a method to distinguish the dynamic pressure change (seismic waves) and the static pressure change (tsunami) in real time.

We examine the contribution to the dynamic pressure change made by the sea-bottom acceleration $\rho_0 h_0 a_z^{bot} = \rho_0 h_0 \ddot{u}_z$ (u_z is the vertical displacement at the sea bottom) relative to the hydrostatic pressure change caused by the sea-surface

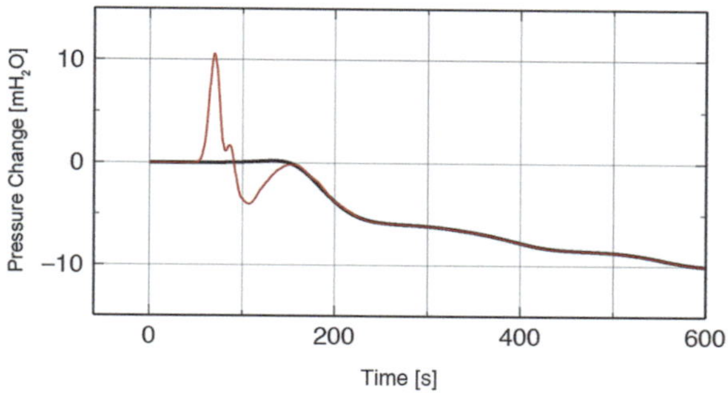

Fig. 5.18 Synthesized ocean-bottom pressure gauge records. Ocean-bottom pressure change with the dynamic pressure change (red) and without the dynamic pressure change (black). A pressure change of 1 mH_2O corresponds to the hydrostatic pressure due to a water depth of 1 m ($1mH_2O = 9.8$ MPa) (Saito and Tsushima 2016, copyright by the American Geophysical Union)

height $\rho_0 g_0 \eta(x, y, t)$. The ratio of the dynamic pressure change to the static pressure change is given by

$$R_D = \frac{h_0 |\ddot{u}_z|}{g_0 |\eta|}. \tag{5.70}$$

If the dominant angular frequency of the vertical displacement at the bottom is ω_c, the acceleration is given by $|\ddot{u}_z| = \omega_C^2 |u_z|$. The sea-surface height is approximately $|\eta| \sim |u_z|$. The ratio is approximately given by

$$R_D \sim \frac{h_0 \omega_C^2 |u_z|}{g_0 |u_z|} = \frac{h_0 \omega_c^2}{g_0}. \tag{5.71}$$

The duration of the sea-bottom deformation t_c is approximately half of the dominant period $t_c \sim \pi / \omega_c$. The ratio is then

$$R_D \sim \frac{\pi^2 h_0}{g_0 t_c^2} \sim \frac{10 h_0}{g_0 t_c^2}. \tag{5.72}$$

This indicates that the contribution of the dynamic pressure change is basically controlled by the duration of the sea-bottom deformation (or this corresponds to the rise time) t_c and decreases rapidly with increasing duration. For example, we consider $R_D = 1/2$ to be a condition that the dynamic pressure change significantly contaminates the static pressure change. The duration t_c is about 90 s if we suppose $h_0 = 4000$ m as an average ocean depth. This indicates that it is inappropriate to assume hydrostatic approximation during ~90 s from the time the sea bottom begins to rise.

5.4 Bridging Generation to Propagation

Sections 5.1, 5.2, and 5.3 have investigated the fundamental mechanism of tsunami generation by supposing a flat sea bottom. It is important to consider the effects of bathymetry when analyzing or simulating realistic records. This section bridges the gap between the analytical solutions derived under a constant sea-depth assumption and tsunami simulations with realistic bathymetry.

5.4.1 Contribution of Sea-Depth Variation

When the flat sea bottom is assumed, the horizontal displacement is allowed to be discontinuous at the boundary between the fluid and solid. Only the vertical component of the displacement should be continuous at the boundary between the fluid and solid (e.g., Ewing et al. 1957; Kennett 2001). Hence, the horizontal displacement at the sea bottom does not excite vertical flow in the sea (note that the vertical displacement at the sea bottom excites horizontal flow in the sea layer during the generation. See Fig. 5.11). When the sea bottom is not flat but is curved, the combination of the horizontal displacement and the sea-bottom topography can excite vertical flow (e.g., Nosov and Kolesov 2011). This mechanism was first taken into account in setting the initial tsunami height distribution by Tanioka and Satake (1996). After the 2011 Tohoku-Oki earthquake, which showed a large horizontal displacement near a deep trench (e.g., Ito et al. 2011), numerous studies pointed out the contribution of the horizontal displacement to the tsunami excitation (e.g., Gusman et al. 2012; Satake et al. 2013; Hooper et al. 2013). Here we introduce the contribution of the sea-bottom horizontal displacement to the tsunami generation process.

Consider the case where the sea bottom is not flat as shown in Fig. 5.19. The displacement normal to the boundary between the sea and crust is continuous at the boundary, whereas the tangential components of the displacement can be discontinuous. When the sea bottom (or top of the crust) is displaced by the displacement vector $\mathbf{u}$, the normal component u_n is continuous on the boundary between the fluid and solid.

The unit vector $\bar{\mathbf{n}}$ normal to the sea bottom $z = -h(x, y)$ is given by

$$\bar{\mathbf{n}} = \frac{1}{\sqrt{(\partial h/\partial x)^2 + (\partial h/\partial y)^2 + 1}} \left(\frac{\partial}{\partial x} h(x, y), \frac{\partial}{\partial y} h(x, y), 1 \right). \tag{5.73}$$

Consider the displacement vector at the top of the crust, $\mathbf{u}(x, y, -h - 0)$. The component of the displacement vector normal to the sea bottom is then given by

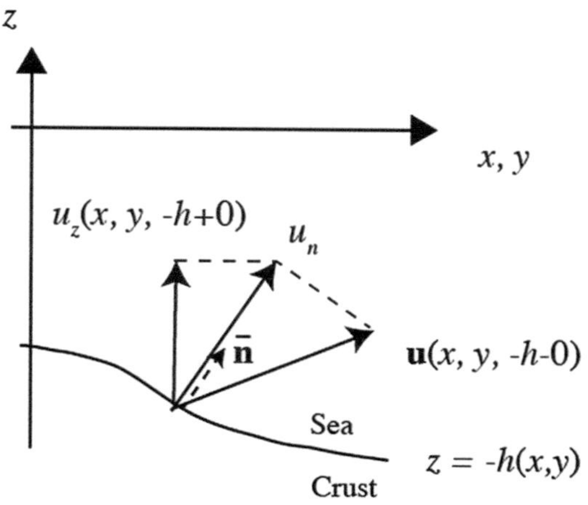

Fig. 5.19 The sea depth is given as a function of the horizontal coordinates (x, y), $z = -h(x, y)$. The bottom of the sea is just above the boundary between the sea and crust, $z = -h(x, y) + 0$, while the top of the crust is just below the boundary $z = -h(x, y) - 0$. The displacement vector at the top of crust is $\mathbf{u} = \mathbf{u}(x, y, -h - 0)$. The bathymetry curve is characterized by the normal vector $\bar{\mathbf{n}}$. The displacement normal to the bathymetry is given by $u_n = \mathbf{u} \cdot \bar{\mathbf{n}}$, which is continuous between the crust and sea. The vertical component at the bottom of the sea is $u_z(x, y, -h + 0)$

$$u_n = \mathbf{u} \cdot \bar{\mathbf{n}} = \frac{1}{\sqrt{(\partial h/\partial x)^2 + (\partial h/\partial y)^2 + 1}} \left(u_x \frac{\partial h}{\partial x} + u_y \frac{\partial h}{\partial y} + u_z \right) \Bigg|_{z=-h-0}. \quad (5.74)$$

Since the normal component is continuous between the boundary, $u_n(x, y, -h + 0) = u_n(x, y, -h - 0)$, the displacement vector at the bottom of the sea is given by

$$\mathbf{u}(x, y, -h + 0) = u_n(x, y, -h - 0)\bar{\mathbf{n}}. \quad (5.75)$$

The vertical component of the sea-bottom displacement $\mathbf{u}(x, y, -h + 0)$ would contribute to the vertical displacement at the sea surface. The projection of the displacement vector $\mathbf{u}(x, y, -h + 0)$ to the vertical axis gives the vertical component at the sea bottom as

$$\begin{aligned} u_z(x, y, -h + 0) &= \mathbf{u}(x, y, -h + 0) \cdot \mathbf{e_z} \\ &= u_n \bar{\mathbf{n}} \cdot \mathbf{e_z} \\ &= \frac{1}{(\partial h/\partial x)^2 + (\partial h/\partial y)^2 + 1} \left(u_x \frac{\partial h}{\partial x} + u_y \frac{\partial h}{\partial y} + u_z \right) \Bigg|_{z=-h-0}. \end{aligned} \quad (5.76)$$

Equation (5.76) represents the vertical displacement at the bottom of the sea $u_z(x, y, -h + 0)$ by using the displacement at the top of the crust $(u_x, u_y, u_z)|_{z=-h-0}$ and the sea depth $z = -h(x, y)$. When the sea depth varies smooth enough as

$$\left(\frac{\partial h}{\partial x}\right)^2 + \left(\frac{\partial h}{\partial y}\right)^2 \ll 1, \tag{5.77}$$

the vertical displacement at the bottom of the water layer is approximately given by

$$u_z(x, y, -h+0) \approx \left(u_x \frac{\partial h}{\partial x} + u_y \frac{\partial h}{\partial y} + u_z \right)\Bigg|_{z=-h-0}. \tag{5.78}$$

We may consider the displacement $(u_x, u_y, u_z)|_{z=-h-0}$ (the right-hand side of Eq. (5.78)) as the displacement vector on the sea bottom. This is often calculated as the sea-bottom displacement with the code of Okada (1985) or other seismic numerical codes. The vertical displacement $u_z(x, y, -h+0)$ on the left-hand side is an effective vertical component of the displacement vector generating the initial tsunami height distribution. If we convolve $u_z(x, y, -h+0)$ with Kajiura's equation (Eq. (5.55)), we obtain a reasonable initial tsunami height distribution for the 2-D tsunami propagation simulations. Equation (5.78) is the same as the equation originally proposed by Tanioka and Satake (1996).

Convincing evidence regarding the contribution of the horizontal displacement to the tsunami excitation was observed during the 2011 Tohoku-Oki earthquake. The black lines in Fig. 5.20a, b show the sea-surface height change observed at the two tide gauges during the Tohoku-Oki earthquake. The blue lines show the theoretically calculated sea-surface height that includes only the contribution from the vertical displacement at the sea bottom. These theoretical calculations can reproduce some parts of the observed records. On the other hand, if the horizontal components are taken into account in setting the initial tsunami height distribution, the theoretically calculated sea-surface height change (red line) successfully explains the overall features of these observations. The two stations used in this comparison are located on the western coast of Honshu, Japan (Fig. 5.20c), where a large tsunami coming from the main tsunami source is blocked by the Japanese mainland. Also, as shown in Fig. 5.20c and d, there were large horizontal displacement compared to vertical displacement. These conditions made it possible to observe convincing evidence indicating the contribution of horizontal displacement to the tsunami excitation.

The fault motion of the 2011 Tohoku-Oki earthquake causes a large horizontal displacement around the deep and steep Japan Trench. Equation (5.78) indicates that these two conditions (large horizontal displacement at a large bathymetry gradient) would efficiently cause the initial tsunami height distribution. By considering the horizontal displacement in calculating the initial tsunami height distribution in the simulation, the tsunami simulation predicts a larger tsunami amplitude (Fig. 5.21).

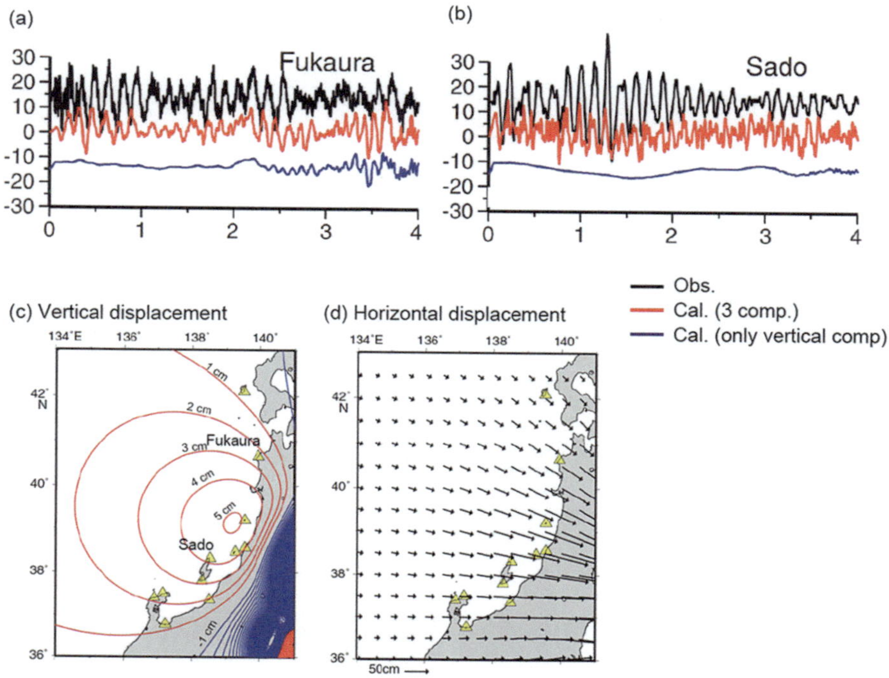

Fig. 5.20 (**a**) Observed and calculated tsunami waveforms at Fukaura (location is plotted in (c)). The black line shows the observed waveform. The red line is the waveform calculated from the total vertical seafloor displacement including the effects of the horizontal displacement. The blue line is the waveform calculated from the vertical displacement. (**b**) Observed and calculated waveforms at Sado. (**c**) Vertical displacement and (**d**) horizontal displacement around Japan calculated from the fault model of the 2011 Tohoku earthquake (Murotani et al. 2015 copyright by Springer)

5.4.2 Synthesis of Ocean-Bottom Pressure Records

We often use the terms "tsunami waveforms" and "tsunami data." Some people may assume that the temporal change of sea-surface displacement is a tsunami waveform, and others may suppose that the sea-bottom pressure change is a tsunami waveform. It is important to recognize what sensor is being used when referring to tsunami data or tsunami waveforms. If one uses tide gauges or GPS sensors, the tsunami waveform corresponds to the vertical sea-surface displacement $u_z(x, y, z_{\mathrm{sur}}, t)$ where $z = z_{\mathrm{sur}}$ is the sea surface. If ocean-bottom pressure gauges are used, the tsunami waveform represents the pressure change at the sea bottom $p_e(x, y, z_{\mathrm{bot}}, t)$ where $z = z_{\mathrm{bot}}$ is the sea bottom. In most cases, after an appropriate low-pass filter is applied, the two values are simply related by a hydrostatic approximation as $p_e(x, y, z_{\mathrm{bot}}, t) = \rho_0 g_0 u_z(x, y, z_{\mathrm{sur}}, t)$ where ρ_0 is the seawater density and g_0 is the gravitational acceleration. However, we cannot use this relation when the sensor is located near the focal area where permanent sea-bottom displacement can occur and the seismic

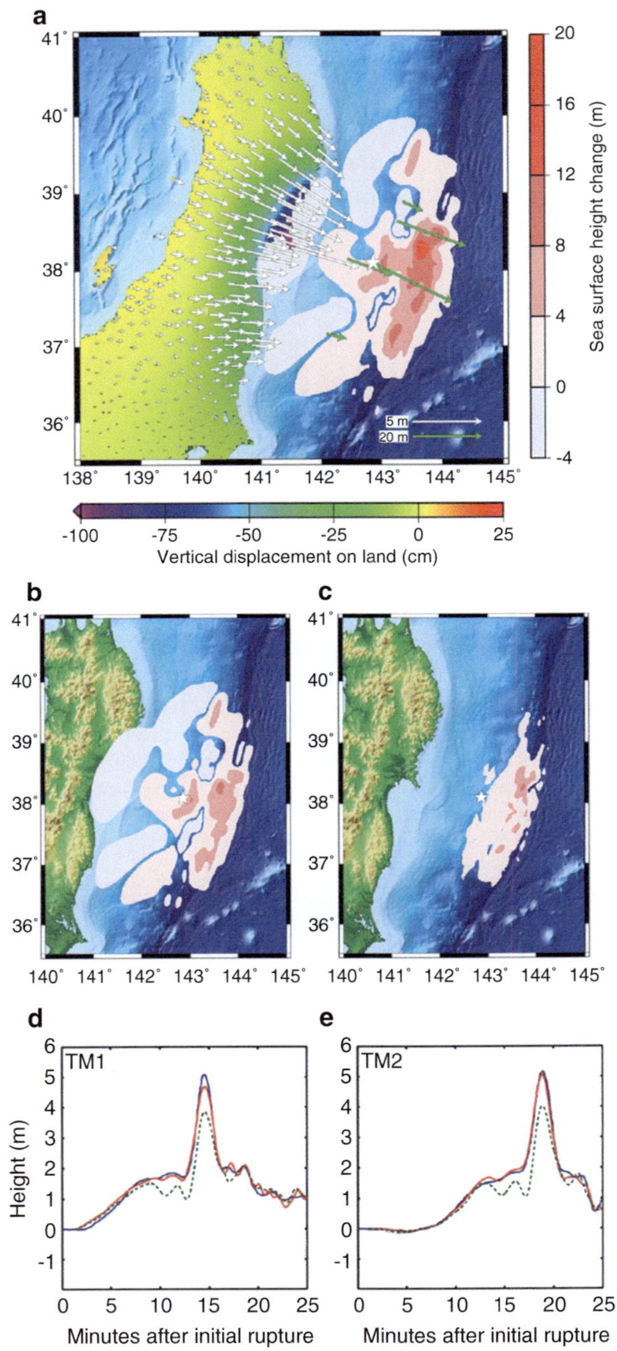

Fig. 5.21 Initial sea-surface height distribution. Red and blue contour colors represent the sea-surface height distribution considering (**a**) the total vertical sea-bottom displacement including both vertical and horizontal displacement, (**b**) the vertical displacement only, and (**c**) the vertical

waves overlap tsunami. This section illustrates a method of synthesizing the ocean-bottom pressure records by taking into account both the seismic waves and tsunami.

There are two methods by which to introduce a source in the tsunami simulations (Fig. 5.22). In Chap. 4, by assuming that the sea layer is compressible ((b) in Fig. 5.22), we synthesized the sea-surface displacement caused by the seismic waves and tsunami. This section synthesizes the ocean-bottom pressure change. We here assume an incompressible sea layer and employ the method shown in (a) in Fig. 5.22. This is because it is reasonable to assume an incompressible sea when calculating the static pressure change.

We propose a two-step method composed of linear and nonlinear simulations (Fig. 5.22a, c). The flowchart is shown in Fig. 5.23. In this method, we decompose the pressure change into two components. One is the pressure change caused by gravity. This contribution is calculated by the tsunami simulation. This corresponds to the static pressure change p_{static}. The other is the pressure change caused by seismic waves, which is excited even when gravity is zero. This is the dynamic pressure change p_{dynamic}. The total pressure change p_{e} is represented as $p_{\mathrm{e}} = p_{\mathrm{dynamic}} + p_{\mathrm{static}}$.

In the first step, we calculate the spatial and temporal distribution of the sea-bottom motion generated by an earthquake. This simulation also calculates the sea-bottom pressure change corresponding to a dynamic pressure change. The particle motion in an elastic medium satisfies the linear equations of motion:

$$\rho \frac{\partial^2 u_i}{\partial t^2} = \tau_{ij,j} + f_i, \tag{5.79}$$

and satisfies a constitutive law,

$$\tau_{ij} = \lambda \delta_{ij} u_{k,k} + \mu \left(u_{i,j} + u_{j,i} \right), \tag{5.80}$$

where τ_{ij} is the stress tensor in the medium. These equations are commonly used in many seismological studies (e.g., Aki and Richards (2002)). The parameters λ and μ are Lamé parameters, which represent the properties of the medium. An equivalent body force of the double couple force system for a point dislocation source is used to represent an earthquake (e.g., Eqs. (4.10) and (4.7) in Chap. 4). We saved the simulation results of the spatial and temporal distribution of the vertical velocity at

Fig. 5.21 (continued) displacement from the horizontal displacement and bathymetry effect. (**d**), (**e**) Tsunami waveforms at two offshore sea-bottom pressure gauges. The blue lines show the observed waveforms. The red lines show the waveforms calculated from the total vertical sea-bottom displacement. The green dashed lines are the waveforms calculated from the vertical displacement (Hooper et al. 2013, copyright by Elsevier)

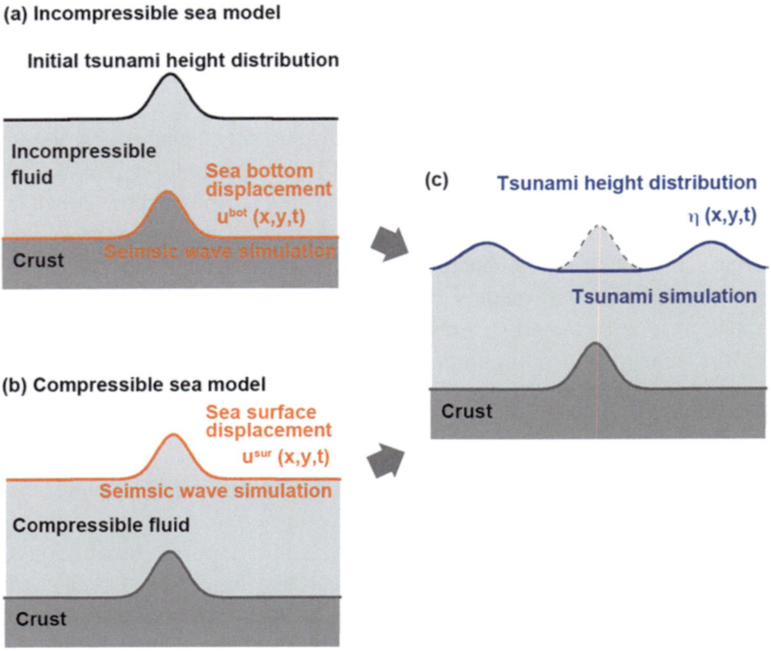

Fig. 5.22 Incompressible sea model and compressible sea model for tsunami simulations. (**a**) The sea is assumed to be an incompressible fluid. (**b**) The sea is assumed to be a compressible fluid. (**c**) Tsunami propagation is calculated based on the tsunami equations in the incompressible sea

Fig. 5.23 Flowchart of the ocean-bottom pressure record synthesis

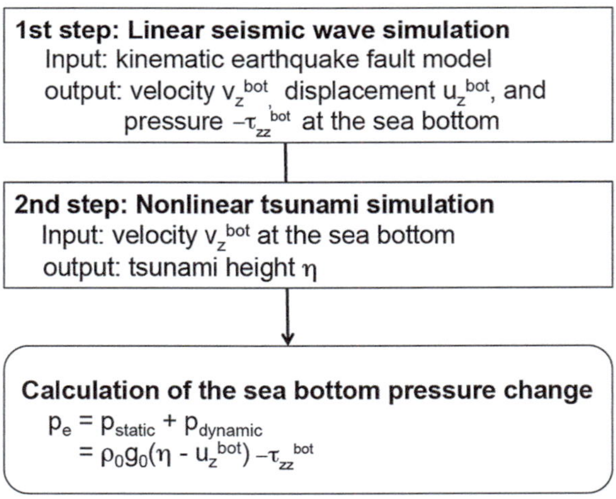

the sea bottom v_z^{bot} and used them as a source in the tsunami simulation in the next step. Additionally, we saved the simulated vertical displacement u_z^{bot} and the stress (pressure) τ_{zz}^{bot} at the sea bottom for the later synthesis of the ocean-bottom pressure records.

In the second step, we calculate the tsunami propagation by using the velocity field at the sea bottom v_z^{bot} as a source. According to the incompressible fluid theory, when the vertical displacement at the sea bottom given by $u_z^{\mathrm{bot}}(x, y)$ occurs instantaneously, the sea-surface height is given by Eq. (5.81) exp(..)

$$\eta_0(x, y) = \frac{1}{(2\pi)^2} \int\int_{-\infty}^{\infty} dk_x dk_y \exp\left[i\left(k_x x + k_y y\right)\right] \frac{\tilde{u}_z^{\mathrm{bot}}\left(k_x, k_y\right)}{\cosh(kh_0)}, \tag{5.81}$$

where $\tilde{u}_z^{\mathrm{bot}}\left(k_x, k_y\right)$ is the 2-D Fourier transform of the vertical displacement distribution at the bottom $u_z^{\mathrm{bot}}(x, y)$ (e.g., Kajiura 1963; Saito and Furumura 2009). Note that when we intend to include the horizontal displacement contribution, we should use Eq. (5.78) for $u_z^{\mathrm{bot}}(x, y)$.

Then, the sea-surface elevation during a fractional duration Δt excited by the time-dependent sea-bottom deformation is given by

$$\Delta\eta_0(x, y, t) = \frac{1}{(2\pi)^2} \int\int_{-\infty}^{\infty} dk_x dk_y \exp\left[i\left(k_x x + k_y y\right)\right] \frac{\tilde{v}_z^{\mathrm{bot}}\left(k_x, k_y, t\right)}{\cosh(kh_0)} \Delta t, \tag{5.82}$$

where $\tilde{v}_z^{\mathrm{bot}}\left(k_x, k_y, t\right)$ is the 2-D Fourier transform of the vertical velocity distribution at the bottom $v_z^{\mathrm{bot}}(x, y, t)$. Substituting the vertical velocity at the sea bottom calculated in the first step of the seismic wave simulation into Eq. (5.82), we obtain the tsunami height increment $\Delta\eta_0(x, y, t)$ for the small time interval Δt at the time t.

The tsunami propagation is assumed to be described by the tsunami equations (see Chap. 6: *Propagation Simulation*). The sea-surface elevation in a fractional duration given by Eq. (5.82) is added to the sea-surface height distribution at each time step. The simulated spatial and temporal distribution of the sea-surface height $\eta(x, y, t)$ is saved to synthesize the ocean-bottom pressure records.

Ocean-bottom pressure records are synthesized using the results of these seismic and tsunami simulations. An ocean-bottom pressure change is considered to be composed of static and dynamic components. The static component of the ocean-bottom pressure change caused by gravity is given by

$$p_{\mathrm{static}} = \rho_0 g_0 \left(\eta - u_z^{\mathrm{bot}}\right) \tag{5.83}$$

where the sea-surface height displacement $\eta(x, y, t)$ and the vertical sea-bottom displacement $u_z^{\mathrm{bot}}(x, y, t)$ are calculated in the tsunami and seismic simulations, respectively. On the other hand, the dynamic ocean-bottom pressure change is given by

$$p_{\mathrm{dynamic}} = -\tau_{zz}^{\mathrm{bot}}(x, y, t). \tag{5.84}$$

The stress at the bottom τ_{zz}^{bot} is calculated in the seismic wave simulation. Note that the sign of the pressure is opposite to that of the stress, and the stress is isotropic as $\tau_{xx} = \tau_{yy} = \tau_{zz}$ in the sea layer. The total pressure change at the sea bottom is given by

$$p_e = p_{\text{static}} + p_{\text{dynamic}}. \qquad (5.85)$$

5.4.2.1 Synthesized Ocean-Bottom Pressure Waveforms

Figure 5.24a shows the ocean-bottom pressure records observed at a station near the focal area of the 2011 Tohoku-Oki earthquake. The record shows the dominant high-frequency component of the pressure change up to an elapsed time of ~500 s. These are mainly ocean acoustic waves. Applying the low-pass filter with a cutoff frequency of 60 s to the record, we remove the ocean acoustic waves and obtain low-frequency component records (Fig. 5.24b). The ocean-bottom pressure change caused by the tsunami is clearly recognized from the elapsed time of 0–1200 s. Moreover, we recognize that the low-frequency wave also appears at the elapsed time of ~100 s. This is the pressure change caused by low-frequency seismic waves.

Figure 5.24c shows the ocean-bottom pressure change simulated by the method described in this chapter. We used a rectangle fault plane with a length of 200 km and a width of 150 km with a uniform slip of 20 m. The depth of the centroid of the fault is set at a depth of 50.5 km. The strike, dip, and rake are set at 200°, 30°, and 90°, respectively. The slip starts simultaneously on the fault, and the rise time is set at $t_c = 30$ s. The simulated waveform indicates the pressure change due to the seismic wave at around 100 s with an amplitude of ~2 mH_2O, and the pressure gradually increases. The maximum value was ~5 mH_2O at ~900 s. The simulated ocean-bottom pressure record roughly reproduced the observed record, even though a uniform slip and simple source time function were assumed in the simulation.

Figure 5.25 compares the synthesized ocean-bottom pressure change for various simulation settings. The light gray line represents the synthesized pressure change including the ocean acoustic waves, low-frequency seismic waves, and tsunami. The synthesized pressure change contains less high-frequency component than the observed records (e.g., see Fig. 5.24a). This is mainly because the earthquake fault model used in this synthesis was characterized by a smoother slip distribution than the actual earthquake. By applying the low-pass filter to the synthesized pressure change (light gray line) and removing the ocean acoustic waves, we obtain a red line that contains low-frequency seismic waves and tsunami. The dark gray line represents the synthesized pressure change including only the tsunami component, i.e., the static pressure change p_{static} in Eq. (5.85). These comparisons clearly indicate that both high- and low-frequency seismic waves significantly contribute to the ocean-bottom pressure waveforms. Even if the low-pass filter is applied, the low-frequency seismic wave still exists in the waveform. This suggests that low-frequency seismic waves can function as noise when ocean-bottom pressure records are analyzed as

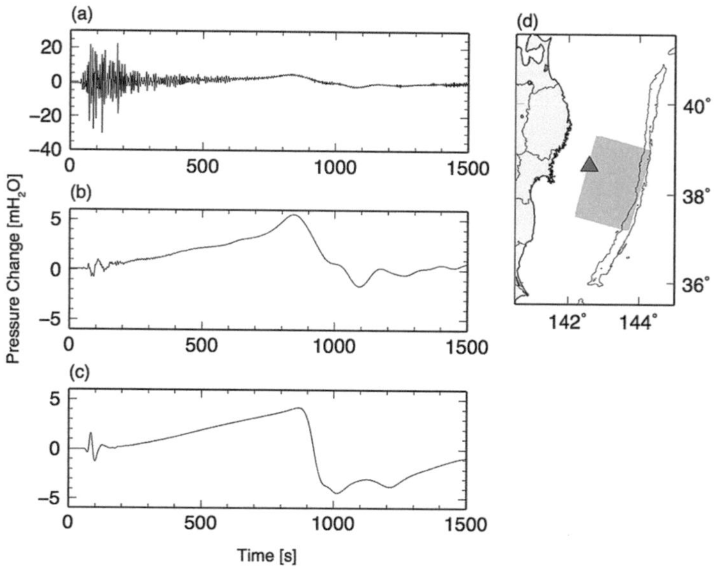

Fig. 5.24 (**a**) Ocean-bottom pressure record of the 2011 Tohoku-Oki earthquake observed at station P06 (a gray triangle in (d)). (**b**) Low-frequency component of the ocean-bottom pressure record obtained by the low-pass filter with a cutoff frequency of 60 s. (**c**) Simulated ocean-bottom pressure change at the station. (**d**) The locations of the station (triangle) and source model of the 2011 Tohoku-Oki earthquake (rectangle) (Saito and Tsushima 2016, copyright by the American Geophysical Union)

tsunami signal. It is important to be especially careful when using ocean-bottom pressure records in an automated tsunami analysis system so as not to mistake low-frequency seismic waves as tsunami.

Appendices

Appendix 5A: Equation (5.28)

We derive Eq. (5.28) from Eq. (5.27). In Eq. (5.27), the residue theorem is used with respect to the integration over the angular frequency. Since the poles are located on the path of integration, we introduce artificial damping parameters to shift the poles from the path and to integrate with the residue theorem. These artificial damping parameters will be taken to be zero after the integration using the residue theorem. An artificial damping parameter ($\varepsilon_1 > 0$) is introduced in Eq. (5.10) as

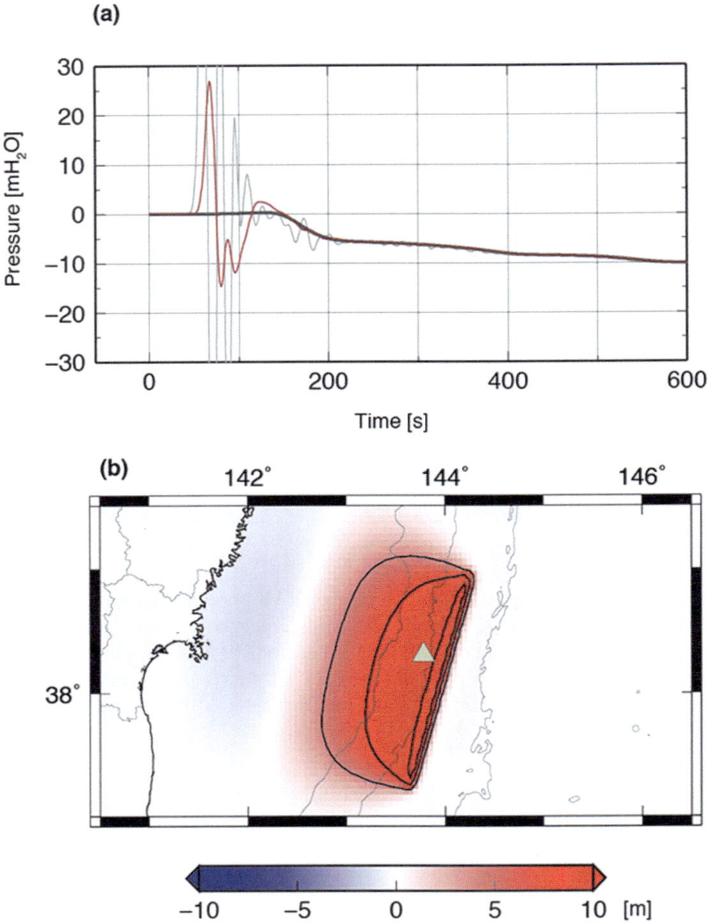

Fig. 5.25 (**a**) Simulated waveforms at a station inside the focal area (the location is indicated in (b)). A synthesized ocean-bottom pressure change including ocean acoustic waves, low-frequency seismic waves, and tsunami (light gray line). The low-pass filter with a cutoff period of 60 s is applied to the simulated waveform (red lines and dark gray lines). The red line is the total pressure change. The dark gray line shows the pressure change due to only the static component. (**b**) The station location and the vertical displacement distribution (Saito and Tsushima 2016, copyright by the American Geophysical Union)

$$\left.\frac{\partial \phi(\mathbf{x}, t)}{\partial t}\right|_{z=0} + g_0 \eta(x, y, t) = -2\varepsilon_1 \phi(\mathbf{x}, t)|_{z=0}. \tag{5.5A.1}$$

Using another surface boundary condition (Eq. (5.9)):

$$\frac{\partial \eta(x,y,t)}{\partial t} = \left.\frac{\partial \phi(\mathbf{x},t)}{\partial z}\right|_{z=0}, \tag{5.5A.2}$$

and (5.5A.1), we obtain

$$\left.\frac{\partial^2 \phi(\mathbf{x},t)}{\partial t^2}\right|_{z=0} + g_0 \left.\frac{\partial \phi(\mathbf{x},t)}{\partial z}\right|_{z=0} = -2\varepsilon_1 \left.\frac{\partial \phi(\mathbf{x},t)}{\partial t}\right|_{z=0}, \tag{5.5A.3}$$

instead of Eq. (5.11) and

$$\left(\frac{d}{dz} - \frac{\omega^2 + 2i\epsilon_1\omega}{g_0}\right)\widehat{\phi}(k_x,k_y,z,\omega)\bigg|_{z=0} = 0, \tag{5.5A.4}$$

instead of Eq. (5.20).

Additionally, another artificial damping parameter ($\varepsilon_2 > 0$) is introduced in Eq. (5.24) as

$$\chi(t) = \frac{1}{T}[H(t) - H(t-T)]e^{-\varepsilon_2 t}. \tag{5.5A.5}$$

Then, we obtain

$$\widehat{\chi}(\omega) = \int_{-\infty}^{\infty} \chi(t)e^{i\omega t}dt = \frac{1}{i(\omega + i\epsilon_2)T}\left(e^{i(\omega+i\epsilon_2)T} - 1\right). \tag{5.5A.6}$$

When Eqs. (5.5A.4) and (5.5A.6) are used, the equation corresponding to Eq. (5.27) is written as

$$\begin{aligned}
&\phi(x,y,z,t)\\
&= \frac{1}{(2\pi)^2}\int_{-\infty}^{\infty}\int_{-\infty}^{\infty} dk_x dk_y\, e^{i(k_x x + k_y y)}\frac{1}{k}\frac{\widehat{d}(k_x,k_y)}{\cosh(kh_0)}\\
&\quad \frac{1}{2\pi T}\frac{i}{}\int_{-\infty}^{\infty} d\omega e^{-i\omega t}\frac{1 - e^{i(\omega+i\epsilon_2)T}}{\omega + i\epsilon_2}\frac{(\omega^2 + i2\epsilon_1\omega)\sinh(kz) + g_0 k\cosh(kz)}{(\omega^2 + i2\epsilon_1\omega) - g_0 k\tanh(kh_0)}\\
&= \frac{1}{(2\pi)^2}\int_{-\infty}^{\infty}\int_{-\infty}^{\infty} dk_x dk_y\, e^{i(k_x x + k_y y)}\frac{1}{k}\frac{\widehat{d}(k_x,k_y)}{\cosh(kh_0)}\\
&\quad \frac{1}{2\pi T}\frac{i}{}\left[\int_{-\infty}^{\infty} d\omega\frac{e^{-i\omega t}}{\omega + i\epsilon_2}\frac{(\omega^2 + i2\epsilon_1\omega)\sinh(kz) + g_0 k\cosh(kz)}{(\omega^2 + i2\epsilon_1\omega) - g_0 k\tanh(kh_0)}\right.\\
&\qquad \left. - \int_{-\infty}^{\infty} d\omega\frac{e^{-i\omega(t-T)}e^{-\epsilon_2 T}}{\omega + i\epsilon_2}\frac{(\omega^2 + i2\epsilon_1\omega)\sinh(kz) + g_0 k\cosh(kz)}{(\omega^2 + i2\epsilon_1\omega) - g_0 k\tanh(kh_0)}\right].
\end{aligned} \tag{5.5A.7}$$

We first consider the integration in the second term:

$$\int_{-\infty}^{\infty} d\omega \frac{e^{-i\omega(t-T)} e^{-\varepsilon_2 T}}{\omega + i\varepsilon_2} \frac{(\omega^2 + i2\varepsilon_1\omega)\sinh(kz) + g_0 k \cosh(kz)}{\omega^2 + i2\varepsilon_1\omega - g_0 k \tanh(kh_0)}. \tag{5.5A.8}$$

Considering that the artificial damping is very small or the artificial damping parameters are much smaller than the angular frequency ω for tsunamis, we employ the residue theorem. The poles of Eq. (5.5A.8) are located at $\omega = -i\varepsilon_1 \pm \omega_0$ and $-i\varepsilon_2$ in the lower half of the ω-plane (Fig. 5.A.1), where

$$\omega_0 \equiv \sqrt{g_0 k \tanh(kh_0)}. \tag{5.5A.9}$$

When $t - T < 0$, we take the path of the integral in the upper half plane (path I in Fig. 5.A.1) including no poles. In this case, then, the integration is zero as

$$\int_{-\infty}^{\infty} d\omega \frac{e^{-i\omega(t-T)} e^{-\varepsilon_2 T}}{\omega + i\varepsilon_2} \frac{(\omega^2 + i2\varepsilon_1\omega)\sinh(kz) + g_0 k \cosh(kz)}{\omega^2 + i2\varepsilon_1\omega - g_0 k \tanh(kh_0)}$$
$$= 0 \quad \text{for } t - T < 0. \tag{5.5A.10}$$

When $t - T > 0$, we take the integral path in the lower half plane including the poles (path II in Fig. 5.A.1). By setting $\varepsilon_1 \to 0$, $\varepsilon_2 \to 0$ and using the residue theorem, we calculate Eq. (5.5A.8) as

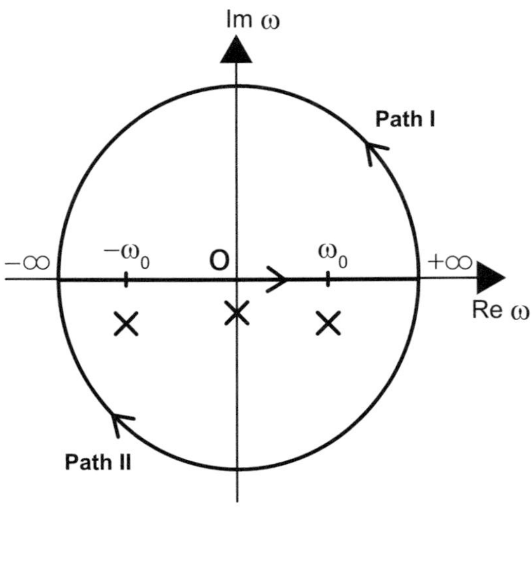

Fig. 5.A.1 Paths of integral in the complex ω-plane for Eq. (5.5A.8). The poles (crosses) are located at $\omega = -i\varepsilon_1 \pm \omega_0$ and $-i\varepsilon_2$ in the lower half plane

$$
\begin{aligned}
&\int_{-\infty}^{\infty} d\omega \frac{e^{-i\omega(t-T)}e^{-\varepsilon_2 T}}{\omega + i\varepsilon_2} \frac{(\omega^2 + i2\varepsilon_1\omega)\sinh(kz) + g_0 k\cosh(kz)}{(\omega^2 + i2\varepsilon_1\omega) - g_0 k\tanh(kh_0)} \\
&\approx \int_{-\infty}^{\infty} d\omega \frac{e^{-i\omega(t-T)}}{\omega + i\varepsilon_2} \frac{\omega^2\sinh(kz) + g_0 k\cosh(kz)}{[\omega - (\omega_0 - i\varepsilon_1)][\omega - (-\omega_0 - i\varepsilon_1)]} \\
&= -2\pi i[\mathrm{Res}(\omega = -i\varepsilon_2) + \mathrm{Res}(\omega = \omega_0 - i\varepsilon_1) + \mathrm{Res}(\omega = -\omega_0 - i\varepsilon_1)] \\
&= -2\pi i\left[\frac{-g_0 k\cosh kz}{\omega_0^2} + \frac{\omega_0^2\sinh kz + g_0 k\cosh kz}{\omega_0^2}\cos\left[\omega_0(t-T)\right]\right]
\end{aligned} \tag{5.5A.11}
$$

Summarizing Eqs. (5.5A.10) and (5.5A.11), we obtain

$$
\begin{aligned}
&\int_{-\infty}^{\infty} d\omega \frac{e^{-i\omega(t-T)}}{\omega} \frac{\omega^2\sinh(kz) + g_0 k\cosh(kz)}{\omega^2 - g_0 k\tanh(kh_0)} \\
&= 2\pi i\left[\frac{g_0 k\cosh kz}{\omega_0^2} - \frac{\omega_0^2\sinh kz + g_0 k\cosh kz}{\omega_0^2}\cos\left[\omega_0(t-T)\right]\right]H(t-T).
\end{aligned}
$$

Using (5.5A.9), we have

$$
\begin{aligned}
&\int_{-\infty}^{\infty} d\omega \frac{e^{-i\omega(t-T)}}{\omega} \frac{\omega^2\sinh(kz) + g_0 k\cosh(kz)}{\omega^2 - g_0 k\tanh(kh_0)} \\
&= 2\pi i\left[\frac{\cosh kz}{\tanh kh_0} - \left(\frac{\cosh kz}{\tanh kh_0} + \sinh kz\right)\cos\left[\omega_0(t-T)\right]\right]H(t-T).
\end{aligned} \tag{5.5A.12}
$$

The first term in the bracket in Eq. (5.5A.7) can also be calculated in a similar way. The velocity potential (5.5A.7) is then given by

$$
\begin{aligned}
&\phi(x, y, z, t) \\
&= -\frac{1}{(2\pi)^2}\int_{-\infty}^{\infty}\int_{-\infty}^{\infty} dk_x dk_y\, e^{i(k_x x + k_y y)} \frac{\widehat{d}(k_x, k_y)}{k\cosh kh_0} \\
&\quad \left[\frac{\cosh kz}{\tanh kh_0}\frac{H(t) - H(t-T)}{T}\right. \\
&\qquad \left. -\left(\frac{\cosh kz}{\tanh kh_0} + \sinh kz\right)\frac{H(t)\cos(\omega_0 t) - H(t-T)\cos(\omega_0(t-T))}{T}\right] \\
&= -\frac{1}{(2\pi)^2}\int_{-\infty}^{\infty}\int_{-\infty}^{\infty} dk_x dk_y\, e^{i(k_x x + k_y y)} \frac{\widehat{d}(k_x, k_y)}{k} \\
&\quad \left[\frac{\cosh kz}{\sinh kh_0}\frac{H(t) - H(t-T)}{T}\right. \\
&\qquad \left. -\left(\frac{\cosh kz}{\sinh kh_0} + \frac{\sinh kz}{\cosh kh_0}\right)\frac{H(t)\cos(\omega_0 t) - H(t-T)\cos(\omega_0(t-T))}{T}\right]
\end{aligned} \tag{5.5A.13}
$$

This equation is identical to Eq. (5.28).

Appendix 5B: Tsunami Propagation from an Initial Height Distribution and Zero Velocity Distribution

Setting the initial conditions as

$$\eta(x, y, t = 0) = \eta_0(x, y), \tag{5.5B.1}$$

and

$$\mathbf{v_H}(\mathbf{x}, t = 0) = 0, \text{and } v_z(\mathbf{x}, t = 0) = 0, \tag{5.5B.2}$$

the tsunami propagation is solved in the Cartesian coordinates (Fig. 5.B.1).

The velocity potential ($\mathbf{v} = \nabla\phi$) satisfies the Laplace equation (Eq. (5.3)):

$$\Delta\phi(\mathbf{x}, t) = 0. \tag{5.5B.3}$$

The velocity potential satisfies the boundary condition at the sea surface (Eq. (5.11)):

$$\left.\frac{\partial^2\phi(\mathbf{x}, t)}{\partial t^2}\right|_{z=0} + g_0 \left.\frac{\partial\phi(\mathbf{x}, t)}{\partial z}\right|_{z=0} = 0, \tag{5.5B.4}$$

and the boundary condition at the sea bottom:

$$\left.\frac{\partial\phi(\mathbf{x}, t)}{\partial z}\right|_{z=-h_0} = 0. \tag{5.5B.5}$$

Note that the initial conditions (Eqs. (5.5B.1) and (5.5B.2)) and the bottom boundary condition (Eq. (5.5B.5)) are different when compared with the tsunami generation problem treated in 5.2.1.

In order to solve this problem, we introduce the Fourier transform as

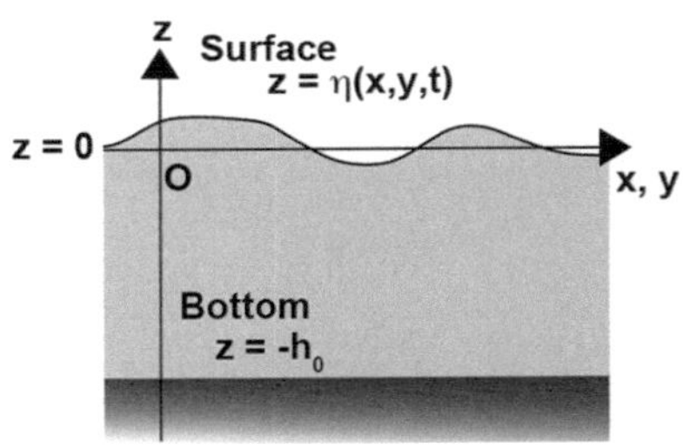

Fig. 5.B.1 Coordinates used in the formulation

$$\widehat{\Phi}(k_x, k_y, z, t) = \int\int_{-\infty}^{\infty} dxdy\, e^{-i(k_x x + k_y y)} \phi(x, y, z, t). \tag{5.5B.6}$$

The corresponding inverse Fourier transform is

$$\phi(x, y, z, t) = \frac{1}{(2\pi)^2} \int\int_{-\infty}^{\infty} dk_x dk_y\, e^{i(k_x x + k_y y)}\, \widehat{\phi}(k_x, k_y, z, t). \tag{5.5B.7}$$

Then, the Laplace equation (Eq. (5.5B.3)) and the boundary conditions (Eqs. (5.5B.4) and (5.5B.5)) are rewritten as

$$\frac{d^2}{dz^2}\widehat{\phi}(k_x, k_y, z, t) = k^2 \widehat{\phi}(k_x, k_y, z, t), \tag{5.5B.8}$$

where $k^2 = {k_x}^2 + {k_y}^2$,

$$\left.\frac{\partial^2 \widehat{\phi}(k_x, k_y, z, t)}{\partial t^2}\right|_{z=0} + g_0 \left.\frac{\partial \widehat{\phi}(k_x, k_y, z, t)}{\partial z}\right|_{z=0} = 0, \tag{5.5B.9}$$

and

$$\left.\frac{\partial \widehat{\phi}(k_x, k_y, z, t)}{\partial z}\right|_{z=-h_0} = 0. \tag{5.5B.10}$$

The general solution of Eq. (5.5B.8) is given by

$$\widehat{\phi}(k_x, k_y, z, t) = A(k_x, k_y, t)\cosh(kz) + B(k_x, k_y, t)\sinh(kz). \tag{5.5B.11}$$

The coefficient $B(k_x, k_y, t)$ is represented by $A(k_x, k_y, t)$ by considering the bottom boundary condition (Eq. (5.5B.10)). Then, Eq. (5.5B.11) is calculated as

$$\widehat{\phi}(k_x, k_y, z, t) = A(k_x, k_y, t) \frac{\cosh[k(z + h_0)]}{\cosh(kh_0)}. \tag{5.5B.12}$$

Substituting (5.5B.12) into (5.5B.9), we obtain

$$\frac{d^2 A(k_x, k_y, t)}{dt^2} + \omega_0^2 A(k_x, k_y, t) = 0, \tag{5.5B.13}$$

where

$$\omega_0^2 = g_0 k \tanh(kh_0). \tag{5.5B.14}$$

Equation (5.5B.14) represents the well-known dispersion relation (Eq. (3.47)). The general solution of Eq. (5.5B.13) is

$$A(k_x, k_y, t) = C(k_x, k_y)\cos(\omega_0 t) + D(k_x, k_y)\sin(\omega_0 t). \tag{5.5B.15}$$

Then, the velocity potential of Eq. (5.5B.11) is given by

$$\widehat{\phi}(k_x, k_y, z, t) = [C\cos(\omega_0 t) + D\sin(\omega_0 t)]\frac{\cosh[k(z+h_0)]}{\cosh(kh_0)}.$$

Considering the initial velocity condition of Eq. (5.5B.2), the coefficient of C needs to be zero:

$$\widehat{\phi}(k_x, k_y, z, t) = D(k_x, k_y)\sin(\omega_0 t)\frac{\cosh[k(z+h_0)]}{\cosh(kh_0)}. \tag{5.5B.16}$$

Then, the velocity potential is represented by the inverse Fourier transform as

$$\phi(x, y, z, t) = \frac{1}{(2\pi)^2}\int\int_{-\infty}^{\infty} dk_x dk_y\, e^{i(k_x x + k_y y)} D(k_x, k_y)\sin(\omega_0 t)\frac{\cosh[k(z+h_0)]}{\cosh(kh_0)}. \tag{5.5B.17}$$

The tsunami height is represented by the velocity potential as

$$\eta(x, y, t) = -\frac{1}{g_0}\frac{\partial \phi}{\partial t}\bigg|_{z=0}. \tag{5.5B.18}$$

By substituting (5.5B.17) into (5.5B.18), we get

$$\eta(x, y, t) = -\frac{1}{(2\pi)^2}\frac{1}{g_0}\int\int_{-\infty}^{\infty} dk_x dk_y\, e^{i(k_x x + k_y y)}\omega_0 D(k_x, k_y)\cos(\omega_0 t). \tag{5.5B.19}$$

Considering the initial tsunami height distribution (5.5B.1), we obtain

$$\widehat{\eta}_0(k_x, k_y) = -\frac{\omega_0 D(k_x, k_y)}{g_0},$$

or

$$D(k_x, k_y) = -\frac{g_0}{\omega_0}\widehat{\eta}_0(k_x, k_y), \tag{5.5B.20}$$

where $\widehat{\eta}_0(k_x, k_y)$ is the Fourier transform of the initial tsunami height distribution $\eta_0(x, y)$. Substituting Eq. (5.5B.20), we obtain the solution of the velocity potential as

$$\phi(x, y, z, t) = \frac{-1}{(2\pi)^2}\int\int_{-\infty}^{\infty} dk_x dk_y\, e^{i(k_x x + k_y y)} \frac{g_0}{\omega_0}\widehat{\eta}_0(k_x, k_y)\sin(\omega_0 t)\frac{\cosh[k(z+h_0)]}{\cosh(kh_0)}.$$

Using the dispersion relation (Eq. (5.5B.14)), we calculate

$$\phi(x, y, z, t) = \frac{-1}{(2\pi)^2} \times \int\int_{-\infty}^{\infty} dk_x dk_y\, e^{i(k_x x + k_y y)}\frac{\omega_0}{k}\widehat{\eta}_0(k_x, k_y)\frac{\cosh[k(z+h_0)]}{\sinh(kh_0)}\sin(\omega_0 t). \tag{5.5B.21}$$

Using the velocity potential of Eq. (5.5B.21), we can represent the velocity, the tsunami height, and the pressure change in the following.

$$\mathbf{v}_H(\mathbf{x}, t) = \nabla_H\phi(\mathbf{x}, t) = \frac{1}{(2\pi)^2}\int_{-\infty}^{\infty}\int_{-\infty}^{\infty} dk_x dk_y\, e^{i(k_x x + k_y y)}\widehat{\eta}_0(k_x, k_y)\left[-i\omega_0\frac{\mathbf{k}_H}{k}\frac{\cosh[k(z+h_0)]}{\sinh kh_0}\sin(\omega_0 t)\right], \tag{5.5B.22}$$

where $\nabla_H = (\partial/\partial x)\mathbf{e_x} + (\partial/\partial y)\mathbf{e_y}$, and $\mathbf{k}_H = k_x\mathbf{e_x} + k_y\mathbf{e_y}$.

$$\mathrm{v}_z(\mathbf{x}, t) = \frac{\partial\phi(\mathbf{x}, t)}{\partial z} = \frac{1}{(2\pi)^2}\int_{-\infty}^{\infty}\int_{-\infty}^{\infty} dk_x dk_y\, e^{i(k_x x + k_y y)}\widehat{\eta}_0(k_x, k_y)\left[-\omega_0\frac{\sinh[k(z+h_0)]}{\sinh kh_0}\sin(\omega_0 t)\right], \tag{5.5B.23}$$

$$\eta(\mathbf{x}, t) = -\frac{1}{g_0}\frac{\partial\phi}{\partial t}\bigg|_{z=0} = \frac{1}{(2\pi)^2}\int_{-\infty}^{\infty}\int_{-\infty}^{\infty} dk_x dk_y\, e^{i(k_x x + k_y y)}\widehat{\eta}_0(k_x, k_y)\cos(\omega_0 t), \tag{5.5B.24}$$

and

$$
\begin{aligned}
p_e(x, y, z, t) &= -\rho_0 \frac{\partial \phi(\mathbf{x}, t)}{\partial t} \\
&= \frac{1}{(2\pi)^2} \int_{-\infty}^{\infty} \int_{-\infty}^{\infty} dk_x dk_y \, e^{i(k_x x + k_y y)} \rho_0 \widehat{\eta}_0(k_x, k_y) \\
&\quad \frac{\omega_0^2}{k} \frac{\cosh(z + h_0)}{\sinh(kh_0)} \cos(\omega_0 t).
\end{aligned}
\tag{5.5B.25}
$$

Using the dispersion relation (Eq. (5.5B.14)), we calculate

$$
\begin{aligned}
p_e(x, y, z, t) = &\frac{1}{(2\pi)^2} \int_{-\infty}^{\infty} \int_{-\infty}^{\infty} dk_x dk_y \, e^{i(k_x x + k_y y)} \\
&\rho_0 g_0 \widehat{\eta}_0(k_x, k_y) \frac{\cosh(z + h_0)}{\cosh(kh_0)} \cos(\omega_0 t).
\end{aligned}
\tag{5.5B.26}
$$

References

Aki K, Richards P (2002) Quantitative seismology. University Science Books, Sausalito

An C, Cai C, Zheng Y, Meng L, Liu P (2017) Theoretical solution and applications of ocean bottom pressure induced by seismic seafloor motion. Geophys Res Lett 44:10,272–10,281. https://doi.org/10.1002/2017GL075137

Ewing WM, Jardetzky WS, Press F (1957) Elastic waves in layered media. McGraw-Hill Book Company, Inc, New York

Filloux JH (1982) Tsunami recorded on the open ocean floor. Geophys Res Lett 9(1):25–28

Fujii Y, Satake K (2008) Tsunami sources of the November 2006 and January 2007 great Kuril earthquakes. Bull Seismol Soc Am 98(3):1559–1571. https://doi.org/10.1785/0120070221

Gill AE (1982) Atmosphere-ocean dynamics, International geophysics series, vol 30. Academic Press, New York

Gradshteyn IS, Ryshik IM (2000) Table of integrals, series and products (6th Ed. In Engl., ed. A. Jeffrey and D. Zwillinger). Academic Press, San Diego

Gusman AR, Tanioka Y, Sakai S, Tsushima H (2012) Source model of the great 2011 Tohoku earthquake estimated from tsunami waveforms and crustal deformation data. Earth Planet Sci Lett 341:234–242. https://doi.org/10.1016/j.epsl.2012.06.006

Hooper A et al (2013) Importance of horizontal seafloor motion on tsunami height for the 2011 Mw= 9.0 Tohoku-Oki earthquake. Earth Planet Sci Lett 361:469–479. https://doi.org/10.1016/j.epsl.2012.11.013

Ito Y et al (2011) Frontal wedge deformation near the source region of the 2011 Tohoku-Oki earthquake. Geophys Res Lett 38(7):L00G05. https://doi.org/10.1029/2011GL048355

Kajiura K (1963) The leading wave of a tsunami. Bull Earthq Res Inst 41:535–571

Kakinuma T, Akiyama M (2007) Numerical analysis of tsunami generation due to seabed deformation. Coast Eng 2006:1490–1502. https://doi.org/10.1142/9789812709554_0127

Kambe T (2007) Elementary fluid mechanics, vol 258. World Scientific, Singapore

Kennett BL (2001) The seismic Wavefield: volume I, introduction and theoretical development. Cambridge University Press, Cambridge

Kervella Y, Dutykh D, Dias F (2007) Comparison between three-dimensional linear and nonlinear tsunami generation models. Theor Comput Fluid Dyn 21(4):245–269. https://doi.org/10.1007/s00162-007-0047-0

Kubota T, Hino R, Inazu D, Ito Y, Iinuma T (2015) Complicated rupture process of the Mw 7.0 intraslab strike-slip earthquake in the Tohoku region on 10 July 2011 revealed by near-field pressure records. Geophys Res Lett 42(22):9733–9739. https://doi.org/10.1002/2015GL066101

Kubota T, Saito T, Suzuki W, Hino R (2017) Estimation of seismic centroid moment tensor using ocean bottom pressure gauges as seismometers. Geophys Res Lett 44:10,907–10,915. https://doi.org/10.1002/2017GL075386

Lotto GC, Nava G, Dunham EM (2017) Should tsunami simulations include a nonzero initial horizontal velocity? Earth Planets Space 69:117. https://doi.org/10.1186/s40623-017-0701-8

Murotani S, Iwai M, Satake K, Shevchenko G, Loskutov A (2015) Tsunami forerunner of the 2011 Tohoku Earthquake observed in the Sea of Japan. Pure Appl Geophys 172(3–4):683–697. https://doi.org/10.1007/s00024-014-1006-5

Nosov MA (1999) Tsunami generation in compressible ocean. Phys Chem Earth (B) 24(5):437–441

Nosov MA, Kolesov SV (2011) Optimal initial conditions for simulation of seismotectonic tsunamis. Pure Appl Geophys 168(6–7):1223–1237. https://doi.org/10.1007/s00024-010-0226-6

Okada Y (1985) Surface deformation due to shear and tensile faults in a half-space. Bull Seismol Soc Am 75(4):1135–1154

Pedlosky J (2013) Waves in the ocean and atmosphere: introduction to wave dynamics. Springer Science & Business Media, New York

Saito T (2013) Dynamic tsunami generation due to sea-bottom deformation: analytical representation based on linear potential theory. Earth Planets Space 65:1411–1423. https://doi.org/10.5047/eps.2013.07.004

Saito T (2017) Tsunami generation: validity and limitations of conventional theories. Geophys J Int 210(3):1888–1900. https://doi.org/10.1093/gji/ggx275

Saito T, Furumura T (2009) Three-dimensional tsunami generation simulation due to sea-bottom deformation and its interpretation based on the linear theory. Geophys J Int 178(2):877–888. https://doi.org/10.1111/j.1365-246X.2009.04206.x

Saito T, Tsushima H (2016) Synthesizing ocean bottom pressure records including seismic wave and tsunami contributions: toward realistic tests of monitoring systems. J Geophys Res Solid Earth 121(11):8175–8195. https://doi.org/10.1002/2016JB013195

Satake K, Kanamori H (1991) Abnormal tsunamis caused by the June 13, 1984, Torishima, Japan, earthquake. J Geophys Res Solid Earth 96(B12):19933–19939. https://doi.org/10.1029/91JB01903

Satake K, Fujii Y, Harada T, Namegaya Y (2013) Time and space distribution of coseismic slip of the 2011 Tohoku earthquake as inferred from tsunami waveform data. Bull Seismol Soc Am 103(2B):1473–1492. https://doi.org/10.1785/0120120122

Song YT, Mohtat A, Yim SC (2017) New insights on tsunami genesis and energy source. J Geophys Res Oceans 122:4238–4256. https://doi.org/10.1002/2016JC012556

Takahashi R (1942) On seismic sea waves caused by deformations of the sea bottom. Bull Earthq Res Inst 20:357–400

Tanioka Y, Satake K (1996) Tsunami generation by horizontal displacement of ocean bottom. Geophys Res Lett 23(8):861–864. https://doi.org/10.1029/96GL00736

Tanioka Y, Seno T (2001) Sediment effect on tsunami generation of the 1896 Sanriku tsunami earthquake. Geophys Res Lett 28(17):3389–3392

Tsushima H, Hino R, Tanioka Y, Imamura F, Fujimoto H (2012) Tsunami waveform inversion incorporating permanent seafloor deformation and its application to tsunami forecasting. J Geophys Res Solid Earth 117(B03311). https://doi.org/10.1029/2011JB008877

Yamamoto T (1982) Gravity waves and acoustic waves generated by submarine earthquakes. Soil Dyn Earthq Eng 1(2):75–82

Chapter 6
Propagation Simulation

Abstract Tsunami propagation simulations are very useful in both theoretical and application studies. Recent improvements in computer performance and the detailed bathymetry surveys in local- and global-scale make numerical simulations more feasible and reliable. This chapter treats the theoretical background and numerical schemes underlying tsunami propagation simulations. Since tsunami wavelength is usually greater than the sea depth, we approximate a 3-D equation of motion using 2-D tsunami equations. There are various kinds of tsunami equations according to the approximations. Hence, it is important to select appropriate tsunami equations depending on the situation and purpose of the simulation. Section 6.1 is an overview of various tsunami equations and introduces some results of the simulations. Section 6.2 derives the 2-D tsunami equations from the 3-D equation of motion by assuming long-wavelength wave propagation. We explain the linear long-wave equations, nonlinear long-wave equations, and linear dispersive equations. Section 6.3 illustrates the finite difference methods for numerically simulating the tsunami propagation across realistic bathymetry.

Keywords Long-wave approximation · Nonlinear long-wave equations · Linear dispersive equations · Nonlinear dispersive equations · Finite difference method

6.1 Nonlinear and Dispersive Tsunami

Assuming that the tsunami wavelength is greater than the sea depth, we approximate the 3-D equation of motion using 2-D tsunami equations. There are various kinds of approximations (e.g., Madsen and Sørensen 1992; Imamura 1996). We usually use different types of 2-D tsunami equations depending on the purpose and situation. In general, if we use more approximations, the equations become simpler, and the simulation cost becomes lower. The optimum balance between the cost and the approximations depends on the problem we work on. The techniques for solving those equations have been developed based on fluid dynamics. For example, Shi et al. (2012) presented a nonlinear Boussinesq model to calculate dispersive waves

T. Saito, *Tsunami Generation and Propagation*, Springer Geophysics,
https://doi.org/10.1007/978-4-431-56850-6_6

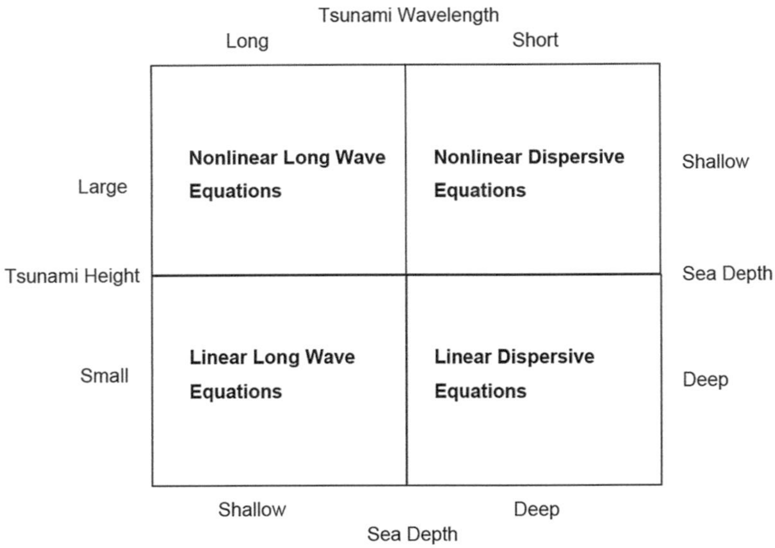

Fig. 6.1 A classification of four tsunami equations (linear long-wave equations, nonlinear long-wave equations, linear dispersive equations, and nonlinear dispersive equations). Linear dispersive equations appropriately reproduce tsunami propagation across a deep ocean. Nonlinear long-wave equations can reproduce tsunamis approaching coasts

including wave breaking and inundation. This book cannot treat all the equations but introduces four basic tsunami equations to obtain a perspective and practical methods for numerical simulation. The equations may be classified as shown in Fig. 6.1 according to the wavelength and height.

First, we consider the tsunami height. When a tsunami propagates in deep ocean, the tsunami height is much smaller than the sea depth. As tsunami approaches coast, the sea depth becomes shallow, but as long as the tsunami height is much smaller than the sea depth, the linear equations can be used, and the superposition principle is valid. In this case, we can use the linear long-wave equations and linear dispersive equations in order to simulate tsunami propagation. When the tsunami height is not much smaller than the sea depth, the nonlinear effects of tsunami become important. Then, the superposition principle is invalid. In order to simulate nonlinear phenomena, the nonlinear long-wave equations and nonlinear dispersive equations are used.

Next, we consider the tsunami wavelength. As shown in the phase velocity of water waves (*Chap. 3: Propagation of Tsunami and Seismic Waves*), the tsunami propagation velocity generally depends on the wavelength (Eq. (3.56)). In order to reproduce frequency-dependent tsunami propagation, we should use nonlinear/linear dispersive tsunami equations. On the other hand, as long as the wavelength is much greater than the sea depth, the tsunami velocity does not show frequency dependence. When the tsunami wavelength is much longer than the sea depth, we can use nonlinear/linear long-wave equations that do not show the dispersion.

During the 2011 Tohoku-Oki earthquake, many higher-quality tsunami waveforms compared to past tsunamis were recorded. These records reliably revealed the earthquake fault kinematics and tsunami source model (e.g., Koketsu et al. 2011; Fujii et al. 2011; Saito et al. 2011). By using a reliable source model and various tsunami equations together with high-quality tsunami records, we show some examples of tsunami propagation simulations.

6.1.1 Nonlinear Waves

6.1.1.1 Energy Absorption

When the tsunami amplitude is not small enough compared to the sea depth, the tsunami shows nonlinear behavior. This behavior is well represented by nonlinear long-wave equation. One of the important nonlinear effects is energy absorption due to bottom friction. This energy absorption is modeled using the nondimensional bottom friction coefficient C_f. Figure 6.2a shows the tsunami waveforms for 8 days (192 h) after the 2011 Tohoku-Oki earthquake occurred. Following the very high leading tsunami, small-amplitude tsunamis continuously arrived at stations KPG1, 51425, 46411, and 32412 (station locations are shown in Fig. 6.2b). Figure 6.2c plots the squared-wave amplitude, which is proportional to tsunami energy density (Eq. (3.93)):

$$E(x, y, t) = \rho_0 g_0 |\eta|^2$$

The squared amplitude is plotted on a logarithmic scale in Fig. 6.2c. The tsunami energy monotonously decreased as the elapsed time increased. Even 7 days after the earthquake, the tsunami energy was significantly larger than the background energy level (compare with the energy before the Tohoku-Oki earthquake). Munk (1963) may be the first to point out that the tsunami energy in the ocean decays in a way similar to sound waves in a closed room. Van Dorn (1984, 1987) confirmed Munk's hypothesis by analyzing the records of many tsunamis across the Pacific Ocean and other oceans. It is interesting to note that such long-lasting waves or wave energy after an earthquake are also recognized in seismic waves, which are referred to as coda waves. Coda waves have been extensively studied in seismology and has provided seismologists with important information about the scattering and energy absorption mechanism in the earth's crust (e.g., Sato et al. 2012). A surprising feature of the coda is that the coda amplitude does not depend on the hypocentral distance but only on the elapsed time. This characteristic seems to be common between seismic waves and tsunami. For example, we cannot identify which station is the closest to the earthquake hypocenter using only the tsunami data with an elapsed time between 96 and 144 h in Fig. 6.2c. In order to reproduce the tsunami coda energy, Saito et al. (2013) conducted numerical simulations by adjusting the bottom friction coefficient C_f. In Fig. 6.2d, the gray symbol indicates the average temporal decay of the

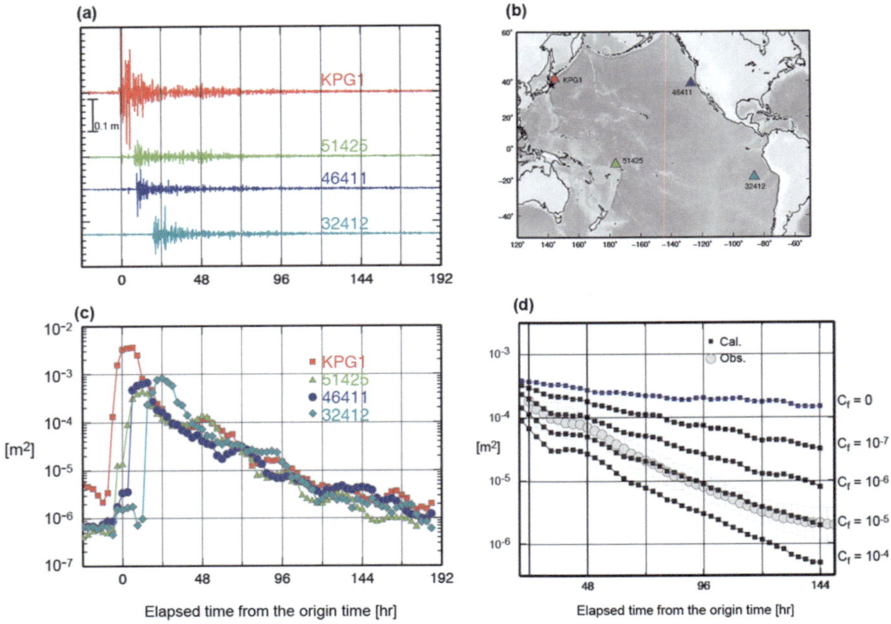

Fig. 6.2 (**a**) Band-pass-filtered waveforms with a passband between 30 and 120 min at stations KPG1 (Japan), 51425 (northwest of Apia, Samoa), 46411 (northwest of San Francisco), and 32412 (southwest of Lima, Peru). The elapsed time is measured from the earthquake origin time. (**b**) The station locations. (**c**) Temporal changes of the squared amplitude at the stations. The time window of 12 h and the time shift of 3 h are used in smoothing the squared amplitude over time. (**d**) Squared amplitude synthesized by numerical tsunami simulations with various nondimensional friction coefficients C_f (blue and black dots) and the observed averaged squared amplitude. (Saito et al. 2013, copyright by the Seismological Society of America)

squared tsunami amplitude in the Pacific Ocean after the Tohoku-Oki earthquake. If we do not consider the bottom friction ($C_f = 0$), the simulation (blue dots) overestimates the coda energy (gray dots). Larger C_f friction coefficients cause the energy decay to occur more rapidly (black dots). If the friction coefficient is set as $C_f = 10^{-5}$, the tsunami energy decay is reproduced well for 6 days following the Tohoku-Oki earthquake (Fig. 6.2d).

The nonlinear effects caused by the advection and the bottom friction are more dominant in shallower seas near coasts. Figure 6.3a shows the observed tsunami waveforms near coasts for the 2011 Tohoku-Oki earthquake (gray) and those calculated by the simulation including energy absorption (red lines by (a) nonlinear dispersive simulations). The simulated waveforms successfully reproduce the observed waveforms. In order to examine the effects of the nonlinearity, Fig. 6.3b shows the simulated waveforms without the nonlinear terms (red lines by (b) linear dispersive simulations). The difference appears only in the later waves that are reflected from the coasts, whereas little difference is recognized in the leading waves. This suggests that the nonlinear effect occurs near the coasts when the tsunami is reflected at the coasts.

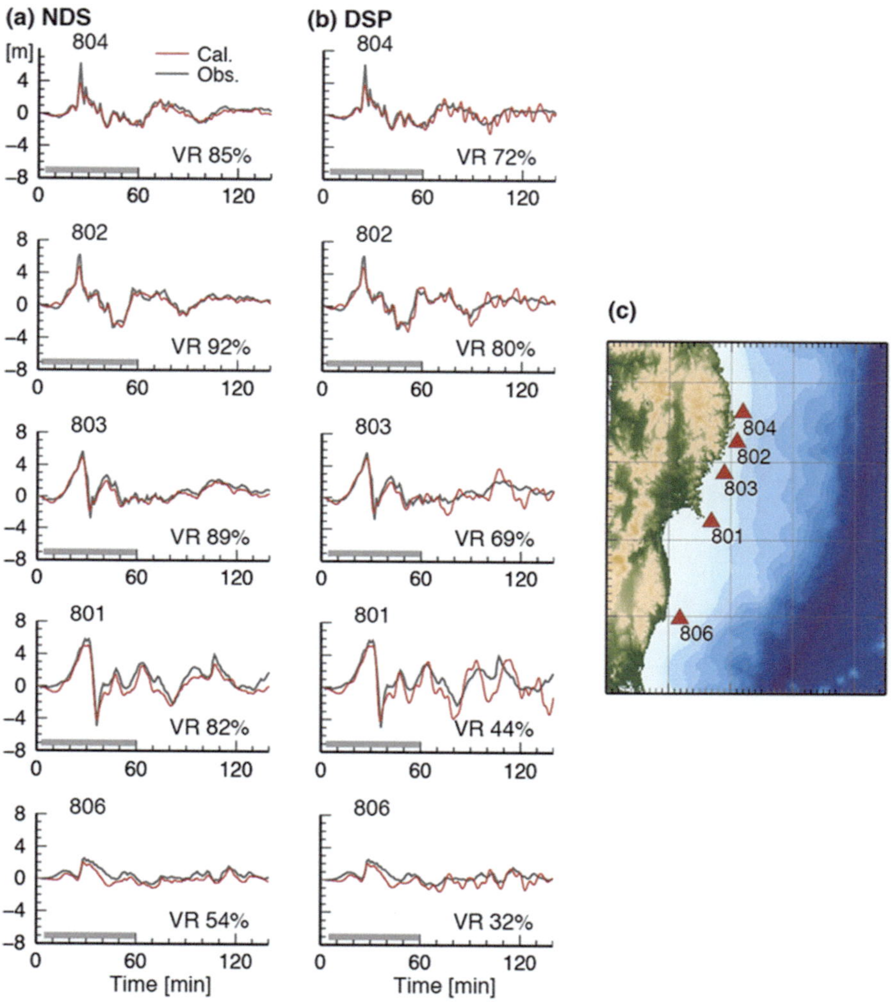

Fig. 6.3 Comparisons of the calculated (red) and observed (black) tsunami waveforms for the station with GPS tsunami gauges (804, 802, 803, 801, and 806). (**a**) The nonlinear dispersive equations (NDS) and (**b**) linear dispersive equations (DSP) were used for the calculation. Gray bars indicate the data used for the tsunami source estimation. The variance reduction (VR) for each calculated waveform is listed in each bin. (**c**) The locations of the GPS tsunami gauges. (Saito et al. 2014, copyright by the American Geophysical Union)

6.1.1.2 Inundation

When a massive amount of water flows onto a coast as a tsunami (tsunami run-up), the tsunami inundates the inland area. The run-up process is also nonlinear. This is unique to tsunami, while the dispersion and energy absorption are also recognized in seismic waves. Tsunami inundation causes devastating damage and leaves geological marks near the coast as tsunami deposits. The inundation process is more complicated than tsunami propagation in deep oceans, but it is very important to understand the run-up mechanism and to reliably assess the inundation area for anticipated huge earthquakes in order to mitigate disasters near coasts. At the same time, by analyzing the tsunami deposits, we can reproduce large historical tsunamis that occurred when no seismographs were available (e.g., Minoura et al. 2001). This is referred to as paleoseismology. This is one of the advantages of tsunami analysis compared to seismogram analysis. Figure 6.4 shows the simulated inundation area of

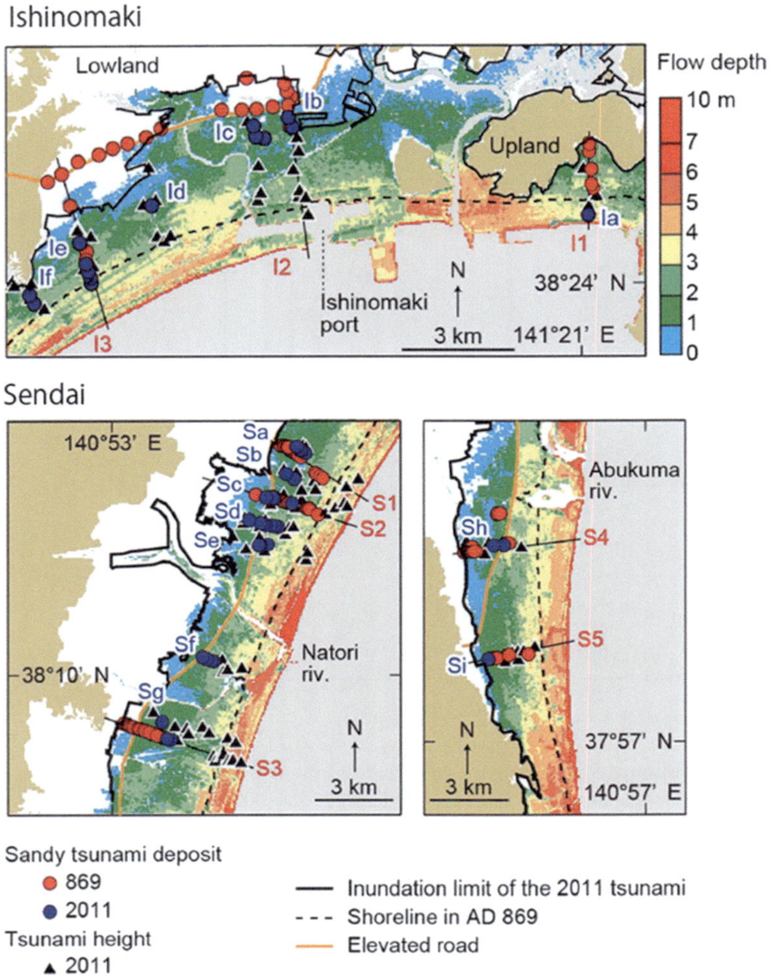

Fig. 6.4 Computed flow depths on Ishinomaki and Sendai Plains using the 2011 Tohoku-Oki earthquake model. Locations of sandy deposits of the A.D. 869 (red circles) and 2011 tsunamis (blue circles) are also shown. Triangles indicate the locations of 2011 tsunami height measurements. Black solid lines indicate the 2011 inundation limit, and dashed lines indicated the shoreline in A.D. 869. Orange lines represent artificially elevated roads in 2011. Ground with an elevation of more than 10 m is defined as "upland" and colored gold. (Namegaya and Satake 2014, copyright by the American Geophysical Union)

the 2011 Tohoku-Oki earthquake together with the distribution of tsunami deposit from the 869 Jogan earthquake (Namegaya and Satake 2014). By comparing the location of tsunami deposit and tsunami flow depth (height above ground), they estimated that the 869 Jogan earthquake had a rupture length of at least 200 km and a moment magnitude larger than 8.6. Note that the maximum magnitude was not delimited by this analysis. It is very important to investigate huge historical tsunami events, because huge earthquakes can occur repeatedly in the future, even though not always by the same rupture process.

To simulate the inundation correctly, high-resolution and reliable bathymetry/topography data and the bottom friction distribution data are necessary. Moreover, the distribution of buildings should be taken into account when we simulate with a spatial resolution of a few meters. Since high-resolution tsunami simulations have high computing costs, it was difficult to conduct an inundation simulation during the short time between the occurrence of a huge earthquake and the inundation of the nearby coasts. However, Oishi et al. (2015) developed a high-performance simulation code on a supercomputer and showed that the inundation can be successfully simulated in a few minutes after the tsunami source is set (~5 min is at least necessary by setting the tsunami source after the earthquake occurs). Recently, new tsunami simulation code has been extensively developed. For example, JAGURS (e.g., Baba et al. 2016), a tsunami simulation code (linear/nonlinear long-wave/Boussinesq equations with/without the effects of elastic deformation of the Earth due to tsunamis) is openly available on the Internet (https://github.com/jagurs-admin/jagurs, accessed 2017-08-24) (Fig. 6.5).

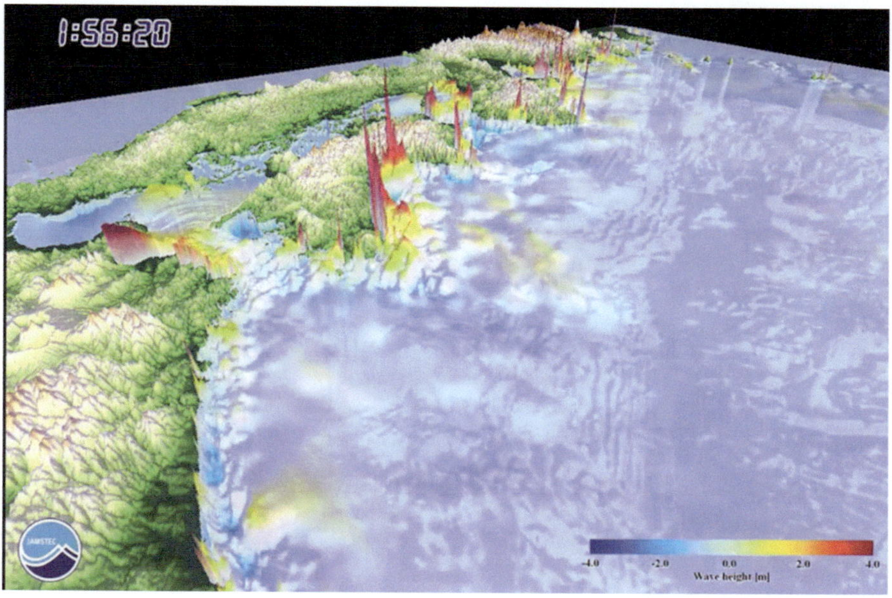

Fig. 6.5 An example of a tsunami simulation (JAGURS code) in the Nankai subduction zone, southwestern Japan. (Courtesy of T. Baba)

6.1.2 Dispersive Waves

It is well known from fluid dynamics that the dispersive wave develops when a tsunami propagates a long distance in a deep ocean. Actually, a dispersive tsunami was observed at a station located far from the source (~6200 km) during the 2011 Tohoku-Oki earthquake (Kirby et al. 2013). On the other hand, recent tsunami observations detected dispersive waves even at stations that were not so far from the source (e.g., Inazu and Saito 2014). Although moderate-sized earthquake faulting usually does not reach the sea bottom and earthquake faults are buried in the subsurface, huge earthquake faulting often reaches the sea bottom. When the rupture reaches the sea bottom, the sea-bottom displacement distribution becomes steep. This is characterized by a rich short-wavelength component in its wavenumber spectrum, which then generates a short-wavelength tsunami. Actually, the Tohoku-Oki earthquake excited a steep initial tsunami height distribution due to a large slip near the trench (e.g., Suzuki et al. 2011; Satake et al. 2013). Figure 6.6a

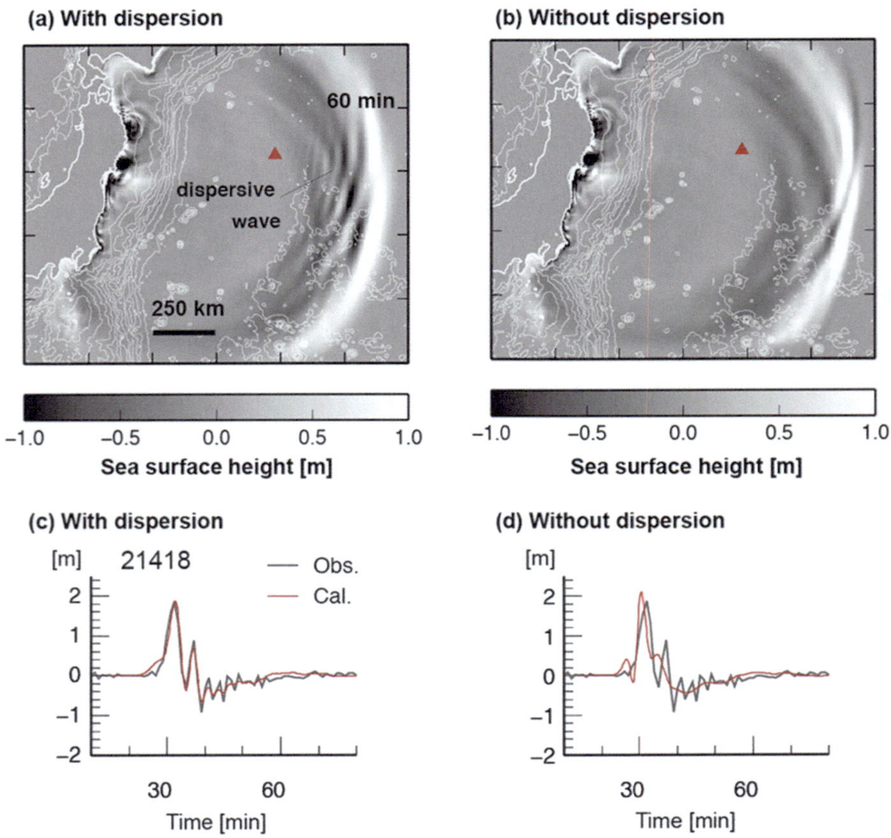

Fig. 6.6 Tsunami height distribution at the elapsed time of 60 min simulated with (**a**) dispersive and (**b**) non-dispersive tsunami equations. The location of the station 21418 is indicated by a red triangle. Sea-surface height changes at the station simulated by (**c**) dispersive (red line) and (**d**) non-dispersive tsunami equations (red line) are plotted together with the observed sea-surface height change (gray line). (Saito et al. 2014, copyright by the American Geophysical Union)

shows the simulated tsunami at the elapsed time of 60 min after the earthquake occurred, which was calculated using the dispersive tsunami equations. Dispersive waves propagate eastward in the deep Pacific Ocean. We recognize these to be dispersive waves by comparing them with the simulation results of the nonlinear long-wave equations (Fig. 6.6b). Figure 6.6c and d shows the tsunami waveform observed at station 21418 (red triangles in Fig. 6.12a and b). Even near station 21418, which is just ~400 km from the source, dispersive waves were clearly recorded in the 2011 Tohoku-Oki earthquake.

For a tsunami caused by an earthquake, the tsunami wavelength usually depends on the fault-strike direction. The wavelength is longer for a tsunami propagating along the fault strike than along the fault dip. Considering that the dispersive wave appears when the tsunami wavelength is not much longer than the sea depth, a tsunami propagating along the fault-dip direction more easily shows dispersion than a tsunami propagating along the fault-strike direction. This character of the dispersive tsunami can be used to determine the earthquake strike direction. Two earthquakes of M7.1 and M7.4 occurred off the Pacific coast of Japan (Fig. 6.7c and d). The earthquakes were both intraplate events in the Philippine Sea Plate and were

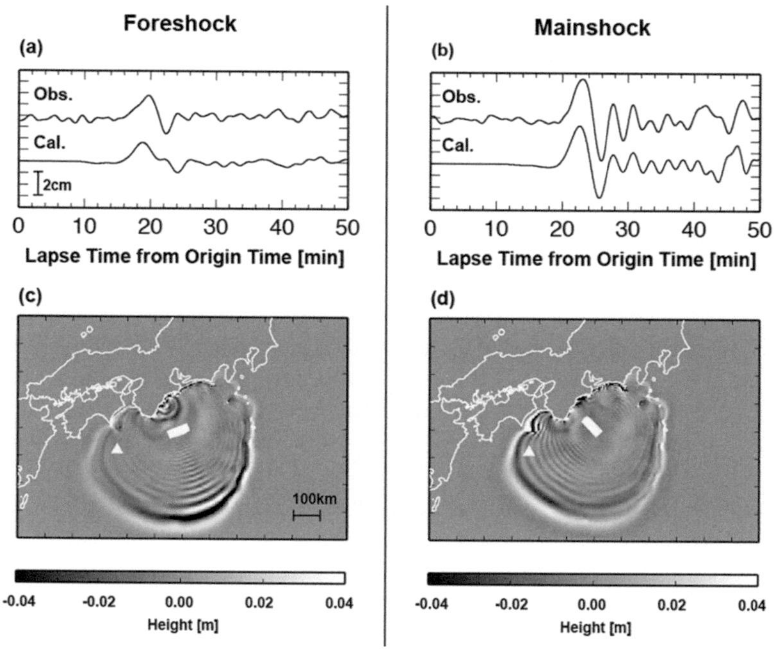

Fig. 6.7 Comparison of the tsunami propagation for the foreshock (M 7.1, 10:07 UTC) and the main shock (M 7.4, 14:57 UTC). Tsunami waveforms were recorded and calculated at the offshore tsunami observation point MPG1 for (**a**) the foreshock and (**b**) the main shock. Snapshots of the tsunami simulation were taken at an elapsed time of 30 min for (**c**) the foreshock and (**d**) the main shock. Triangles indicate the location of MPG1. Rectangles indicate the source region of the foreshock and the main shock. (Saito et al. 2010, copyright by the American Geophysical Union)

characterized by reverse faults. However, because the focal areas were outside the inland seismic network, it was difficult to determine the fault plane by seismogram analysis or aftershock distributions. Although the fault direction information was lost in the far-field seismograms, the short-wavelength dispersive tsunami contains the earthquake fault information. Figure 6.7a and c shows that a dispersive wave was not clearly recognized at the station located westward from the earthquake fault for the foreshock because the fault strike lays in the east-west direction. On the other hand, the dispersive waves are clearly recognized for the main shock (Fig. 6.7b and d). This is because the fault strike ran in the NW-SE direction. By analyzing this directional dependence based on the dispersive tsunami equations, Saito et al. (2010) determined the fault directions of the two earthquakes and concluded that the directions of the foreshock and the main shock were different.

6.2 Tsunami Equations Derived from Equations of Motion

This section illustrates the derivation of the 2-D tsunami equations (linear/nonlinear and dispersive/non-dispersive equations) from the 3-D equation of motion (e.g., Japan Society of Civil Engineers 1994). This outline of the derivation will be helpful when a new type of tsunami equations is being proposed.

6.2.1 Integration with Respect to Sea Depth: 3-D to 2-D Equations

In order to describe tsunami propagation, we usually use the equations describing the velocity and the sea-surface height in the 2-D horizontal space (x, y). The 2-D equations are referred to as 2-D tsunami equations or just tsunami equations in this book. In the derivation of the tsunami equations, the elasticity of the media is neglected; incompressible fluid over a rigid seafloor is assumed. Gravity works as the restoring force of the tsunami. The tsunami equations are obtained by an integration of the 3-D equation of motion along the vertical direction from the sea bottom to the sea surface.

We use the Cartesian coordinates shown in Fig. 6.8, where the x- and y-axes lie in the horizontal plane and the z-axis is taken as positive in the upward direction. The 3-D equation of motion in an incompressible sea is given by

$$\frac{Dv_x(\mathbf{x}, t)}{Dt} = -\frac{1}{\rho_0}\frac{\partial p(\mathbf{x}, t)}{\partial x}, \tag{6.1}$$

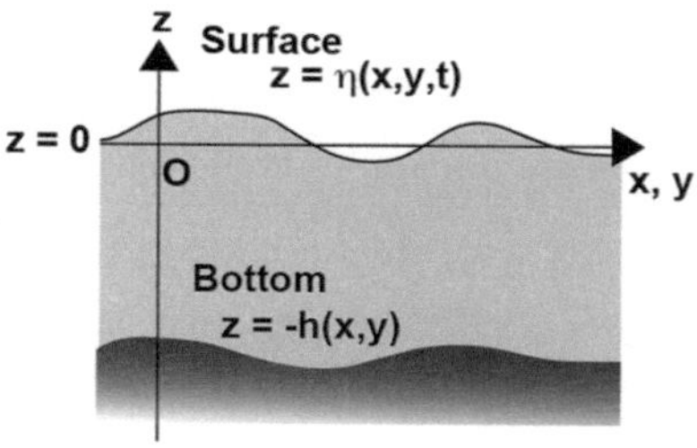

Fig. 6.8 Coordinates used in deriving tsunami equations

$$\frac{Dv_y(\mathbf{x},t)}{Dt} = -\frac{1}{\rho_0}\frac{\partial p(\mathbf{x},t)}{\partial y}, \tag{6.2}$$

$$\frac{Dv_z(\mathbf{x},t)}{Dt} = -\frac{1}{\rho_0}\frac{\partial p(\mathbf{x},t)}{\partial z} - g_0, \tag{6.3}$$

where v_x, v_y, v_z are the velocities of the water particle, p is the pressure, ρ_0 is the water density, and g_0 is the gravitational acceleration. The total time derivative D/Dt is represented by the partial time derivative as

$$\frac{D}{Dt} = \frac{\partial}{\partial t} + v_x\frac{\partial}{\partial x} + v_y\frac{\partial}{\partial y} + v_z\frac{\partial}{\partial z}. \tag{6.4}$$

Assuming that the seawater is incompressible, the conservation of mass gives the conservation of the water volume as

$$\frac{\partial v_x}{\partial x} + \frac{\partial v_y}{\partial y} + \frac{\partial v_z}{\partial z} = 0. \tag{6.5}$$

The motion and the shape at a boundary cannot be independent of each other in a continuum medium when wave breaking is not included. We consider a kinematic boundary condition, which defines the relation between the motion and the shape of the boundary. When the sea surface at a time t is given by $z = \eta(x, y, t)$ as shown in Fig. 6.9, a function $f(x, y, z, t)$ is introduced as

$$f(x,y,z,t) = z - \eta(x,y,t). \tag{6.6}$$

The function:

$$f(x,y,z,t) = 0 \tag{6.7}$$

represents a surface in 3-D space. A particle is located at the surface satisfies $f(x, y, z, t) = 0$ for any time t. We consider that a particle on the surface located at (x, y, z) at time t moves to $\left(x + v_x^0\Delta t, y + v_y^0\Delta t, z + v_z^0\Delta t\right)$ after the time Δt with the velocity $\left(v_x^0, v_y^0, v_z^0\right)$ at the boundary (Fig. 6.9). At time $t + \Delta t$, the particle is also located on the surface. Then, the function should satisfy

$$f\left(x + v_x^0\Delta t, y + v_y^0\Delta t, z + v_z^0\Delta t, t + \Delta t\right) = 0. \tag{6.8}$$

By taking the Taylor series and setting Δt at an infinitesimally small value, we obtain

$$\frac{\partial f(x, y, z, t)}{\partial t} + v_x^0\frac{\partial f(x, y, z, t)}{\partial x} + v_y^0\frac{\partial f(x, y, z, t)}{\partial y} + v_z^0\frac{\partial f(x, y, z, t)}{\partial z} = 0. \tag{6.9}$$

Substituting Eq. (6.6) into Eq. (6.9), we obtain

$$v_z^0 = \frac{\partial \eta(x, y, t)}{\partial t} + v_x^0\frac{\partial \eta(x, y, t)}{\partial x} + v_y^0\frac{\partial \eta(x, y, t)}{\partial y}, \tag{6.10}$$

or

$$v_z = \frac{\partial \eta(x, y, t)}{\partial t} + v_x\frac{\partial \eta(x, y, t)}{\partial x} + v_y\frac{\partial \eta(x, y, t)}{\partial y} \text{ on } z = \eta(x, y, t). \tag{6.11}$$

This is a kinematic boundary condition at the sea surface $z = \eta(x, y, t)$. This governs the relation between the motion at the surface (v_x, v_y, v_z) and the shape at the surface η. In a similar way, we obtain the kinematic boundary condition at the sea bottom $z = -h(x, y)$ as

$$v_z = -v_x\frac{\partial h(x, y)}{\partial x} - v_y\frac{\partial h(x, y)}{\partial y} \text{ on } z = -h(x, y). \tag{6.12}$$

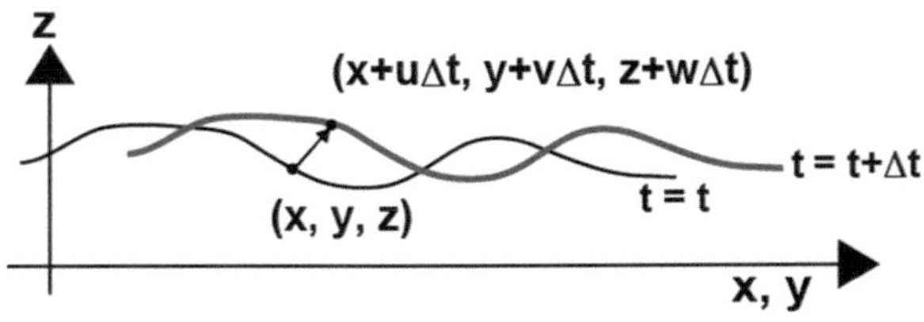

Fig. 6.9 A boundary for the coordinates (x, y, z)

Moreover, on the sea surface $z = \eta(x, y, t)$, the pressure is constant and should satisfy

$$p(\mathbf{x}, t) = 0 \quad \text{on} \quad z = \eta(x, y, t). \tag{6.13}$$

This is a boundary condition with respect to force on the moving boundary $z = \eta(x, y, t)$. This is referred to as a dynamic boundary condition. Equations (6.1), (6.2), (6.3), (6.5), (6.11), (6.12), and (6.13) govern the propagation of water waves or tsunamis. We derive the 2-D tsunami equations from these equations.

We are going to describe the fluid motion in 2-D (x, y) space. We do not directly use the horizontal velocity $v_x(x, y, z, t)$ or $v_y(x, y, z, t)$, but the horizontal velocity is integrated from the sea bottom $z = -h(x, y)$ to the sea surface $z = \eta(x, y, t)$, for example, as

$$\int_{-h(x,y)}^{\eta(x,y,t)} v_x(x, y, z, t) dz. \tag{6.14}$$

Also, instead of $\partial v_x/\partial x$, we use the partial derivative of Eq. (6.14) as

$$\frac{\partial}{\partial x} \int_{-h(x,y)}^{\eta(x,y,t)} v_x(x, y, z, t) dz. \tag{6.15}$$

Because an independent variable x is also included in the integration range through $\eta(x, y, t)$ or $-h(x, y)$, we carefully differentiate Eq. (6.15) as

$$\begin{aligned} &\frac{\partial}{\partial x} \int_{-h(x,y)}^{\eta(x,y,t)} v_x(x, y, z, t) dz \\ &= \lim_{\delta x \to 0} \frac{1}{\delta x} \left[\int_{-h(x+\delta x,y)}^{\eta(x+\delta x,y,t)} v_x(x + \delta x, y, z, t) dz - \int_{-h(x,y)}^{\eta(x,y,t)} v_x(x, y, z, t) dz \right]. \end{aligned} \tag{6.16}$$

The first term on the right-hand side is expanded using Taylor series with respect to x as

$$\begin{aligned} &\int_{-h(x+\delta x,y)}^{\eta(x+\delta x,y,t)} v_x(x + \delta x, y, z, t) dz \\ &\approx \int_{-h-\partial h/\partial x\, \delta x}^{\eta+\partial \eta/\partial x\, \delta x} \left[v_x(x, y, z, t) + \frac{\partial v_x}{\partial x} \delta x \right] dz \\ &\approx \int_{-h-\partial h/\partial x\, \delta x}^{-h} v_x(x, y, z, t) dz + \int_{-h}^{\eta} \left[v_x(x, y, z, t) + \frac{\partial v_x}{\partial x} \delta x \right] dz + \int_{\eta}^{\eta+\partial \eta/\partial x\, \delta x} v_x(x, y, z, t) dz \\ &\approx \frac{\partial h}{\partial x} \delta x\, v_x(x, y, -h, t) + \int_{-h}^{\eta} v_x(x, y, z, t) dz + \int_{-h}^{\eta} \frac{\partial v_x}{\partial x} \delta x\, dz + \frac{\partial \eta}{\partial x} \delta x\, v_x(x, y, \eta, t), \end{aligned}$$

where we omit the higher terms of δx but keep only the first order. Using Eq. (6.16), we represent Eq. (6.15) as

$$\frac{\partial}{\partial x}\int_{-h(x,y)}^{\eta(x,y,t)} v_x(x,y,z,t)dz = \int_{-h}^{\eta} \frac{\partial v_x(x,y,z,t)}{\partial x} dz + v_x(x,y,\eta,t)\frac{\partial \eta(x,y)}{\partial x} + v_x(x,y,-h,t)\frac{\partial h(x,y)}{\partial x}. \tag{6.17}$$

This derivative rule is also known as Leibnitz' rule. By using Leibnitz' rule, we obtain

$$\int_{-h(x,y)}^{\eta(x,y,t)} \frac{\partial v_x}{\partial x} dz = \frac{\partial}{\partial x}\int_{-h(x,y)}^{\eta(x,y,t)} v_x(x,y,z,t)dz - v_x(x,y,\eta,t)\frac{\partial \eta}{\partial x} - v_x(x,y,-h,t)\frac{\partial h}{\partial x}. \tag{6.18}$$

By integrating the conservation of the volume $\nabla \cdot \mathbf{v} = 0$. (Eq. (6.5)) from the sea bottom to the sea surface:

$$0 = \int_{-h}^{\eta} \frac{\partial v_x}{\partial x} dz + \int_{-h}^{\eta} \frac{\partial v_y}{\partial y} dz + v_z(x,y,\eta) - v_z(x,y,-h). \tag{6.19}$$

Using Eq. (6.18) and a similar equation for v_y, Eq. (6.19) is rewritten as

$$\begin{aligned}
0 = &\frac{\partial}{\partial x}\int_{-h(x,y)}^{\eta(x,y,t)} v_x(x,y,z,t)dz - v_x(x,y,\eta,t)\frac{\partial \eta}{\partial x} - v_x(x,y,-h,t)\frac{\partial h}{\partial x}\\
&+\frac{\partial}{\partial y}\int_{-h(x,y)}^{\eta(x,y,t)} v_y(x,y,z,t)dz - v_y(x,y,\eta,t)\frac{\partial \eta}{\partial y} - v_y(x,y,-h,t)\frac{\partial h}{\partial y}\\
&+v_z(x,y,\eta) - v_z(x,y,-h)\\
= &\frac{\partial}{\partial x}\int_{-h(x,y)}^{\eta(x,y,t)} v_x(x,y,z,t)dz + \frac{\partial}{\partial y}\int_{-h(x,y)}^{\eta(x,y,t)} v_y(x,y,z,t)dz\\
&+v_z(x,y,\eta) - v_x(x,y,\eta,t)\frac{\partial \eta}{\partial x} - v_y(x,y,\eta,t)\frac{\partial \eta}{\partial y}\\
&-v_z(x,y,-h) - v_x(x,y,-h,t)\frac{\partial h}{\partial x} - v_y(x,y,-h,t)\frac{\partial h}{\partial y}.
\end{aligned} \tag{6.20}$$

Using the kinematic boundary condition at the surface (Eq. (6.11)) and the bottom (Eq. (6.12)), Eq. (6.20) gives the conservation of the volume in the 2-D ($x - y$ space) form as

$$\frac{\partial \eta}{\partial t} + \frac{\partial}{\partial x}\int_{-h(x,y)}^{\eta(x,y,t)} v_x(x,y,z,t)dz + \frac{\partial}{\partial y}\int_{-h(x,y)}^{\eta(x,y,t)} v_y(x,y,z,t)dz = 0. \tag{6.21}$$

Similarly, integrating Eq. (6.5) from the sea bottom to the depth z as

$$\int_{-h}^{z} \frac{\partial v_z}{\partial z} dz = -\int_{-h}^{z} \frac{\partial v_x}{\partial x} dz - \int_{-h}^{z} \frac{\partial v_y}{\partial y} dz,$$

and using Leibnitz' rule, we calculate

$$v_z(z) - v_z(-h) = -\frac{\partial}{\partial x}\int_{-h}^{z} v_x dz + v_x(-h)\frac{\partial h}{\partial x} - \frac{\partial}{\partial y}\int_{-h}^{z} v_y dz + v_y(-h)\frac{\partial h}{\partial y}.$$

Using the boundary condition at the bottom (Eq. (6.12)), we represent the vertical velocity $v_z(x, y, z, t)$ at a depth of z by using the horizontal velocities as

$$v_z(x,y,z,t) = -\frac{\partial}{\partial x}\int_{-h(x,y)}^{z} v_x(x,y,z,t)dz - \frac{\partial}{\partial y}\int_{-h(x,y)}^{z} v_y(x,y,z,t)dz. \tag{6.22}$$

Let us interpret this equation visually. We consider a 1-D case for the interpretation as

$$v_z(x,z,t) = -\frac{\partial}{\partial x}\int_{-h(x)}^{z} v_x(x,z',t)dz' = -\frac{\partial}{\partial x}F(x,z,t) \tag{6.22$'$}$$

where the function $F(x, z)$ is defined as

$$F(x,z,t) = \int_{-h(x)}^{z} v_x(x,z',t)dz'.$$

The function $F(x, z, t)$ indicates the total water volume flowing along the x-axis across the line from the sea bottom $z = -h(x)$ to z as shown in the bold gray vector in Fig. 6.10. We approximately represent Eq. (6.22′) in a finite difference form or an integration form as

$$F(x,z,t) = v_z(x,z,t)\Delta x + F(x+\Delta x, z, t).$$

This represents the conservation of water volume for a region from $x = x$ to $x + \Delta x$ and $z = -h(x)$ to $z = z$ (a region shown in white in Fig. 6.10). The incoming water

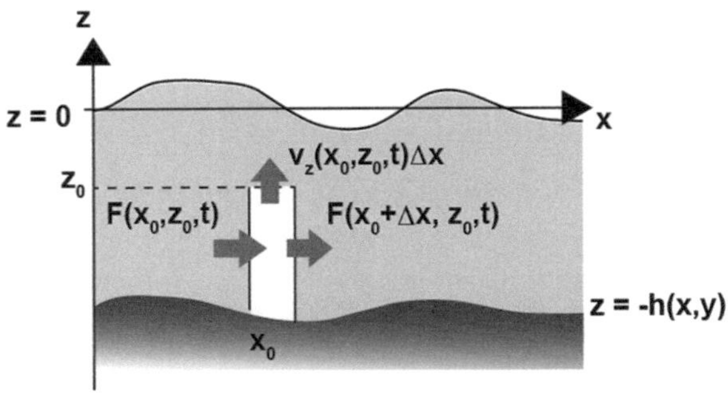

Fig. 6.10 Volume conservation in a region from $x = x_0$ to $x_0 + \Delta x$ and $z = -h(x)$ to z_0 (the region shown in white). The water volume coming into the region per unit time is given by $F(x_0, z_0, t) = \int_{-h(x_0)}^{z_0} v_x(x, z', t)dz'$. The outlaying water volume in the horizontal direction is $F(x_0 + \Delta x, z_0, t)$, and in the vertical direction, it is $v_z(x_0, z_0, t)\Delta x$. The incoming and outlaying water volumes are equal as $F(x_0, z_0, t) = v_z(x_0, z_0, t)\Delta x + F(x_0 + \Delta x, z_0, t)$

volume per unit time is given by $F(x, z, t)$. The outlaying water volume in the horizontal direction is $F(x + \Delta x, z, t)$, and in the vertical direction, it is $v_z(x, z, t)\Delta x$. This equation shows that the incoming and outlaying water volumes are equal. We may consider Eq. (6.21) as a special case of Eq. (6.22) by setting $z = \eta$.

Integrating the equation of motion for the vertical velocity (Eq. (6.3)) from the sea surface ($z = \eta(x, y, t)$) to the depth z and using the dynamic boundary condition at the surface ($p(x, y, \eta, t) = 0$), we obtain the pressure at depth z as

$$p(x, y, z, t) = \rho_0 g_0 [\eta(x, y, t) - z] - \rho_0 \int_{\eta(x,y,t)}^{z} \frac{Dv_z(x, y, z, t)}{Dt} dz. \tag{6.23}$$

Integrating the equation of motion of the x component (Eq. (6.1)) from the sea bottom to the sea surface gives

$$\int_{-h}^{\eta} \frac{\partial v_x}{\partial t} dz + \int_{-h}^{\eta} v_x \frac{\partial v_x}{\partial x} dz + \int_{-h}^{\eta} v_y \frac{\partial v_x}{\partial y} dz + \int_{-h}^{\eta} v_z \frac{\partial v_x}{\partial z} dz + \frac{1}{\rho_0} \int_{-h}^{\eta} \frac{\partial p}{\partial x} dz = 0. \tag{6.24}$$

Substituting Eq. (6.22) and Eq. (6.23) for v_z and p, respectively, in Eq. (6.24) and using Leibnitz' rule with respect to the time derivative, we calculate

$$
\begin{aligned}
0 = & \frac{\partial}{\partial t}\int_{-h}^{\eta} v_x dz - \frac{\partial \eta}{\partial t} v_x(\eta) + \int_{-h}^{\eta} v_x \frac{\partial v_x}{\partial x} dz + \int_{-h}^{\eta} v_y \frac{\partial v_x}{\partial y} dz \\
& - \int_{-h}^{\eta} \frac{\partial v_x}{\partial z}\left[\frac{\partial}{\partial x}\int_{-h}^{z} v_x(x,y,z',t)dz' + \frac{\partial}{\partial y}\int_{-h}^{z} v_y(x,y,z',t)dz'\right] dz \\
& + \int_{-h}^{\eta} \frac{\partial}{\partial x}\left\{ g_0[\eta(x,y,t) - z] - \int_{-h}^{z} \frac{Dv_z(x,y,z',t)}{Dt} dz' \right\} dz \\
= & \frac{\partial}{\partial t}\int_{-h}^{\eta} v_x dz - \frac{\partial \eta}{\partial t} v_x(\eta) + \int_{-h}^{\eta} v_x \frac{\partial v_x}{\partial x} dz + \int_{-h}^{\eta} v_y \frac{\partial v_x}{\partial y} dz \\
& - \int_{-h}^{\eta} \frac{\partial v_x}{\partial z}\left[\int_{-h}^{z} \frac{\partial v_x(x,y,z',t)}{\partial x} dz' + \int_{-h}^{z} \frac{\partial v_y(x,y,z',t)}{\partial y} dz' + v_x(-h)\frac{\partial h}{\partial x} + v_y(-h)\frac{\partial h}{\partial y}\right] dz \\
& + g_0 \int_{-h}^{\eta} dz \frac{\partial \eta(x,y,t)}{\partial x} - \int_{-h}^{\eta} dz \frac{\partial}{\partial x}\int_{\eta}^{z} \frac{Dv_z(x,y,z',t)}{Dt} dz' \\
= & \frac{\partial}{\partial t}\int_{-h}^{\eta} v_x dz - \frac{\partial \eta}{\partial t} v_x(\eta) + \int_{-h}^{\eta} v_x \frac{\partial v_x}{\partial x} dz + \int_{-h}^{\eta} v_y \frac{\partial v_x}{\partial y} dz \\
& - \int_{-h}^{\eta} \frac{\partial v_x}{\partial z}\left[\int_{-h}^{z} \frac{\partial v_x(x,y,z',t)}{\partial x} dz' + \int_{-h}^{z} \frac{\partial v_y(x,y,z',t)}{\partial y} dz' + v_x(-h)\frac{\partial h}{\partial x} + v_y(-h)\frac{\partial h}{\partial y}\right] dz \\
& + g_0 \int_{-h}^{\eta} dz \frac{\partial \eta(x,y,t)}{\partial x} - \frac{\partial}{\partial x}\int_{-h}^{\eta} dz \int_{\eta}^{z} \frac{Dv_z(x,y,z',t)}{Dt} dz' - \frac{\partial h}{\partial x}\int_{-h}^{\eta} \frac{Dv_z(x,y,z',t)}{Dt} dz' \\
= & \frac{\partial}{\partial t}\int_{-h}^{\eta} v_x dz - \frac{\partial \eta}{\partial t} v_x(\eta) + \int_{-h}^{\eta} v_x \frac{\partial v_x}{\partial x} dz + \int_{-h}^{\eta} v_y \frac{\partial v_x}{\partial y} dz \\
& - \int_{-h}^{\eta} \frac{\partial v_x}{\partial z}\left[\int_{-h}^{z} \frac{\partial v_x(x,y,z',t)}{\partial x} dz' + \int_{-h}^{z} \frac{\partial v_y(x,y,z',t)}{\partial y} dz' + v_x(-h)\frac{\partial h}{\partial x} + v_y(-h)\frac{\partial h}{\partial y}\right] dz \\
& + g_0(\eta + h)\frac{\partial \eta(x,y,t)}{\partial x} - \frac{\partial}{\partial x}\int_{-h}^{\eta} dz \int_{\eta}^{z} \frac{Dv_z(x,y,z',t)}{Dt} dz' - \frac{\partial h}{\partial x}\int_{-h}^{\eta} \frac{Dv_z(x,y,z',t)}{Dt} dz'.
\end{aligned}
\tag{6.25}
$$

Note that one of the terms in Eq. (6.25) is calculated with partial integration as

$$
\begin{aligned}
& \int_{-h}^{\eta} dz \frac{\partial v_x}{\partial z} \int_{-h}^{z} \frac{\partial v_x(x,y,z',t)}{\partial x} dz' \\
& = \left[v_x \int_{-h}^{z} \frac{\partial v_x(x,y,z',t)}{\partial x} dz' \right]_{z=-h}^{\eta} - \int_{-h}^{\eta} dz v_x \frac{\partial}{\partial z}\int_{-h}^{z} \frac{\partial v_x(x,y,z',t)}{\partial x} dz' \qquad (6.26) \\
& = v_x(\eta) \int_{-h}^{\eta} \frac{\partial v_x(x,y,z',t)}{\partial x} dz' - \int_{-h}^{\eta} dz\, v_x \frac{\partial v_x(x,y,z,t)}{\partial x}.
\end{aligned}
$$

Similarly, we calculate

$$
\int_{-h}^{\eta} dz \frac{\partial v_x}{\partial z} \int_{-h}^{z} \frac{\partial v_y(x,y,z',t)}{\partial y} dz' = v_x(\eta) \int_{-h}^{\eta} \frac{\partial v_y(x,y,z',t)}{\partial y} dz' - \int_{-h}^{\eta} dz\, v_x \frac{\partial v_y}{\partial y}.
\tag{6.27}
$$

Substituting Eqs. (6.26) and (6.27) into Eq. (6.25), we obtain

$$
\begin{aligned}
&\frac{\partial}{\partial t}\int_{-h}^{\eta} v_x dz - v_x(\eta)\frac{\partial \eta}{\partial t} + \int_{-h}^{\eta} v_x \frac{\partial v_x}{\partial x} dz + \int_{-h}^{\eta} v_y \frac{\partial v_x}{\partial y} dz - v_x(\eta)\int_{-h}^{\eta} \frac{\partial v_x}{\partial x} dz \\
&\quad + \int_{-h}^{\eta} v_x \frac{\partial v_x}{\partial x} dz - v_x(\eta)\int_{-h}^{\eta} \frac{\partial v_y}{\partial y} dz + \int_{-h}^{\eta} v_x \frac{\partial v_y}{\partial y} dz \\
&\quad + v_x(-h)\frac{\partial h}{\partial x}[v_x(\eta) - v_x(-h)] \\
&\quad + v_y(-h)\frac{\partial h}{\partial y}[v_x(\eta) - v_x(-h)] + g_0(\eta + h)\frac{\partial \eta(x,y,t)}{\partial x} \\
&\quad - \frac{\partial}{\partial x}\int_{-h}^{\eta} dz \int_{\eta}^{z} \frac{Dv_z(x,y,z',t)}{Dt} dz' - \frac{\partial h}{\partial x}\int_{-h}^{\eta} \frac{Dv_z(x,y,z',t)}{Dt} dz' = 0.
\end{aligned}
\tag{6.28}
$$

This is rewritten as

$$
\begin{aligned}
0 =& \frac{\partial}{\partial t}\int_{-h}^{\eta} v_x dz - v_x(\eta)\frac{\partial \eta}{\partial t} + \int_{-h}^{\eta} 2v_x \frac{\partial v_x}{\partial x} dz + \int_{-h}^{\eta}\left(v_y \frac{\partial v_x}{\partial y} + v_x \frac{\partial v_y}{\partial y}\right) dz \\
& - v_x(\eta)\left[\frac{\partial}{\partial x}\int_{-h}^{\eta} v_x dz - v_x(\eta)\frac{\partial \eta}{\partial x} - v_x(-h)\frac{\partial h}{\partial x}\right] \\
& - v_x(\eta)\left[\frac{\partial}{\partial y}\int_{-h}^{\eta} v_y dz - v_y(\eta)\frac{\partial \eta}{\partial y} - v_y(-h)\frac{\partial h}{\partial y}\right] \\
& - v_x(-h)\frac{\partial h}{\partial x}[v_x(\eta) - v_x(-h)] \\
& - v_y(-h)\frac{\partial h}{\partial y}[v_x(\eta) - v_x(-h)] + g_0(\eta + h)\frac{\partial \eta(x,y,t)}{\partial x} \\
& - \frac{\partial}{\partial x}\int_{-h}^{\eta} dz \int_{\eta}^{z} \frac{Dv_z(x,y,z',t)}{Dt} dz' - \frac{\partial h}{\partial x}\int_{-h}^{\eta} \frac{Dv_z(x,y,z',t)}{Dt} dz' \\
=& \frac{\partial}{\partial t}\int_{-h}^{\eta} v_x dz - v_x(\eta)\frac{\partial \eta}{\partial t} + \int_{-h}^{\eta} \frac{\partial}{\partial x}(v_x)^2 dz + \int_{-h}^{\eta} \frac{\partial}{\partial y}(v_x v_y) dz \\
& - v_x(\eta)\left(\frac{\partial}{\partial x}\int_{-h}^{\eta} v_x dz + \frac{\partial}{\partial y}\int_{-h}^{\eta} v_y dz\right) + (v_x(\eta))^2\frac{\partial \eta}{\partial x} \\
& + v_x(\eta)v_x(-h)\frac{\partial h}{\partial x} + v_x(\eta)v_y(\eta)\frac{\partial \eta}{\partial y} + v_x(\eta)v_y(-h)\frac{\partial h}{\partial y} \\
& - v_x(-h)\frac{\partial h}{\partial x}[v_x(\eta) - v_x(-h)] \\
& - v_y(-h)\frac{\partial h}{\partial y}[v_x(\eta) - v_x(-h)] + g_0(\eta + h)\frac{\partial \eta(x,y,t)}{\partial x} \\
& - \frac{\partial}{\partial x}\int_{-h}^{\eta} dz \int_{\eta}^{z} \frac{Dv_z(x,y,z',t)}{Dt} dz' - \frac{\partial h}{\partial x}\int_{-h}^{\eta} \frac{Dv_z(x,y,z',t)}{Dt} dz'.
\end{aligned}
\tag{6.29}
$$

Using the volume conservation (Eq. (6.21)), Eq. (6.29) becomes

$$\begin{aligned}&\frac{\partial}{\partial t}\int_{-h}^{\eta} v_x dz + \int_{-h}^{\eta}\frac{\partial}{\partial x}(v_x)^2 dz + (v_x(\eta))^2\frac{\partial \eta}{\partial x} + (v_x(-h))^2\frac{\partial h}{\partial x} + \int_{-h}^{\eta}\frac{\partial}{\partial y}(v_x v_y)dz \\ &+v_x(\eta)v_y(\eta)\frac{\partial \eta}{\partial y} + v_x(-h)v_y(-h)\frac{\partial h}{\partial y} + g_0(\eta+h)\frac{\partial \eta(x,y,t)}{\partial x} - \\ &\frac{\partial}{\partial x}\int_{-h}^{\eta} dz\int_{\eta}^{z}\frac{Dv_z(x,y,z',t)}{Dt}dz' - \frac{\partial h}{\partial x}\int_{-h}^{\eta}\frac{Dv_z(x,y,z',t)}{Dt}dz' = 0\end{aligned} \quad (6.30)$$

Using Leibnitz' rule again, we obtain

$$\begin{aligned}&\frac{\partial}{\partial t}\int_{-h}^{\eta} v_x dz + \frac{\partial}{\partial x}\int_{-h}^{\eta}(v_x)^2 dz + \frac{\partial}{\partial y}\int_{-h}^{\eta}(v_x v_y)dz + g_0(\eta+h)\frac{\partial \eta}{\partial x} \\ &= \frac{\partial}{\partial x}\int_{-h}^{\eta} dz\int_{\eta}^{z}\frac{Dv_z(x,y,z',t)}{Dt}dz' + \frac{\partial h}{\partial x}\int_{-h}^{\eta}\frac{Dv_z(x,y,z',t)}{Dt}dz'.\end{aligned} \quad (6.31)$$

This equation describes the time evolution of $\int_{-h}^{\eta} v_x dz$.

We also obtain an equation with respect to the velocity in the y direction as

$$\begin{aligned}&\frac{\partial}{\partial t}\int_{-h}^{\eta} v_y dz + \frac{\partial}{\partial x}\int_{-h}^{\eta}(v_x v_y)dz + \frac{\partial}{\partial y}\int_{-h}^{\eta}(v_y)^2 dz + g_0(\eta+h)\frac{\partial \eta}{\partial y} \\ &= \frac{\partial}{\partial y}\int_{-h}^{\eta} dz\int_{\eta}^{z}\frac{Dv_z(x,y,z',t)}{Dt}dz' + \frac{\partial h}{\partial y}\int_{-h}^{\eta} dz'\frac{Dv_z(x,y,z',t)}{Dt}.\end{aligned} \quad (6.32)$$

Equations (6.31) and (6.32) are considered to be the equations of motion with respect to the horizontal flow integrated over the sea depth ($-h \le z \le \eta$) given by

$$\int_{-h(x,y)}^{\eta(x,y,t)} v_x(x,y,z,t)dz \text{ and } \int_{-h(x,y)}^{\eta(x,y,t)} v_y(x,y,z,t)dz.$$

The equations of motion (6.31) and (6.32) and the conservation of the volume (Eq. (6.21)) play an important role in the tsunami propagation in the 2-D form.

6.2.1.1 Normalization

It is quite useful to use nondimensional variables in representing the equations because they give essential dimensional values that govern tsunami phenomena. In order to use nondimensional variables, we introduce the characteristic horizontal length L_0, the sea depth H_0, and the sea-surface height η_0. All these values are constant. Then, the variables in the equations are represented by nondimensional values such as $\bar{x}$, $\bar{t}$, and $\bar{v}_x$:

$$\begin{aligned} &x = L_0\bar{x}, y = L_0\bar{y}, z = H_0\bar{z}, t = \frac{L_0}{\sqrt{g_0 H_0}}\bar{t}, \\ &\eta = \eta_0\bar{\eta}, v_x = \sqrt{g_0 H_0}\frac{\eta_0}{H_0}\bar{v}_x, v_y = \sqrt{g_0 H_0}\frac{\eta_0}{H_0}\bar{v}_y, v_z = \sqrt{g_0 H_0}\frac{\eta_0}{L_0}\bar{v}_z \\ &p = \rho_0 g_0 H_0 \bar{p}, \quad h = H_0\bar{h}. \end{aligned} \tag{6.33}$$

We rewrote Eqs. (6.21) and (6.22) by using the nondimensional variables (e.g., $\bar{x}, \bar{y}$, and $\bar{t}$) defined in Eq. (6.33) as

$$\frac{\partial\bar{\eta}(\bar{x},\bar{y},\bar{t})}{\partial\bar{t}} = -\frac{\partial}{\partial\bar{x}}\int_{-\bar{h}}^{\epsilon\bar{\eta}}\bar{v}_x(\bar{x},\bar{y},\bar{z},\bar{t})d\bar{z} - \frac{\partial}{\partial\bar{y}}\int_{-\bar{h}}^{\epsilon\bar{\eta}}\bar{v}_y(\bar{x},\bar{y},\bar{z},\bar{t})d\bar{z}, \tag{6.34}$$

and

$$\bar{v}_z(\bar{x},\bar{y},\bar{z},\bar{t}) = -\frac{\partial}{\partial\bar{x}}\int_{-\bar{h}}^{\bar{z}}\bar{v}_x d\bar{z} - \frac{\partial}{\partial\bar{y}}\int_{-\bar{h}}^{\bar{z}}\bar{v}_y d\bar{z}, \tag{6.35}$$

where we introduced the following two nondimensional parameters:

$$\sigma = \frac{H_0}{L_0} \text{ and } \epsilon = \frac{\eta_0}{H_0}. \tag{6.36}$$

Then, the total time derivative for v_z is rewritten using normalized variables as

$$\frac{Dv_z(x,y,z,t)}{Dt} = \frac{\partial v_z}{\partial t} + v_x\frac{\partial v_z}{\partial x} + v_y\frac{\partial v_z}{\partial y} + v_z\frac{\partial v_z}{\partial z} = g_0\sigma^2\epsilon\frac{D\overline{v_z}(\bar{x},\bar{y},\bar{z},\bar{t})}{D\bar{t}}, \tag{6.37}$$

where we define

$$\frac{D\overline{v_z}(\bar{x},\bar{y},\bar{z},\bar{t})}{D\bar{t}} = \frac{\partial\overline{v_z}}{\partial\bar{t}} + \epsilon\left(\overline{v_x}\frac{\partial\overline{v_z}}{\partial\bar{x}} + \overline{v_y}\frac{\partial\overline{v_z}}{\partial\bar{y}} + \overline{v_z}\frac{\partial\overline{v_z}}{\partial\bar{z}}\right).$$

From Eqs. (6.31) and (6.32), the normalized equations of motion for the integrated horizontal flows are

$$\begin{aligned} &\frac{\partial}{\partial\bar{t}}\int_{-\bar{h}}^{\epsilon\bar{\eta}}\bar{v}_x(\bar{x},\bar{y},\bar{z},\bar{t})d\bar{z} + \epsilon\frac{\partial}{\partial\bar{x}}\int_{-\bar{h}}^{\epsilon\bar{\eta}}(\bar{v}_x)^2 d\bar{z} + \epsilon\frac{\partial}{\partial\bar{y}}\int_{-\bar{h}}^{\epsilon\bar{\eta}}\bar{v}_x\bar{v}_y d\bar{z} + (\epsilon\bar{\eta}+\bar{h})\frac{\partial\bar{\eta}}{\partial\bar{x}} \\ &= \sigma^2\frac{\partial}{\partial\bar{x}}\int_{-\bar{h}}^{\epsilon\bar{\eta}}d\bar{z}\int_{\epsilon\bar{\eta}}^{\bar{z}}d\bar{z}'\frac{D\bar{v}_z(\bar{x},\bar{y},\bar{z}',\bar{t})}{D\bar{t}} + \sigma^2\frac{\partial\bar{h}}{\partial\bar{x}}\int_{-\bar{h}}^{\epsilon\bar{\eta}}d\bar{z}'\frac{D\bar{v}_z(\bar{x},\bar{y},\bar{z}',\bar{t})}{D\bar{t}}, \end{aligned} \tag{6.38}$$

and

$$\frac{\partial}{\partial \bar{t}}\int_{-\bar{h}}^{\epsilon\bar{\eta}} \bar{v}_y(\bar{x},\bar{y},\bar{z},\bar{t})d\bar{z} + \epsilon\frac{\partial}{\partial \bar{x}}\int_{-\bar{h}}^{\epsilon\bar{\eta}} \bar{v}_x\bar{v}_y d\bar{z} + \epsilon\frac{\partial}{\partial \bar{y}}\int_{-\bar{h}}^{\epsilon\bar{\eta}} (\bar{v}_x)^2 d\bar{z} + (\epsilon\bar{\eta}+\bar{h})\frac{\partial \bar{\eta}}{\partial \bar{y}}$$
$$= \sigma^2\frac{\partial}{\partial \bar{y}}\int_{-\bar{h}}^{\epsilon\bar{\eta}} d\bar{z}\int_{\epsilon\bar{\eta}}^{\bar{z}} d\bar{z}'\frac{D\bar{v}_z(\bar{x},\bar{y},\bar{z}',\bar{t})}{D\bar{t}} + \sigma^2\frac{\partial \bar{h}}{\partial \bar{y}}\int_{-\bar{h}}^{\epsilon\bar{\eta}} d\bar{z}'\frac{D\bar{v}_z(\bar{x},\bar{y},\bar{z}',\bar{t})}{D\bar{t}}. \quad (6.39)$$

It should be noted that we did not use any approximations in deriving these equations from the original governing equations (Eqs. (6.1)–(6.5), (6.11), (6.12), and (6.13)).

6.2.2 Long-Wave Approximations

We will consider the case when the horizontal scale L_0 is much longer than the order of the sea depth H_0, i.e., $\sigma \ll 1$ for tsunami propagation. This is a long-wave approximation.

We assume that the flow is irrotational or the rotation of the velocity vector is zero:

$$\frac{\partial v_z}{\partial y} - \frac{\partial v_y}{\partial z} = 0, \quad \frac{\partial v_x}{\partial z} - \frac{\partial v_z}{\partial x} = 0, \quad \frac{\partial v_y}{\partial x} - \frac{\partial v_x}{\partial y} = 0. \quad (6.40)$$

From the first and the second equations in (6.40), we obtain

$$\frac{\partial \bar{v}_y}{\partial \bar{z}} = \sigma^2\frac{\partial \bar{v}_z}{\partial \bar{y}}, \quad \frac{\partial \bar{v}_x}{\partial \bar{z}} = \sigma^2\frac{\partial \bar{v}_z}{\partial \bar{x}}. \quad (6.41)$$

These equations indicate that the vertical variation of the horizontal velocity is on the order of σ^2, ($O(\sigma^2)$).

The average horizontal velocity over depth is introduced as

$$\bar{v}_x^{\mathrm{av}}(\bar{x},\bar{y},\bar{t}) = \frac{1}{\epsilon\bar{\eta}+\bar{h}}\int_{-\bar{h}}^{\epsilon\bar{\eta}} \bar{v}_x(\bar{x},\bar{y},\bar{z},\bar{t})d\bar{z}, \quad (6.42)$$

and

$$\bar{v}_y^{\mathrm{av}}(\bar{x},\bar{y},\bar{t}) = \frac{1}{\epsilon\bar{\eta}+\bar{h}}\int_{-\bar{h}}^{\epsilon\bar{\eta}} \bar{v}_y(\bar{x},\bar{y},\bar{z},\bar{t})d\bar{z}. \quad (6.43)$$

Considering that the vertical variation of the horizontal velocity is on the order of σ^2 (Eq. (6.41)), we set the horizontal velocity as the sum of the average horizontal velocity and its fluctuation as

$$\bar{v}_x(\bar{x},\bar{y},\bar{z},\bar{t}) = \bar{v}_x^{\mathrm{av}}(\bar{x},\bar{y},\bar{t}) + \sigma^2 \bar{v}_x'(\bar{x},\bar{y},\bar{z},\bar{t}), \tag{6.44}$$

and

$$\bar{v}_y(\bar{x},\bar{y},\bar{z},\bar{t}) = \bar{v}_y^{\mathrm{av}}(\bar{x},\bar{y},\bar{t}) + \sigma^2 \bar{v}_y'(\bar{x},\bar{y},\bar{z},\bar{t}). \tag{6.45}$$

The horizontal fluctuation averaged over depth becomes zero as

$$\int_{-h}^{\epsilon\bar{\eta}} \bar{v}_x'(\bar{x},\bar{y},\bar{z},\bar{t})d\bar{z} = 0, \tag{6.46}$$

and

$$\int_{-h}^{\epsilon\bar{\eta}} \bar{v}_y'(\bar{x},\bar{y},\bar{z},\bar{t})d\bar{z} = 0. \tag{6.47}$$

Substituting Eqs. (6.44) and (6.45) into Eqs. (6.34) and (6.35), we obtain

$$\frac{\partial\bar{\eta}}{\partial\bar{t}} + \frac{\partial}{\partial\bar{x}}\left[(\epsilon\bar{\eta}+\bar{h})\bar{v}_x^{\mathrm{av}}(\bar{x},\bar{y},\bar{t})\right] + \frac{\partial}{\partial\bar{y}}\left[(\epsilon\bar{\eta}+\bar{h})\bar{v}_y^{\mathrm{av}}(\bar{x},\bar{y},\bar{t})\right] = 0, \tag{6.48}$$

and

$$\begin{aligned}\bar{v}_z(\bar{x},\bar{y},\bar{z},\bar{t}) = &-\frac{\partial}{\partial\bar{x}}\left[(\bar{z}+\bar{h})\bar{v}_x^{\mathrm{av}}(\bar{x},\bar{y},\bar{t})\right] - \sigma^2\frac{\partial}{\partial\bar{x}}\int_{-\bar{h}}^{\bar{z}} \bar{v}_x'(\bar{x},\bar{y},\bar{z}',\bar{t})d\bar{z}' \\ &-\frac{\partial}{\partial\bar{y}}\left[(\bar{z}+\bar{h})\bar{v}_y^{\mathrm{av}}(\bar{x},\bar{y},\bar{t})\right] - \sigma^2\frac{\partial}{\partial\bar{y}}\int_{-\bar{h}}^{\bar{z}} \bar{v}_y'(\bar{x},\bar{y},\bar{z}',\bar{t})d\bar{z}'.\end{aligned} \tag{6.49}$$

Also, substituting Eqs. (6.44) and (6.45) into Eqs. (6.38) and (6.39), we obtain

$$\begin{aligned}&\frac{\partial}{\partial\bar{t}}\left[(\epsilon\bar{\eta}+\bar{h})\bar{v}_x^{\mathrm{av}}(\bar{x},\bar{y},\bar{t})\right] + \epsilon\frac{\partial}{\partial\bar{x}}\left[(\epsilon\bar{\eta}+\bar{h})\left(\bar{v}_x^{\mathrm{av}}\right)^2\right] + \epsilon\sigma^4\frac{\partial}{\partial\bar{x}}\int_{-\bar{h}}^{\epsilon\bar{\eta}}\left(\bar{v}_x'\right)^2 d\bar{z} \\ &\quad + \epsilon\frac{\partial}{\partial\bar{y}}\left[(\epsilon\bar{\eta}+\bar{h})\bar{v}_x^{\mathrm{av}}\,\bar{v}_y^{\mathrm{av}}\right] + \epsilon\sigma^4\frac{\partial}{\partial\bar{y}}\int_{-\bar{h}}^{\epsilon\bar{\eta}} \bar{v}_x'\bar{v}_y' d\bar{z} + (\epsilon\bar{\eta}+\bar{h})\frac{\partial\bar{\eta}}{\partial\bar{x}} \\ &= \sigma^2\frac{\partial}{\partial\bar{x}}\int_{-\bar{h}}^{\epsilon\bar{\eta}} d\bar{z}\int_{\epsilon\bar{\eta}}^{\bar{z}} d\bar{z}'\frac{D\bar{v}_z(\bar{x},\bar{y},\bar{z}',\bar{t})}{D\bar{t}} + \sigma^2\frac{\partial\bar{h}}{\partial\bar{x}}\int_{-\bar{h}}^{\epsilon\bar{\eta}} d\bar{z}'\frac{D\bar{v}_z(\bar{x},\bar{y},\bar{z}',\bar{t})}{D\bar{t}}.\end{aligned} \tag{6.50}$$

and

$$\frac{\partial}{\partial \bar{t}}\left[(\epsilon\bar{\eta}+\bar{h})\bar{v}_y^{\mathrm{av}}(\bar{x},\bar{y},\bar{t})\right]+\epsilon\frac{\partial}{\partial \bar{x}}\left[(\epsilon\bar{\eta}+\bar{h})\bar{v}_x^{\mathrm{av}}\ \bar{v}_y^{\mathrm{av}}\right]+\epsilon\sigma^4\frac{\partial}{\partial \bar{x}}\int_{-\bar{h}}^{\epsilon\bar{\eta}}\bar{v}_x{}'\bar{v}_y{}'d\bar{z}$$
$$+\epsilon\frac{\partial}{\partial \bar{y}}\left[(\epsilon\bar{\eta}+\bar{h})\left(\bar{v}_y^{\mathrm{av}}\right)^2\right]+\epsilon\sigma^4\frac{\partial}{\partial \bar{y}}\int_{-\bar{h}}^{\epsilon\bar{\eta}}\left(\bar{v}_y{}'\right)^2d\bar{z}+(\epsilon\bar{\eta}+\bar{h})\frac{\partial\bar{\eta}}{\partial\bar{y}} \tag{6.51}$$
$$=\sigma^2\frac{\partial}{\partial \bar{y}}\int_{-\bar{h}}^{\epsilon\bar{\eta}}d\bar{z}\int_{\epsilon\bar{\eta}}^{\bar{z}}d\bar{z}'\frac{D\bar{v}_z(\bar{x},\bar{y},\bar{z}',\bar{t})}{D\bar{t}}+\sigma^2\frac{\partial\bar{h}}{\partial\bar{y}}\int_{-\bar{h}}^{\epsilon\bar{\eta}}d\bar{z}'\frac{D\bar{v}_z(\bar{x},\bar{y},\bar{z}',\bar{t})}{D\bar{t}}.$$

The common parts of

$$\int_{-\bar{h}}^{\epsilon\bar{\eta}}d\bar{z}\int_{\epsilon\bar{\eta}}^{\bar{z}}d\bar{z}'\frac{D\bar{v}_z(\bar{x},\bar{y},\bar{z}',\bar{t})}{D\bar{t}}$$

and

$$\int_{-\bar{h}}^{\epsilon\bar{\eta}}d\bar{z}'\frac{D\bar{v}_z(\bar{x},\bar{y},\bar{z}',\bar{t})}{D\bar{t}}$$

appear in (6.50) and (6.51). Using Eqs. (6.37b) and (6.49), we calculate these as

$$\begin{aligned}
&\int_{-\bar{h}}^{\epsilon\bar{\eta}}d\bar{z}\int_{\epsilon\bar{\eta}}^{\bar{z}}d\bar{z}'\frac{D\bar{v}_z(\bar{x},\bar{y},\bar{z}',\bar{t})}{D\bar{t}}\\
&=\int_{-\bar{h}}^{\epsilon\bar{\eta}}d\bar{z}\int_{\epsilon\bar{\eta}}^{\bar{z}}d\bar{z}'\left[\frac{\partial\bar{v}_z(\bar{x},\bar{y},\bar{z}',\bar{t})}{\partial\bar{t}}+\mathrm{O}(\epsilon)\right]\\
&=-\int_{-\bar{h}}^{\epsilon\bar{\eta}}d\bar{z}\int_{\epsilon\bar{\eta}}^{\bar{z}}d\bar{z}'\left[\frac{\partial}{\partial\bar{t}}\left\{\frac{\partial}{\partial\bar{x}}\left[(\bar{z}+\bar{h})\bar{v}_x^{\mathrm{av}}(\bar{x},\bar{y},\bar{t})\right]+\frac{\partial}{\partial\bar{y}}\left[(\bar{z}+\bar{h})\bar{v}_y^{\mathrm{av}}(\bar{x},\bar{y},\bar{t})\right]+\mathrm{O}(\sigma^2)\right\}+\mathrm{O}(\epsilon)\right]\\
&\approx-\int_{-\bar{h}}^{0}d\bar{z}\int_{0}^{\bar{z}}d\bar{z}'\left[\frac{\partial}{\partial\bar{t}}\left\{\frac{\partial}{\partial\bar{x}}\left[(\bar{z}'+\bar{h})\bar{v}_x^{\mathrm{av}}(\bar{x},\bar{y},\bar{t})\right]+\frac{\partial}{\partial\bar{y}}\left[(\bar{z}'+\bar{h})\bar{v}_y^{\mathrm{av}}(\bar{x},\bar{y},\bar{t})\right]\right\}\right]\\
&=-\int_{-\bar{h}}^{0}d\bar{z}\int_{0}^{\bar{z}}d\bar{z}'\left[\frac{\partial}{\partial\bar{t}}\left(\frac{\partial\bar{v}_x^{\mathrm{av}}}{\partial\bar{x}}+\frac{\partial\bar{v}_y^{\mathrm{av}}}{\partial\bar{y}}\right)\bar{z}'+\frac{\partial}{\partial\bar{t}}\left(\frac{\partial\bar{v}_x^{\mathrm{av}}}{\partial\bar{x}}+\frac{\partial\bar{v}_y^{\mathrm{av}}}{\partial\bar{y}}\right)\bar{h}+\frac{\partial}{\partial\bar{t}}\left(\frac{\partial\bar{h}}{\partial\bar{x}}\bar{v}_x^{\mathrm{av}}+\frac{\partial\bar{h}}{\partial\bar{y}}\bar{v}_y^{\mathrm{av}}\right)\right]\\
&=-\frac{1}{6}\bar{h}^3\frac{\partial}{\partial\bar{t}}\left(\frac{\partial\bar{v}_x^{\mathrm{av}}}{\partial\bar{x}}+\frac{\partial\bar{v}_y^{\mathrm{av}}}{\partial\bar{y}}\right)+\frac{1}{2}\bar{h}^3\frac{\partial}{\partial\bar{t}}\left(\frac{\partial\bar{v}_x^{\mathrm{av}}}{\partial\bar{x}}+\frac{\partial\bar{v}_y^{\mathrm{av}}}{\partial\bar{y}}\right)+\frac{1}{2}\bar{h}^2\frac{\partial}{\partial\bar{t}}\left(\frac{\partial\bar{h}}{\partial\bar{x}}\bar{v}_x^{\mathrm{av}}+\frac{\partial\bar{h}}{\partial\bar{y}}\bar{v}_y^{\mathrm{av}}\right)\\
&=\frac{1}{3}\bar{h}^3\frac{\partial}{\partial\bar{t}}\left(\frac{\partial\bar{v}_x^{\mathrm{av}}}{\partial\bar{x}}+\frac{\partial\bar{v}_y^{\mathrm{av}}}{\partial\bar{y}}\right)+\frac{1}{2}\bar{h}^2\frac{\partial}{\partial\bar{t}}\left(\frac{\partial\bar{h}}{\partial\bar{x}}\bar{v}_x^{\mathrm{av}}+\frac{\partial\bar{h}}{\partial\bar{y}}\bar{v}_y^{\mathrm{av}}\right),
\end{aligned} \tag{6.52}$$

and

$$\begin{aligned}
&\int_{-\bar{h}}^{\epsilon\bar{\eta}} d\bar{z}' \frac{D\bar{v}_z(\bar{x},\bar{y},\bar{z}',\bar{t})}{D\bar{t}} \\
&= \int_{-\bar{h}}^{\epsilon\bar{\eta}} d\bar{z}' \left[\frac{\partial \bar{v}_z(\bar{x},\bar{y},\bar{z}',\bar{t})}{\partial \bar{t}} + \mathrm{O}(\epsilon)\right] \\
&= -\int_{-\bar{h}}^{\epsilon\bar{\eta}} d\bar{z}' \left[\frac{\partial}{\partial \bar{t}}\left\{\frac{\partial}{\partial \bar{x}}\left[(\bar{z}+\bar{h})\bar{v}_x^{\mathrm{av}}(\bar{x},\bar{y},\bar{t})\right] + \frac{\partial}{\partial \bar{y}}\left[(\bar{z}+\bar{h})\bar{v}_y^{\mathrm{av}}(\bar{x},\bar{y},\bar{t})\right] + \mathrm{O}(\sigma^2)\right\} + \mathrm{O}(\epsilon)\right] \\
&\approx -\int_{-\bar{h}}^{0} d\bar{z}' \left[\frac{\partial}{\partial \bar{t}}\left\{\frac{\partial}{\partial \bar{x}}\left[(\bar{z}'+\bar{h})\bar{v}_x^{\mathrm{av}}(\bar{x},\bar{y},\bar{t})\right] + \frac{\partial}{\partial \bar{y}}\left[(\bar{z}'+\bar{h})\bar{v}_y^{\mathrm{av}}(\bar{x},\bar{y},\bar{t})\right]\right\}\right] \\
&= -\int_{-\bar{h}}^{0} d\bar{z}' \left[\frac{\partial}{\partial \bar{t}}\left(\frac{\partial \bar{v}_x^{\mathrm{av}}}{\partial \bar{x}} + \frac{\partial \bar{v}_y^{\mathrm{av}}}{\partial \bar{y}}\right)\bar{z}' + \frac{\partial}{\partial \bar{t}}\left(\frac{\partial \bar{v}_x^{\mathrm{av}}}{\partial \bar{x}} + \frac{\partial \bar{v}_y^{\mathrm{av}}}{\partial \bar{y}}\right)\bar{h} + \frac{\partial}{\partial \bar{t}}\left(\frac{\partial \bar{h}}{\partial \bar{x}}\bar{v}_x^{\mathrm{av}} + \frac{\partial \bar{h}}{\partial \bar{y}}\bar{v}_y^{\mathrm{av}}\right)\right] \\
&= \frac{1}{2}\bar{h}^2\frac{\partial}{\partial \bar{t}}\left(\frac{\partial \bar{v}_x^{\mathrm{av}}}{\partial \bar{x}} + \frac{\partial \bar{v}_y^{\mathrm{av}}}{\partial \bar{y}}\right) - \bar{h}^2\frac{\partial}{\partial \bar{t}}\left(\frac{\partial \bar{v}_x^{\mathrm{av}}}{\partial \bar{x}} + \frac{\partial \bar{v}_y^{\mathrm{av}}}{\partial \bar{y}}\right) - \bar{h}\frac{\partial}{\partial \bar{t}}\left(\frac{\partial \bar{h}}{\partial \bar{x}}\bar{v}_x^{\mathrm{av}} + \frac{\partial \bar{h}}{\partial \bar{y}}\bar{v}_y^{\mathrm{av}}\right) \\
&= -\frac{1}{2}\bar{h}^2\frac{\partial}{\partial \bar{t}}\left(\frac{\partial \bar{v}_x^{\mathrm{av}}}{\partial \bar{x}} + \frac{\partial \bar{v}_y^{\mathrm{av}}}{\partial \bar{y}}\right) - \bar{h}\frac{\partial}{\partial \bar{t}}\left(\frac{\partial \bar{h}}{\partial \bar{x}}\bar{v}_x^{\mathrm{av}} + \frac{\partial \bar{h}}{\partial \bar{y}}\bar{v}_y^{\mathrm{av}}\right).
\end{aligned} \tag{6.53}$$

Then, these parts are approximately given by

$$\begin{aligned}
&\int_{-\bar{h}}^{\epsilon\bar{\eta}} d\bar{z} \int_{\epsilon\bar{\eta}}^{\bar{z}} d\bar{z}' \frac{D\bar{v}_z(\bar{x},\bar{y},\bar{z}',\bar{t})}{D\bar{t}} \\
&= \frac{1}{3}\bar{h}^3\frac{\partial}{\partial \bar{t}}\left(\frac{\partial \bar{v}_x^{\mathrm{av}}}{\partial \bar{x}} + \frac{\partial \bar{v}_y^{\mathrm{av}}}{\partial \bar{y}}\right) + \frac{1}{2}\bar{h}^2\frac{\partial}{\partial \bar{t}}\left(\bar{v}_x^{\mathrm{av}}\frac{\partial \bar{h}}{\partial x} + \bar{v}_y^{\mathrm{av}}\frac{\partial \bar{h}}{\partial y}\right) + \mathrm{O}(\epsilon) + \mathrm{O}(\sigma^2) \\
&\approx \frac{1}{3}\bar{h}^3\frac{\partial}{\partial \bar{t}}\left(\frac{\partial \bar{v}_x^{\mathrm{av}}}{\partial \bar{x}} + \frac{\partial \bar{v}_y^{\mathrm{av}}}{\partial \bar{y}}\right) + \frac{1}{2}\bar{h}^2\frac{\partial}{\partial \bar{t}}\left(\bar{v}_x^{\mathrm{av}}\frac{\partial \bar{h}}{\partial \bar{x}} + \bar{v}_y^{\mathrm{av}}\frac{\partial \bar{h}}{\partial \bar{y}}\right)
\end{aligned} \tag{6.54}$$

and

$$\begin{aligned}
&\int_{-\bar{h}}^{\epsilon\bar{\eta}} d\bar{z}' \frac{D\bar{v}_z(\bar{x},\bar{y},\bar{z}',\bar{t})}{D\bar{t}} \\
&= -\frac{1}{2}\bar{h}^2\frac{\partial}{\partial \bar{t}}\left(\frac{\partial \bar{v}_x^{\mathrm{av}}}{\partial \bar{x}} + \frac{\partial \bar{v}_y^{\mathrm{av}}}{\partial \bar{y}}\right) - \bar{h}\frac{\partial}{\partial \bar{t}}\left(\bar{v}_x^{\mathrm{av}}\frac{\partial \bar{h}}{\partial \bar{x}} + \bar{v}_y^{\mathrm{av}}\frac{\partial \bar{h}}{\partial \bar{y}}\right) + \mathrm{O}(\epsilon) + \mathrm{O}(\sigma^2) \\
&\approx -\frac{1}{2}\bar{h}^2\frac{\partial}{\partial \bar{t}}\left(\frac{\partial \bar{v}_x^{\mathrm{av}}}{\partial \bar{x}} + \frac{\partial \bar{v}_y^{\mathrm{av}}}{\partial \bar{y}}\right) - \bar{h}\frac{\partial}{\partial \bar{t}}\left(\bar{v}_x^{\mathrm{av}}\frac{\partial \bar{h}}{\partial \bar{x}} + \bar{v}_y^{\mathrm{av}}\frac{\partial \bar{h}}{\partial \bar{y}}\right).
\end{aligned} \tag{6.55}$$

Substituting Eqs. (6.54) and (6.55) into Eqs. (6.50) and (6.51), we obtain

$$
\begin{aligned}
&\frac{\partial}{\partial \bar{t}}\left[(\epsilon\bar{\eta}+\bar{h})\bar{v}_x^{\mathrm{av}}(\bar{x},\bar{y},\bar{t})\right]+\epsilon\frac{\partial}{\partial \bar{x}}\left[(\epsilon\bar{\eta}+\bar{h})\left(\bar{v}_x^{\mathrm{av}}\right)^2\right]+\epsilon\sigma^4\frac{\partial}{\partial \bar{x}}\int_{-\bar{h}}^{\epsilon\bar{\eta}}\left(\bar{v}_x{}'\right)^2 d\bar{z}\\
&\quad+\epsilon\frac{\partial}{\partial \bar{y}}\left[(\epsilon\bar{\eta}+\bar{h})\bar{v}_x^{\mathrm{av}}\,\bar{v}_y^{\mathrm{av}}\right]+\epsilon\sigma^4\frac{\partial}{\partial \bar{y}}\int_{-\bar{h}}^{\epsilon\bar{\eta}}\bar{v}_x{}'\bar{v}_y{}' d\bar{z}+(\epsilon\bar{\eta}+\bar{h})\frac{\partial\bar{\eta}}{\partial\bar{x}}\\
&\approx\sigma^2\frac{\partial}{\partial\bar{x}}\left[\frac{1}{3}\bar{h}^3\frac{\partial}{\partial\bar{t}}\left(\frac{\partial\bar{v}_x^{\mathrm{av}}}{\partial\bar{x}}+\frac{\partial\bar{v}_y^{\mathrm{av}}}{\partial\bar{y}}\right)+\frac{1}{2}\bar{h}^2\frac{\partial}{\partial\bar{t}}\left(\bar{v}_x^{\mathrm{av}}\frac{\partial\bar{h}}{\partial\bar{x}}+\bar{v}_y^{\mathrm{av}}\frac{\partial\bar{h}}{\partial\bar{y}}\right)\right]\\
&\quad-\sigma^2\frac{\partial\bar{h}}{\partial\bar{x}}\left[\frac{1}{2}\bar{h}^2\frac{\partial}{\partial\bar{t}}\left(\frac{\partial\bar{v}_x^{\mathrm{av}}}{\partial\bar{x}}+\frac{\partial\bar{v}_y^{\mathrm{av}}}{\partial\bar{y}}\right)+\bar{h}\frac{\partial}{\partial\bar{t}}\left(\bar{v}_x^{\mathrm{av}}\frac{\partial\bar{h}}{\partial\bar{x}}+\bar{v}_y^{\mathrm{av}}\frac{\partial\bar{h}}{\partial\bar{y}}\right)\right]
\end{aligned}
\tag{6.56}
$$

and

$$
\begin{aligned}
&\frac{\partial}{\partial \bar{t}}\left[(\epsilon\bar{\eta}+\bar{h})\bar{v}_y^{\mathrm{av}}(\bar{x},\bar{y},\bar{t})\right]+\epsilon\frac{\partial}{\partial \bar{x}}\left[(\epsilon\bar{\eta}+\bar{h})\bar{v}_x^{\mathrm{av}}\,\bar{v}_y^{\mathrm{av}}\right]+\epsilon\sigma^4\frac{\partial}{\partial \bar{x}}\int_{-\bar{h}}^{\epsilon\bar{\eta}}\bar{v}_x{}'\bar{v}_y{}' d\bar{z}\\
&\quad+\epsilon\frac{\partial}{\partial \bar{y}}\left[(\epsilon\bar{\eta}+\bar{h})\left(\bar{v}_y^{\mathrm{av}}\right)^2\right]+\epsilon\sigma^4\frac{\partial}{\partial \bar{y}}\int_{-\bar{h}}^{\epsilon\bar{\eta}}\left(\bar{v}_y{}'\right)^2 d\bar{z}+(\epsilon\bar{\eta}+\bar{h})\frac{\partial\bar{\eta}}{\partial\bar{y}}\\
&\approx\sigma^2\frac{\partial}{\partial\bar{y}}\left[\frac{1}{3}\bar{h}^3\frac{\partial}{\partial\bar{t}}\left(\frac{\partial\bar{v}_x^{\mathrm{av}}}{\partial\bar{x}}+\frac{\partial\bar{v}_y^{\mathrm{av}}}{\partial\bar{y}}\right)+\frac{1}{2}\bar{h}^2\frac{\partial}{\partial\bar{t}}\left(\bar{v}_x^{\mathrm{av}}\frac{\partial\bar{h}}{\partial\bar{x}}+\bar{v}_y^{\mathrm{av}}\frac{\partial\bar{h}}{\partial\bar{y}}\right)\right]\\
&\quad-\sigma^2\frac{\partial\bar{h}}{\partial\bar{y}}\left[\frac{1}{2}\bar{h}^2\frac{\partial}{\partial\bar{t}}\left(\frac{\partial\bar{v}_x^{\mathrm{av}}}{\partial\bar{x}}+\frac{\partial\bar{v}_y^{\mathrm{av}}}{\partial\bar{y}}\right)+\bar{h}\frac{\partial}{\partial\bar{t}}\left(\bar{v}_x^{\mathrm{av}}\frac{\partial\bar{h}}{\partial\bar{x}}+\bar{v}_y^{\mathrm{av}}\frac{\partial\bar{h}}{\partial\bar{y}}\right)\right].
\end{aligned}
\tag{6.57}
$$

The equations of the average horizontal velocities (Eqs. (6.56) and (6.57)) and the conservation of the water volume (Eq. (6.48)) constitute the 2-D tsunami equations. The equations can be further approximated according to the situation.

6.2.2.1 Linear Long-Wave Equations

When the tsunami height η_0 is much smaller than the sea depth H_0, that is, $\epsilon = \eta_0/H_0 \ll 1$ (see Eq. (6.36)), and the tsunami wavelength is much longer than the sea depth $\sigma = H_0/L_0 \ll 1$ (see Eq. (6.36)), we obtain the linear long-wave equations by setting $\epsilon = 0$ and $\sigma = 0$ in Eqs. (6.48), (6.56), and (6.57) and noting that $\bar{h}$ is independent of time as

$$
\frac{\partial\bar{\eta}}{\partial\bar{t}}+\frac{\partial}{\partial\bar{x}}\left[\bar{h}\bar{v}_x^{\mathrm{av}}(\bar{x},\bar{y},\bar{t})\right]+\frac{\partial}{\partial\bar{y}}\left[\bar{h}\bar{v}_y^{\mathrm{av}}(\bar{x},\bar{y},\bar{t})\right]=0, \tag{6.58}
$$

$$
\frac{\partial}{\partial\bar{t}}\left[\bar{v}_x^{\mathrm{av}}(\bar{x},\bar{y},\bar{t})\right]+\frac{\partial\bar{\eta}}{\partial\bar{x}}=0, \tag{6.59}
$$

and

$$\frac{\partial}{\partial \bar{t}}\left[\bar{v}_y^{\mathrm{av}}(\bar{x}, \bar{y}, \bar{t})\right] + \frac{\partial \bar{\eta}}{\partial \bar{y}} = 0. \tag{6.60}$$

By using the values of physical dimensions (Eq. (6.33)), we rewrite Eqs. (6.58), (6.59), and (6.60) as

$$\frac{\partial \eta(x, y, t)}{\partial t} + \frac{\partial}{\partial x}\left[h v_x^{\mathrm{av}}(x, y, t)\right] + \frac{\partial}{\partial y}\left[h v_y^{\mathrm{av}}(x, y, t)\right] = 0, \tag{6.61}$$

$$\frac{\partial v_x^{\mathrm{av}}(x, y, t)}{\partial t} + g_0 \frac{\partial \eta}{\partial x} = 0, \tag{6.62}$$

and

$$\frac{\partial v_y^{\mathrm{av}}(x, y, t)}{\partial t} + g_0 \frac{\partial \eta}{\partial y} = 0. \tag{6.63}$$

where

$$v_x^{\mathrm{av}}(x, y, t) = \frac{1}{\eta + h}\int_{-h}^{\eta} v_x(x, y, z, t)dz \text{ and } v_y^{\mathrm{av}}(x, y, t) = \frac{1}{\eta + h}\int_{-h}^{\eta} v_y(x, y, z, t)dz. \tag{6.64}$$

This system of Eqs. (6.61)–(6.63), referred to as linear long-wave equations, is the simplest form for representing tsunami propagation. They are often used in calculating the propagation process for tsunami source estimation because these are linear equations with respect to the tsunami height η and are suitable for the superposition of the waves from each element of the source (e.g., Satake et al. 2013).

When the sea depth is constant, $h = h_0$, by differentiating Eq. (6.61) with respect to time and substituting Eqs. (6.62) and (6.63) into it, we rewrite the equations as the wave equation

$$\frac{\partial^2 \eta(x, y, t)}{\partial t^2} = (c_0)^2 \left(\frac{\partial^2 \eta}{\partial x^2} + \frac{\partial^2 \eta}{\partial y^2}\right)$$

where the phase velocity c_0 is given by $c_0 = \sqrt{g_0 h_0}$.

6.2.2.2 Nonlinear Long-Wave Equations

When the tsunami wavelength is much longer than the sea depth $\sigma = H_0/L_0 \ll 1$, but the tsunami height is not much smaller than the sea depth, we only set $\sigma = 0$ and do not set $\epsilon = 0$. We calculate by setting $\sigma = 0$ in Eqs. (6.56) and (6.57) as follows:

$$\frac{\partial}{\partial \bar{t}}\left[(\epsilon\bar{\eta}+\bar{h})\bar{v}_x^{\mathrm{av}}(\bar{x},\bar{y},\bar{t})\right]+\epsilon\frac{\partial}{\partial \bar{x}}\left[(\epsilon\bar{\eta}+\bar{h})\left(\bar{v}_x^{\mathrm{av}}\right)^2\right]$$
$$+\epsilon\frac{\partial}{\partial \bar{y}}\left[(\epsilon\bar{\eta}+\bar{h})\bar{v}_x^{\mathrm{av}}\ \bar{v}_y^{\mathrm{av}}\right]+(\epsilon\bar{\eta}+\bar{h})\frac{\partial\bar{\eta}}{\partial\bar{x}}=0,$$

and

$$\frac{\partial}{\partial \bar{t}}\left[(\epsilon\bar{\eta}+\bar{h})\bar{v}_y^{\mathrm{av}}(\bar{x},\bar{y},\bar{t})\right]+\epsilon\frac{\partial}{\partial \bar{x}}\left[(\epsilon\bar{\eta}+\bar{h})\bar{v}_x^{\mathrm{av}}\ \bar{v}_y^{\mathrm{av}}\right]$$
$$+\epsilon\frac{\partial}{\partial \bar{y}}\left[(\epsilon\bar{\eta}+\bar{h})\left(\bar{v}_y^{\mathrm{av}}\right)^2\right]+(\epsilon\bar{\eta}+\bar{h})\frac{\partial\bar{\eta}}{\partial\bar{y}}=0. \quad (6.65)$$

These equations are rewritten with the values of the physical dimensions as

$$\frac{\partial}{\partial t}\left[(\eta+h)v_x^{\mathrm{av}}\right]+\frac{\partial}{\partial x}\left[(\eta+h)\left(v_x^{\mathrm{av}}\right)^2\right]+\frac{\partial}{\partial y}\left[(\eta+h)v_x^{\mathrm{av}}v_y^{\mathrm{av}}\right]+g_0(\eta+h)\frac{\partial\eta}{\partial x}=0, \quad (6.66)$$

and

$$\frac{\partial}{\partial t}\left[(\eta+h)v_y^{\mathrm{av}}\right]+\frac{\partial}{\partial x}\left[(\eta+h)v_x^{\mathrm{av}}v_y^{\mathrm{av}}\right]+\frac{\partial}{\partial y}\left[(\eta+h)\left(v_y^{\mathrm{av}}\right)^2\right]+g_0(\eta+h)\frac{\partial\eta}{\partial y}=0. \quad (6.67)$$

The volume conservation Eq. (6.48) is given by

$$\frac{\partial\eta}{\partial t}+\frac{\partial}{\partial x}\left[(\eta+h)v_x^{\mathrm{av}}\right]+\frac{\partial}{\partial y}\left[(\eta+h)v_y^{\mathrm{av}}\right]=0. \quad (6.68)$$

By calculating Eqs. (6.66) and (6.67) with Eq. (6.68), we obtain

$$\frac{\partial v_x^{\mathrm{av}}}{\partial t}+v_x^{\mathrm{av}}\frac{\partial v_x^{\mathrm{av}}}{\partial x}+v_y^{\mathrm{av}}\frac{\partial v_x^{\mathrm{av}}}{\partial y}+g_0\frac{\partial\eta}{\partial x}=0, \quad (6.69)$$

and

$$\frac{\partial v_y^{\mathrm{av}}}{\partial t}+v_x^{\mathrm{av}}\frac{\partial v_y^{\mathrm{av}}}{\partial x}+v_y^{\mathrm{av}}\frac{\partial v_y^{\mathrm{av}}}{\partial y}+g_0\frac{\partial\eta}{\partial y}=0. \quad (6.70)$$

Equations (6.68), (6.69), and (6.70) are referred to as nonlinear long-wave equations. This system of equations is not linear due to the term $\partial/\partial x_i\left[(\eta+h)v_i^{\mathrm{av}}\right]$ in Eq. (6.68) and the term $v_j^{\mathrm{av}}\partial v_i^{\mathrm{av}}/\partial x_j$ (advection term) in

Eqs. (6.69) and (6.70). When the sea depth h is constant $h = h_0$, the phase velocity is approximately given by (Appendix 2B)

$$c \approx \sqrt{g_0 h_0}\left(1 + \frac{3}{2}\frac{\eta}{h_0}\right). \tag{6.71}$$

This indicates that a tsunami propagates faster when the tsunami height is higher, which causes the tsunami shape to become deformed with increasing travel distance (see Fig. 2.19).

The nonlinear long-wave equations are usually used in evaluating tsunami propagation in very shallow seas and near the coast by incorporating bottom friction. If an appropriate moving boundary condition between the sea and land is included, these equations can be used to calculate the inundation process (e.g., Oishi et al. 2015; Baba et al. 2016).

6.2.2.3 Linear Dispersive Equations

When the tsunami height is much smaller than the sea depth $\epsilon = \eta_0/H_0 \ll 1$ but the tsunami wavelength is not much longer than the sea depth, we assume dispersive equations by setting $\epsilon = 0$ but not $\sigma = 0$ in Eqs. (6.48), (6.56), and (6.57):

$$\frac{\partial \bar{\eta}}{\partial \bar{t}} + \frac{\partial}{\partial \bar{x}}\left[\bar{h}\bar{v}_x^{\text{av}}(\bar{x},\bar{y},\bar{t})\right] + \frac{\partial}{\partial \bar{y}}\left[\bar{h}\bar{v}_y^{\text{av}}(\bar{x},\bar{y},\bar{t})\right] = 0, \tag{6.72}$$

$$\begin{aligned}
&\frac{\partial}{\partial \bar{t}}\left[\bar{h}\bar{v}_x^{\text{av}}(\bar{x},\bar{y},\bar{t})\right] + \bar{h}\frac{\partial \bar{\eta}}{\partial \bar{x}} \\
&= \sigma^2\frac{\partial}{\partial \bar{x}}\left[\frac{1}{3}\bar{h}^2\frac{\partial}{\partial \bar{t}}\left(\frac{\partial \bar{v}_x^{\text{av}}}{\partial \bar{x}} + \frac{\partial \bar{v}_y^{\text{av}}}{\partial \bar{y}}\right) + \frac{1}{2}\bar{h}^2\frac{\partial}{\partial \bar{t}}\left(\bar{v}_x^{\text{av}}\frac{\partial \bar{h}}{\partial \bar{x}} + \bar{v}_y^{\text{av}}\frac{\partial \bar{h}}{\partial \bar{y}}\right)\right] \\
&\quad -\sigma^2\frac{\partial \bar{h}}{\partial \bar{x}}\left[\frac{1}{2}\bar{h}^2\frac{\partial}{\partial \bar{t}}\left(\frac{\partial \bar{v}_x^{\text{av}}}{\partial \bar{x}} + \frac{\partial \bar{v}_y^{\text{av}}}{\partial \bar{y}}\right) + \bar{h}\frac{\partial}{\partial \bar{t}}\left(\bar{v}_x^{\text{av}}\frac{\partial \bar{h}}{\partial \bar{x}} + \bar{v}_y^{\text{av}}\frac{\partial \bar{h}}{\partial \bar{y}}\right)\right],
\end{aligned} \tag{6.73}$$

and

$$\begin{aligned}
&\frac{\partial}{\partial \bar{t}}\left[\bar{h}\bar{v}_y^{\text{av}}(\bar{x},\bar{y},\bar{t})\right] + \bar{h}\frac{\partial \bar{\eta}}{\partial \bar{y}} \\
&= \sigma^2\frac{\partial}{\partial \bar{y}}\left[\frac{1}{3}\bar{h}^2\frac{\partial}{\partial \bar{t}}\left(\frac{\partial \bar{v}_x^{\text{av}}}{\partial \bar{x}} + \frac{\partial \bar{v}_y^{\text{av}}}{\partial \bar{y}}\right) + \frac{1}{2}\bar{h}^2\frac{\partial}{\partial \bar{t}}\left(\bar{v}_x^{\text{av}}\frac{\partial \bar{h}}{\partial \bar{x}} + \bar{v}_y^{\text{av}}\frac{\partial \bar{h}}{\partial \bar{y}}\right)\right] \\
&\quad -\sigma^2\frac{\partial \bar{h}}{\partial \bar{y}}\left[\frac{1}{2}\bar{h}^2\frac{\partial}{\partial \bar{t}}\left(\frac{\partial \bar{v}_x^{\text{av}}}{\partial \bar{x}} + \frac{\partial \bar{v}_y^{\text{av}}}{\partial \bar{y}}\right) + \bar{h}\frac{\partial}{\partial \bar{t}}\left(\bar{v}_x^{\text{av}}\frac{\partial \bar{h}}{\partial \bar{x}} + \bar{v}_y^{\text{av}}\frac{\partial \bar{h}}{\partial \bar{y}}\right)\right].
\end{aligned} \tag{6.74}$$

These equations are rewritten with the values of the physical dimensions (6.33) as

$$\frac{\partial \eta(x,y,t)}{\partial t} + \frac{\partial}{\partial x}\left[h v_x^{\mathrm{av}}(x,y,t)\right] + \frac{\partial}{\partial \bar{y}}\left[h v_y^{\mathrm{av}}(x,y,t)\right] = 0, \tag{6.75}$$

$$\begin{aligned}
&\frac{\partial \left(h v_x^{\mathrm{av}}\right)}{\partial t} + g_0 h \frac{\partial \eta}{\partial x} \\
&= \frac{\partial}{\partial x}\left[\frac{1}{3}h^3 \frac{\partial}{\partial t}\left(\frac{\partial v_x^{\mathrm{av}}}{\partial x} + \frac{\partial v_y^{\mathrm{av}}}{\partial y}\right) + \frac{1}{2}h^2 \frac{\partial}{\partial t}\left(v_x^{\mathrm{av}}\frac{\partial h}{\partial x} + v_y^{\mathrm{av}}\frac{\partial h}{\partial y}\right)\right] \\
&\quad - \frac{\partial h}{\partial x}\left[\frac{1}{2}h^2 \frac{\partial}{\partial t}\left(\frac{\partial v_x^{\mathrm{av}}}{\partial x} + \frac{\partial v_y^{\mathrm{av}}}{\partial y}\right) + h\frac{\partial}{\partial t}\left(v_x^{\mathrm{av}}\frac{\partial h}{\partial x} + v_y^{\mathrm{av}}\frac{\partial h}{\partial y}\right)\right],
\end{aligned} \tag{6.76}$$

and

$$\begin{aligned}
&\frac{\partial \left(h v_y^{\mathrm{av}}\right)}{\partial t} + g_0 h \frac{\partial \eta}{\partial y} \\
&= \frac{\partial}{\partial y}\left[\frac{1}{3}h^3 \frac{\partial}{\partial t}\left(\frac{\partial v_x^{\mathrm{av}}}{\partial x} + \frac{\partial v_y^{\mathrm{av}}}{\partial y}\right) + \frac{1}{2}h^2 \frac{\partial}{\partial t}\left(v_x^{\mathrm{av}}\frac{\partial h}{\partial x} + v_y^{\mathrm{av}}\frac{\partial h}{\partial y}\right)\right] \\
&\quad - \frac{\partial h}{\partial y}\left[\frac{1}{2}h^2 \frac{\partial}{\partial t}\left(\frac{\partial v_x^{\mathrm{av}}}{\partial x} + \frac{\partial v_y^{\mathrm{av}}}{\partial y}\right) + h\frac{\partial}{\partial t}\left(v_x^{\mathrm{av}}\frac{\partial h}{\partial x} + v_y^{\mathrm{av}}\frac{\partial h}{\partial y}\right)\right].
\end{aligned} \tag{6.77}$$

When the tsunami wavelength is much longer than the horizontal scale of the bathymetry fluctuation, we neglect the spatial derivative of the sea depth on the right-hand side of Eqs. (6.76) and (6.77) as

$$\frac{\partial v_x^{\mathrm{av}}(x,y,t)}{\partial t} + g_0 \frac{\partial \eta}{\partial x} = \frac{h}{3}\frac{\partial}{\partial t}\frac{\partial}{\partial x}\left(\frac{\partial}{\partial x}\left(h v_x^{\mathrm{av}}\right) + \frac{\partial}{\partial y}\left(h v_y^{\mathrm{av}}\right)\right) \tag{6.78}$$

and

$$\frac{\partial v_y^{\mathrm{av}}(x,y,t)}{\partial t} + g_0 \frac{\partial \eta}{\partial y} = \frac{h}{3}\frac{\partial}{\partial t}\frac{\partial}{\partial y}\left(\frac{\partial}{\partial x}\left(h v_x^{\mathrm{av}}\right) + \frac{\partial}{\partial y}\left(h v_y^{\mathrm{av}}\right)\right). \tag{6.79}$$

We often use Eqs. (6.75), (6.78), and (6.79) when a tsunami propagates in deep ocean. When the sea depth h is constant, $h = h_0$, and a plane wave $\eta = \exp\left[i(kx - \omega t)\right]$ propagating in the x-direction is substituted into Eqs. (6.75) and (6.78), we obtain the relation between the wavenumber k and the angular frequency ω (dispersion relation) as

$$\omega^2 = g_0 h_0 k^2 \left(1 + \frac{h_0^2 k^2}{3}\right)^{-1}. \tag{6.80}$$

Then, the phase velocity $c = \omega/k$ is given by

$$c = \frac{\omega}{k} = c_0 \left(1 + \frac{h_0^2 k^2}{3}\right)^{-\frac{1}{2}} \tag{6.81}$$

where $c_0 = \sqrt{g_0 h_0}$ is the phase velocity of the wave calculated by the linear long-wave equations (Eqs. (6.61)–(6.63)). Equation (6.81) indicates that the phase velocity decreases as the wavenumber k increases, indicating that shorter-wavelength tsunamis propagate more slowly.

6.3 Numerical Simulations: Finite Difference Method

Numerical simulations are very powerful for the investigation of tsunami propagation with realistic bathymetry. This section illustrates various finite difference methods for numerically calculating tsunami propagation. Section 6.3.1 explains the fundamental framework of the numerical simulation by taking the linear long-wave equations as an example. Section 6.3.2 explains nonlinear long-wave equations. A moving boundary condition and energy dissipation are also incorporated in the simulation. These are often used for numerically simulating tsunami inundation into coastal areas. Section 6.3.3 shows an implicit scheme for stably solving the linear dispersive equations, which are suitable for calculating short-wavelength tsunamis in deep oceans. Section 6.3.4 illustrates a two-step method using explicit and implicit schemes for solving the nonlinear dispersive equations. In all these numerical schemes, we consider two-dimensional Cartesian coordinates (x, y) in the horizontal plane for the propagation of tsunami height $\eta(x, y, t)$ and horizontal velocity, $v_x^{\mathrm{av}}(x, y, t)$ and $v_y^{\mathrm{av}}(x, y, t)$, on the x- and y- axes.

6.3.1 *Linear Long-Wave Equations: Simplest Equations*

There are different kinds of finite difference numerical schemes according to the tsunami equations. First, taking the simplest tsunami equation, the linear long-wave equations, as an example, we illustrate a basic procedure for performing the numerical simulation.

When the wavelength λ is much longer than the sea depth h ($\lambda \gg h$), and the sea-surface height η is much smaller than the sea depth ($\eta \ll h$), the tsunami is well represented by the linear long-wave equations (Eqs. (6.61)–(6.63)):

$$\frac{\partial \eta(x,y,t)}{\partial t} + \frac{\partial}{\partial x}\left[hv_x^{\mathrm{av}}(x,y,t)\right] + \frac{\partial}{\partial y}\left[hv_y^{\mathrm{av}}(x,y,t)\right] = 0, \qquad (6.61)$$

$$\frac{\partial v_x^{\mathrm{av}}(x,y,t)}{\partial t} + g_0 \frac{\partial \eta}{\partial x} = 0, \qquad (6.62)$$

and

$$\frac{\partial v_y^{\mathrm{av}}(x,y,t)}{\partial t} + g_0 \frac{\partial \eta}{\partial y} = 0. \qquad (6.63)$$

We numerically solve the system of the equations with a finite difference scheme. A cell and grids are distributed in the 2-D horizontal space (x, y) as shown in Fig. 6.11. The variables of η, v_x^{av}, and v_y^{av} are defined at different grid points. The grid distribution is referred to as a staggered grid.

The center of the (i, j)th cell is located at $x = x_i = \Delta x/2 + (i - 1)\,\Delta x$, $y = y_j = \Delta y/2 + (j - 1)\,\Delta y$, where Δx and Δy represent the side lengths of the cell on the x- and y-axes, respectively. The i and j are integer. Also, Δx and Δy represent the grid spacing on the x- and y- axes, respectively. The tsunami height, $\eta_{ij} = \eta(x_i, y_j, t)$, and sea depth, $h_{ij} = h(x_i, y_j)$, are given at a grid point located at the center of the cell (x_i, y_i). The horizontal velocity in the x component $v_{x,ij}^{\mathrm{av}} = v_x^{\mathrm{av}}\left(x_i + \Delta x/2, y_j, t\right)$ is given at the grid point on the boundary between the cells of (i, j) and $(i + 1, j)$. Similarly, the

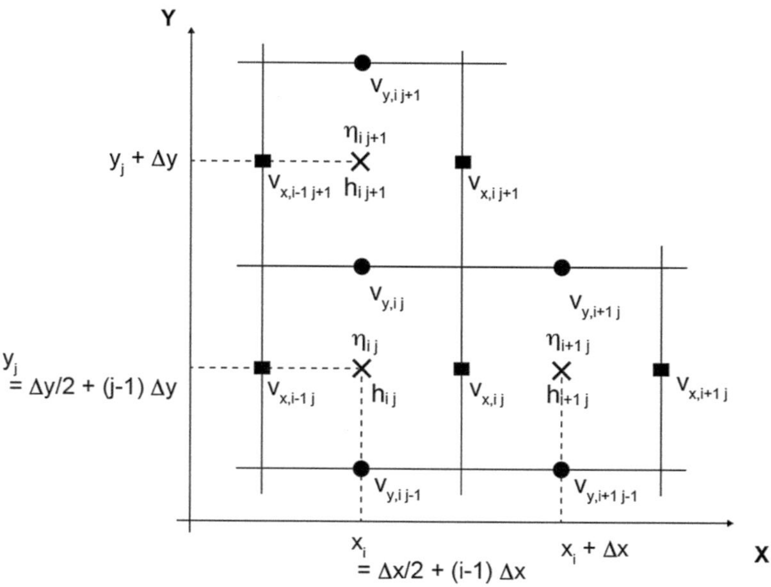

Fig. 6.11 Cell and grids used in the finite difference scheme. The center of the (i, j)th cell is located at $x = x_i = \Delta x/2 + (i - 1)\,\Delta x$, $y = y_j = \Delta y/2 + (j - 1)\,\Delta y$, where Δx and Δy are the grid spacing on the x- and y- axes, respectively

horizontal particle velocity $v_{y,ij}^{\text{av}} = v_y^{\text{av}}\left(x_i, y_j + \Delta y/2, t\right)$ is given at the grid point on the boundary between the cells of (i,j) and $(i, j + 1)$. We define the sea-surface height $\eta_{ij}^k = \eta\left(x_i, y_j, k\Delta t\right)$ at the time of $t = k\Delta t$ (k is integer, $k = 0, 1, \ldots$), while we define the horizontal velocities $v_{x,ij}^{k+1/2} = v_x^{\text{av}}\left(x_i + \Delta x/2, y_j, (k - 1/2)\Delta t\right)$ and $v_{y,ij}^{k+1/2} = v_x^{\text{av}}\left(x_i, y_j + \Delta y/2, (k - 1/2)\Delta t\right)$ at the time of $t = (k - 1/2)\Delta t$.

Equation (6.61) is discretized at the grid point (x_i, y_j) with the adjacent grid points as

$$\frac{\eta_{ij}^{k+1} - \eta_{ij}^k}{\Delta t} + \frac{h_{i+\frac{1}{2}j} v_{x,ij}^{k+\frac{1}{2}} - h_{i-\frac{1}{2}j} v_{x,i-1\,j}^{k+\frac{1}{2}}}{\Delta x} + \frac{h_{i\,j+\frac{1}{2}} v_{y,ij}^{k+\frac{1}{2}} - h_{i\,j-\frac{1}{2}} v_{y,i\,j-1}^{k+\frac{1}{2}}}{\Delta y} = 0. \tag{6.82}$$

Equation (6.62) is discretized at the grid point $(x_i + \Delta x/2, y_j)$ as

$$\frac{v_{x,ij}^{k+1/2} - v_{x,ij}^{k-1/2}}{\Delta t} + g_0 \frac{\eta_{i+1\,j}^k - \eta_{ij}^k}{\Delta x} = 0, \tag{6.83}$$

and Eq. (6.63) is discretized at the grid point $(x_i, y_j + \Delta y/2)$ as

$$\frac{v_{y,ij}^{k+1/2} - v_{y,ij}^{k-1/2}}{\Delta t} + g_0 \frac{\eta_{i\,j+1}^k - \eta_{ij}^k}{\Delta y} = 0. \tag{6.84}$$

The grid points given by non-integer values are given by an interpolation. For example, the value of $h_{i\,+\,1/2\,j}$ is given by

$$h_{i+\frac{1}{2}j} = \frac{1}{2}\left(h_{ij} + h_{i+1\,j}\right). \tag{6.85}$$

We rewrite Eqs. (6.82)–(6.84) as

$$\begin{aligned} \eta_{ij}^{k+1} = \eta_{ij}^k &- \frac{\Delta t}{\Delta x}\left(h_{i+\frac{1}{2}j} v_{x,ij}^{k+\frac{1}{2}} - h_{i-\frac{1}{2}j} v_{x,i-1\,j}^{k+\frac{1}{2}}\right) \\ &- \frac{\Delta t}{\Delta y}\left(h_{i\,j+\frac{1}{2}} v_{y,ij}^{k+\frac{1}{2}} - h_{i\,j-\frac{1}{2}} v_{y,i\,j-1}^{k+\frac{1}{2}}\right), \end{aligned} \tag{6.86}$$

$$v_{x,ij}^{k+1/2} = v_{x,ij}^{k-1/2} - g_0 \frac{\Delta t}{\Delta x}\left(\eta_{i+1\,j}^k - \eta_{ij}^k\right), \tag{6.87}$$

and

$$v_{y,ij}^{k+1/2} = v_{y,ij}^{k-1/2} - g_0 \frac{\Delta t}{\Delta y}\left(\eta_{i\,j+1}^k - \eta_{ij}^k\right). \tag{6.88}$$

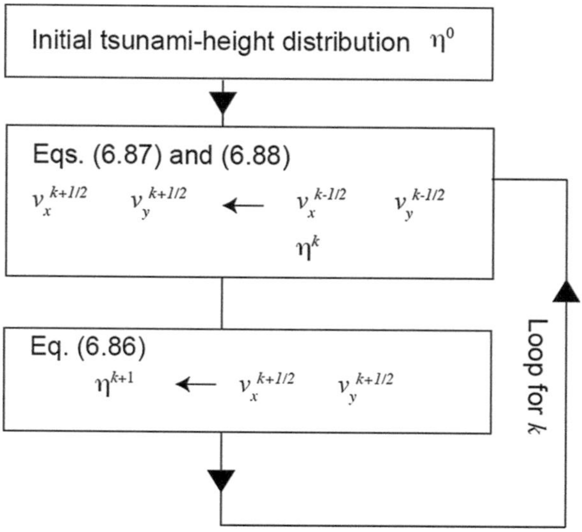

Fig. 6.12 Flowchart of the finite difference simulation using the linear long-wave equations

Figure 6.12 indicates a flowchart for calculating the finite difference equations (Eqs. (6.86)–(6.88)). The finite difference calculation starts with the tsunami height distribution at the time of $t = 0$ or the initial tsunami height distribution, $\eta_{ij}^0 = \eta(x_i, y_j, 0)$. Although sometimes the initial tsunami height distribution is assumed to be identical to the vertical displacement at the sea bottom, this assumption can be inappropriate when a small tsunami source in deep ocean. The previous chapter (Chap. 5) discussed how the initial tsunami height distribution $\eta_{ij}^0 = \eta(x_i, y_j, 0)$ is given by the tsunami generation theory. Considering $k = 0$ in Eqs. (6.87) and (6.88), we then calculate the vertically averaged horizontal velocities $v_{x,ij}^{1/2}$ and $v_{y,ij}^{1/2}$ by substituting $\eta_{ij}^0 = \eta(x_i, y_j, 0)$ and $v_{x,ij}^{-1/2} = v_{y,ij}^{-1/2} = 0$ on the right-hand side of Eqs. (6.87) and (6.88). Note that the analytical solution of the linear incompressible tsunami generation theory (Saito 2013) and the numerical simulations of seismic-tsunami waves with a sloping bathymetry (Lotto et al. 2017) support that the initial horizontal velocity distribution should be zero, $v_{x,ij}^{-1/2} = v_{y,ij}^{-1/2} = 0$. By using these calculated values of $v_{x,ij}^{1/2}$ and $v_{y,ij}^{1/2}$, we calculate the tsunami height at the time of $t = \Delta t$, $\eta_{ij}^1 = \eta(x_i, y_j, \Delta t)$ considering $k = 0$ in Eq. (6.86). Using a similar procedure for $k = 1$ in Eqs. (6.87) and (6.88) and successively in Eq. (6.86), we can calculate the tsunami height at the time of $t = 2\Delta t$ from the tsunami height at the time of $t = \Delta t$. Taking this procedure in sequence, we obtain the tsunami height for the time of $t = 0, \Delta t, 2\Delta t, \ldots$.

We need to set the boundary condition between land and sea. The simplest boundary condition is that the horizontal velocity $(v_x, \ v_y)$ at the grid point between the land and the sea is zero. When the cell located at $(x_i, \ y_j)$ is considered to be land, $h_{ij} < 0$, the horizontal velocities at the grid points attached to this cell are zero:

$$v_{x,i-1j} = 0, v_{x,ij} = 0, v_{y,ij-1} = 0, v_{y,ij} = 0 \,\text{when}\, h_{ij} < 0. \tag{6.89}$$

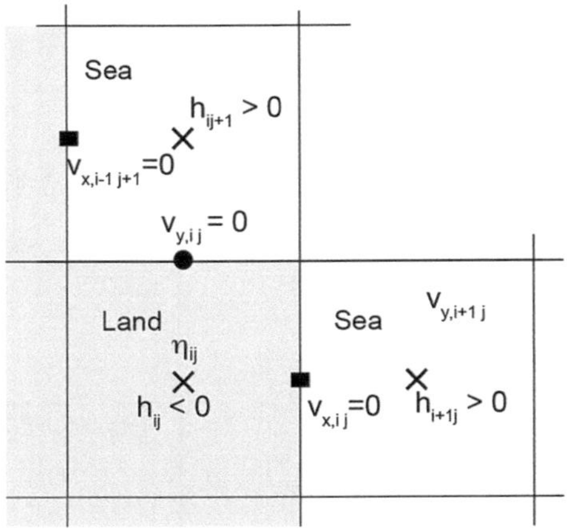

Fig. 6.13 An example of the boundary condition between sea and land. When the cell located at $(x_i, \ y_j)$ has a negative sea depth as $h_{ij} < 0$, the cell is considered to be land. The horizontal velocities between the land and sea are set at zero

Figure 6.13 shows an example. The horizontal velocities between the land and sea are set at zero.

The tsunami is reflected at the edge boundaries of the simulation regions. The reflected waves are artificial waves that we do not want to have in the simulations. There are various ways to damp the artificial waves, including open-boundary condition (e.g., Hwang et al. 1972) and the perfectly matching layer method (e.g., Maeda et al. 2016). The simplest method is to set an absorbing buffer region (e.g., Cerjan et al. 1985). The method is explained in *4.3 Seismic Wave Simulation*.

6.3.2 Nonlinear Long-Wave Equations: Inundation

When the tsunami wavelength is much longer than the sea depth, $\lambda \gg h_0$, but the tsunami height is not much smaller than the sea depth, the tsunami propagation is well described by the nonlinear long-wave equations as

$$\frac{\partial \eta(x,y,t)}{\partial t} + \frac{\partial}{\partial x}\left[(h+\eta)v_x^{\mathrm{av}}\right] + \frac{\partial}{\partial y}\left[(h+\eta)v_y^{\mathrm{av}}\right] = 0, \tag{6.68}$$

$$\frac{\partial v_x^{\mathrm{av}}(x,y,t)}{\partial t} + v_x^{\mathrm{av}}\frac{\partial v_x^{\mathrm{av}}}{\partial x} + v_y^{\mathrm{av}}\frac{\partial v_x^{\mathrm{av}}}{\partial y} + g_0\frac{\partial \eta}{\partial x} = 0, \tag{6.69}$$

and

$$\frac{\partial v_y^{\mathrm{av}}(x,y,t)}{\partial t} + v_x^{\mathrm{av}}\frac{\partial v_y^{\mathrm{av}}}{\partial x} + v_y^{\mathrm{av}}\frac{\partial v_y^{\mathrm{av}}}{\partial y} + g_0\frac{\partial \eta}{\partial x} = 0. \tag{6.70}$$

In actual tsunami propagation, the horizontal velocities of v_x^{av} and v_y^{av} may decrease due to the friction at the sea bottom and small-scale turbulence in the sea. The dissipation would increase with increasing horizontal velocity. We include this dissipation mechanism phenomenologically into Eqs. (6.69) and (6.70) as

$$\frac{\partial v_x^{\mathrm{av}}}{\partial t} + v_x^{\mathrm{av}} \frac{\partial v_x^{\mathrm{av}}}{\partial x} + v_y^{\mathrm{av}} \frac{\partial v_x^{\mathrm{av}}}{\partial y} + g_0 \frac{\partial \eta}{\partial x} = -C_f \frac{v_x^{\mathrm{av}} \sqrt{\left(v_x^{\mathrm{av}}\right)^2 + \left(v_y^{\mathrm{av}}\right)^2}}{h+\eta}, \tag{6.90}$$

and

$$\frac{\partial v_y^{\mathrm{av}}}{\partial t} + v_x^{\mathrm{av}} \frac{\partial v_y^{\mathrm{av}}}{\partial x} + v_y^{\mathrm{av}} \frac{\partial v_y^{\mathrm{av}}}{\partial y} + g_0 \frac{\partial \eta}{\partial x} = -C_f \frac{v_y^{\mathrm{av}} \sqrt{\left(v_x^{\mathrm{av}}\right)^2 + \left(v_y^{\mathrm{av}}\right)^2}}{h+\eta}, \tag{6.91}$$

where C_f is referred to as the nondimensional bottom friction coefficient.

If C_f is constant, the horizontal velocity decreases proportionally to the square of the horizontal velocity and inversely proportional to the total sea depth (sum of the sea depth and the tsunami height). In many tsunami applications, the C_f is represented as a function of sea depth as

$$C_f = \frac{g_0 n_0^2}{(h+\eta)^{1/3}}, \tag{6.92}$$

where n_0 is referred to as Manning's roughness parameter. The value of n_0 can depend on the condition of the sea bottom. The value of $n_0 = 0.03$ [m$^{-1/3}$ s] is often used in tsunami simulations (e.g., Satake 1995).

By using the grid points shown in Fig. 6.11, we represent Eqs. (6.68), (6.90), and (6.91) in the finite difference form. We also use a numerical technique referred to as an upwind scheme to realize a stable calculation of the advection terms in Eqs. (6.90) and (6.91). The finite difference forms are given as follows. By introducing the total sea depth from the sea bottom to the sea surface as

$$d_{ij}^k = \eta_{ij}^k + h_{ij} \tag{6.93}$$

we represent Eqs. (6.68), (6.90), and (6.91) as

$$\frac{\eta_{ij}^{k+1} - \eta_{ij}^k}{\Delta t} + \frac{d_{i+\frac{1}{2}j}^k v_{x,ij}^{k+\frac{1}{2}} - d_{i-\frac{1}{2}j}^k v_{x,i-1\,j}^{k+\frac{1}{2}}}{\Delta x} + \frac{d_{i\,j+\frac{1}{2}}^k v_{y,ij}^{k+\frac{1}{2}} - d_{i,j-\frac{1}{2}}^k v_{y,i\,j-1}^{k+\frac{1}{2}}}{\Delta y} = 0, \tag{6.94}$$

$$\frac{v_{x,ij}^{k+1/2}-v_{x,ij}^{k-1/2}}{\Delta t}+v_{x,ij}^{k-1/2}\frac{\lambda_{11}v_{x,i+1\,j}^{k-1/2}+\lambda_{21}v_{x,ij}^{k-1/2}+\lambda_{31}v_{x,i-1\,j}^{k-1/2}}{\Delta x}$$
$$+v_{y,i+1/2\,j-1/2}^{k-1/2}\frac{\mu_{11}v_{x,i\,j+1}^{k-1/2}+\mu_{21}v_{x,ij}^{k-1/2}+\mu_{31}v_{x,i\,j-1}^{k-1/2}}{\Delta y}+g_0\frac{\eta_{i+1\,j}^{k}-\eta_{i\,j}^{k}}{\Delta x} \tag{6.95}$$
$$=-C_f\frac{v_{x,i\,j}^{k+1/2}+v_{x,i\,j}^{k-1/2}}{2}\frac{\sqrt{\left(v_{x,ij}^{k-1/2}\right)^2+\left(v_{y,i+1/2\,j-1/2}^{k-1/2}\right)^2}}{d_{i+\frac{1}{2}j}^{k}},$$

$$\frac{v_{y,ij}^{k+1/2}-v_{y,ij}^{k-1/2}}{\Delta t}+v_{x,i-1/2\,j+1/2}^{k-1/2}\frac{\lambda_{12}v_{y,i+1\,j}^{k-1/2}+\lambda_{22}v_{y,ij}^{k-1/2}+\lambda_{32}v_{y,i-1\,j}^{k-1/2}}{\Delta x}$$
$$+v_{y,ij}^{k-1/2}\frac{\mu_{12}v_{y,i\,j+1}^{k-1/2}+\mu_{22}v_{y,i\,j}^{k-1/2}+\mu_{32}v_{y,i\,j-1}^{k-1/2}}{\Delta y}+g_0\frac{\eta_{i\,j+1}^{k}-\eta_{ij}^{k}}{\Delta x} \tag{6.96}$$
$$=-C_f\frac{v_{y,ij}^{k+1/2}+v_{y,ij}^{k-1/2}}{2}\frac{\sqrt{\left(v_{x,i-1/2\,j+1/2}^{k-1/2}\right)^2+\left(v_{y,ij}^{k-1/2}\right)^2}}{d_{i\,j+\frac{1}{2}}^{k}}.$$

We used an upwind scheme for the finite difference from of the advection terms. The coefficients λ_{ij} and μ_{ij} for the upwind scheme are chosen depending on the direction of the velocity as

$$\begin{aligned}
&\lambda_{11}=0,\ \lambda_{21}=1,\ \lambda_{31}=-1 \quad \text{for} \quad v_{x,ij}^{k-\frac{1}{2}}\geq 0\\
&\lambda_{11}=1,\ \lambda_{21}=-1,\ \lambda_{31}=0 \quad \text{for} \quad v_{x,ij}^{k-\frac{1}{2}}< 0\\
&\mu_{11}=0,\ \mu_{21}=1,\ \mu_{31}=-1 \quad \text{for} \quad v_{y,i+\frac{1}{2}j-\frac{1}{2}}^{k-\frac{1}{2}}\geq 0\\
&\mu_{11}=1,\ \mu_{21}=-1,\ \mu_{31}=0 \quad \text{for} \quad v_{y,i+\frac{1}{2}j-\frac{1}{2}}^{k-\frac{1}{2}}< 0\\
&\lambda_{12}=0,\ \lambda_{22}=1,\ \lambda_{32}=-1 \quad \text{for} \quad v_{x,i-\frac{1}{2}j+\frac{1}{2}}^{k-\frac{1}{2}}\geq 0\\
&\lambda_{12}=1,\ \lambda_{22}=-1,\ \lambda_{32}=0 \quad \text{for} \quad v_{x,i-\frac{1}{2}j+\frac{1}{2}}^{k-\frac{1}{2}}< 0\\
&\mu_{12}=0,\ \mu_{22}=1,\ \mu_{32}=-1 \quad \text{for} \quad v_{y,ij}^{k-\frac{1}{2}}\geq 0\\
&\mu_{12}=1,\ \mu_{22}=-1,\ \mu_{32}=0 \quad \text{for} \quad v_{y,ij}^{k-\frac{1}{2}}< 0.
\end{aligned} \tag{6.97}$$

Equations (6.94), (6.95), and (6.96) are then rewritten as follows.

$$\eta_{ij}^{k+1} = \eta_{ij}^{k} - \frac{\Delta t}{\Delta x}\left(d_{i+\frac{1}{2}j}^{k} v_{x,ij}^{k+\frac{1}{2}} - d_{i-\frac{1}{2}j}^{k} v_{x,i-1j}^{k+\frac{1}{2}}\right) - \frac{\Delta t}{\Delta y}\left(d_{ij+\frac{1}{2}}^{k} v_{y,ij}^{k+\frac{1}{2}} - d_{ij-\frac{1}{2}}^{k} v_{y,ij-1}^{k+\frac{1}{2}}\right), \tag{6.98}$$

$$\begin{aligned} v_{x,ij}^{k+1/2} &= \frac{1}{1+Q_{ij}^{k}\Delta t/2}\left[\left(1-\frac{Q_{ij}^{k}\Delta t}{2}\right)v_{x,ij}^{k-1/2} - \frac{g_0 \Delta t}{\Delta x}\left(\eta_{i+1\,j}^{k} - \eta_{ij}^{k}\right)\right. \\ &\quad + \frac{\Delta t}{\Delta x} v_{x,ij}^{k-1/2}\left(\lambda_{11} v_{x,i+1\,j}^{k-1/2} + \lambda_{21} v_{x,ij}^{k-1/2} + \lambda_{31} v_{x,i-1\,j}^{k-1/2}\right) \\ &\quad \left. + \frac{\Delta t}{\Delta y} v_{y,i+1/2\,j-1/2}^{k-1/2}\left(\mu_{11} v_{x,i\,j+1}^{k-1/2} + \mu_{21} v_{x,ij}^{k-1/2} + \mu_{31} v_{x,i\,j-1}^{k-1/2}\right)\right], \end{aligned} \tag{6.99}$$

and

$$\begin{aligned} v_{y,ij}^{k+1/2} &= \frac{1}{1+R_{ij}^{k}\Delta t/2}\left[\left(1-\frac{R_{ij}^{k}\Delta t}{2}\right)v_{y,ij}^{k-1/2} - \frac{g_0 \Delta t}{\Delta y}\left(\eta_{i\,j+1}^{k} - \eta_{ij}^{k}\right)\right. \\ &\quad + \frac{\Delta t}{\Delta x} v_{x,i-1/2\,j+1/2}^{k-1/2}\left(\lambda_{12} v_{y,i+1\,j}^{k-1/2} + \lambda_{22} v_{y,ij}^{k-1/2} + \lambda_{32} v_{y,i-1\,j}^{k-1/2}\right) \\ &\quad \left. + \frac{\Delta t}{\Delta y} v_{y,ij}^{k-1/2}\left(\mu_{12} v_{y,i\,j+1}^{k-1/2} + \mu_{22} v_{y,ij}^{k-1/2} + \mu_{32} v_{y,i\,j-1}^{k-1/2}\right)\right]. \end{aligned} \tag{6.100}$$

where $Q_{i,j}^{k}$ and $R_{i,j}^{k}$ are defined as

$$\begin{aligned} Q_{ij}^{k} &= C_f \frac{\sqrt{\left(v_{x,ij}^{k-1/2}\right)^2 + \left(v_{y,i+1/2\,j-1/2}^{k-1/2}\right)^2}}{d_{i+\frac{1}{2}j}^{k}}, \\ R_{ij}^{k} &= C_f \frac{\sqrt{\left(v_{x,i-1/2\,j+1/2}^{k-1/2}\right)^2 + \left(v_{y,ij}^{k-1/2}\right)^2}}{d_{i\,j+1/2}^{k}}. \end{aligned} \tag{6.101}$$

The finite difference forms of Eqs. (6.98), (6.99), and (6.100) numerically describe the tsunami propagation.

6.3.2.1 Inundation: Moving Boundary Between Land and Sea

In order to simulate tsunami inundation, we consider that the boundary between dry area and wet area varies with time (e.g., Iwasaki and Mano 1979; Saito et al. 2014). As shown in Fig. 6.14, we consider the boundary in 1-D space (x-axis), for simplicity. The extension to 2-D space (x-y coordinates) is straightforward. Depending on the sign of the total depth d_i (or the water column height) at the cell located at x_i, we assign each cell as wet or dry. If $d_i > 0$, the cell is assigned to be wet. If $d_i \leq 0$, the cell is assigned to be dry.

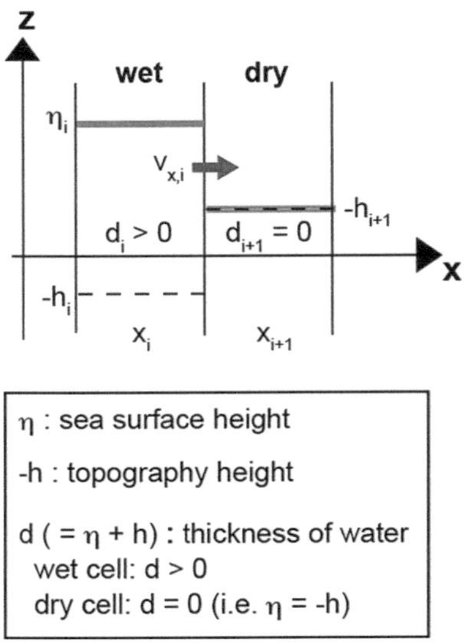

Fig. 6.14 A boundary between dry and wet cells. If the water column height at x_i is positive $d_i > 0$, the cell is assigned to be wet. If $d_i \leq 0$, the cell is assigned to be dry

In the example shown in Fig. 6.14, a wet cell is located at $x = x_i$ and a dry cell is located at $x = x_{i+1}$ next to the wet cell. Inundation occurs from the wet cell to the dry cell for each time step only if the sea-surface height η_i at the wet cell is higher than the topography height of the dry cell at $x = x_{i+1}$.The topography height of the dry cell at $x = x_{i+1}$ is given by $-h_{i+1}$ (because h_{i+1} represents the sea depth). The condition of inundation is then represented as $\eta_i > -h_{i+1}$. In that case, the water flows from the wet cell at x_i into the dry cell at x_{i+1} with the velocity of $v_{x,i}$. For the sake of calculation, we reset the total depth as $d_{i+1} = 0$; when the total depth is negative, $d_{i+1} \leq 0$. By using a total sea depth of zero, $d_{i+1} = 0$, or, in other words, by setting a virtual sea-surface height at the dry cell $\eta_{i+1} = -h_{i+1}$, we calculate the horizontal flow based on Eq. (6.99) as

$$v_{x,i}^{k+1/2} = \frac{1}{1 + Q_i^k \Delta t/2} \left[\left(1 - \frac{Q_i^k \Delta t}{2}\right) v_{x,i}^{k-1/2} - \frac{g_0 \Delta t}{\Delta x}\left(\eta_{i+1}^k - \eta_i^k\right) + \frac{\Delta t}{\Delta x} v_{x,i}^{k-1/2}\left(v_{x,i}^{k-1/2} - v_{x,i-1}^{k-1/2}\right)\right] \tag{6.102}$$

where

$$Q_i^k = C_f \frac{\left|v_{x,i}^{k-1/2}\right|}{d_{i+1/2}^k}. \tag{6.103}$$

When the water flows into the dry cell ($v_{x,i}^{k+1/2} > 0$), the dry cell then turns into a wet cell.

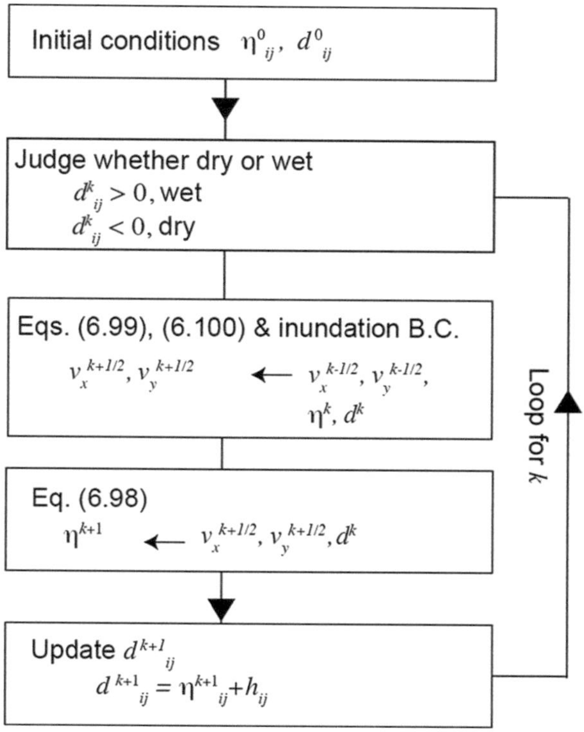

Fig. 6.15 Flowchart for tsunami inundation simulation based on the nonlinear long-wave equations

If the sea-surface height η_i at the wet cell is smaller than the topographical height of the dry cell $\eta_i < -h_{i+1}$, the water does not flow into the dry cell. In this case, the velocity is set at zero, $v_{x,\,i} = 0$.

We numerically solve the finite difference equations Eqs. (6.98), (6.99), and (6.100) based on the flowchart shown in Fig. 6.15. Before starting the time step iteration with respect to k, the initial condition of d_{ij}^0 is set as $d_{ij}^0 = \eta_{ij}^0 + h_{ij}$ where η_{ij}^0 is the initial tsunami height distribution. In the loop of the time step increment with respect to $k = 0$, the (i, j)th cell located at $(x, y) = (x_i, y_j)$ is judged as to whether the cell is wet or dry according to the value of the water column height d_{ij}^0. We consider that when d_{ij}^0 is larger than the very small positive value ϵ (e.g.,= 0.01 m), the cell is wet. Otherwise, when $d_{ij}^0 < \epsilon$, the cell is dry. We then set $d_{ij}^0 = 0$; in other words, we set the tsunami height $\eta_{ij}^0 = -h_{ij}$ at the dry sells to use Eqs. (6.99) and (6.100) at the boundary between the dry and wet cells. For the grids of wet cells, the horizontal velocities are calculated based on Eqs. (6.99) and (6.100). Then, the tsunami height η_{ij}^1 at the time step of $k = 0$ is calculated based on Eq. (6.98). At the end of the loop, the water column height d_{ij}^1 is updated as $d_{ij}^1 = \eta_{ij}^1 + h_{ij}$. We then go back to the beginning of the loop. We set $k = 1$ and judge whether each cell is wet or dry according to the value of d_{ij}^1. Using a similar procedure, we calculate η_{ij}^2. By iterating the loop for k, we obtain the time evolution of the tsunami height distribution as η_{ij}^3, η_{ij}^4, ...

6.3.3 Linear Dispersive Equations: Deep Ocean

Ocean-bottom pressure gauges enable us to observe tsunamis outside bays and harbors. These gauges have greatly contributed to fundamental tsunami research and to the development of early warning systems. The linear dispersive equations (or linear Boussinesq equation (e.g., Peregrine 1972)) are suitable for simulating tsunami records observed in the deep ocean where the sea depth is not much shallower than the tsunami wavelength.

When the tsunami height is much smaller than the sea depth $\eta \ll h_0$ but the tsunami wavelength is not much longer than the sea depth, the tsunami propagation is well described by the linear dispersive equations ((6.75), (6.78), and (6.79)) as

$$\frac{\partial \eta}{\partial t} + \frac{\partial}{\partial x}\left[h v_x^{\mathrm{av}}(x,y,t)\right] + \frac{\partial}{\partial y}\left[h v_y^{\mathrm{av}}(x,y,t)\right] = 0, \tag{6.75}$$

$$\frac{\partial v_x^{\mathrm{av}}}{\partial t} + g_0 \frac{\partial \eta}{\partial x} = \frac{h}{3}\frac{\partial}{\partial t}\frac{\partial}{\partial x}\left(\frac{\partial}{\partial x}\left(h v_x^{\mathrm{av}}\right) + \frac{\partial}{\partial y}\left(h v_y^{\mathrm{av}}\right)\right), \tag{6.78}$$

and

$$\frac{\partial v_y^{\mathrm{av}}}{\partial t} + g_0 \frac{\partial \eta}{\partial y} = \frac{h}{3}\frac{\partial}{\partial t}\frac{\partial}{\partial y}\left(\frac{\partial}{\partial x}\left(h v_x^{\mathrm{av}}\right) + \frac{\partial}{\partial y}\left(h v_y^{\mathrm{av}}\right)\right). \tag{6.79}$$

The equations are written with a finite difference form with the grid points shown in Fig. 6.11 as

$$\frac{\eta_{ij}^{k+1} - \eta_{ij}^{k}}{\Delta t} + \frac{h_{i+\frac{1}{2}\,j} v_{x,ij}^{k+\frac{1}{2}} - h_{i-\frac{1}{2}\,j} v_{x,i-1\,j}^{k+\frac{1}{2}}}{\Delta x} + \frac{h_{i\,j+\frac{1}{2}} v_{y,ij}^{k+\frac{1}{2}} - h_{ij-\frac{1}{2}} v_{y,i\,j-1}^{k+\frac{1}{2}}}{\Delta y} = 0, \tag{6.104}$$

$$\begin{aligned}
&\frac{v_{x,ij}^{k+1/2} - v_{x,ij}^{k-1/2}}{\Delta t} + g_0 \frac{\eta_{i+1\,j}^{k} - \eta_{ij}^{k}}{\Delta x} \\
&= \frac{h_{i+\frac{1}{2}\,j}}{3\Delta t}\left[\frac{1}{(\Delta x)^2}\left(h_{i+\frac{3}{2}\,j} v_{x,i+1\,j}^{k+1/2} - 2h_{i+\frac{1}{2}\,j} v_{x,ij}^{k+1/2} + h_{i-\frac{1}{2},j} v_{x,i-1\,j}^{k+1/2}\right)\right. \\
&\quad - \frac{1}{(\Delta x)^2}\left(h_{i+\frac{3}{2}\,j} v_{x,i+1\,j}^{k+1/2} - 2h_{i+\frac{1}{2}\,j} v_{x,ij}^{k+1/2} + h_{i-\frac{1}{2},j} v_{x,i-1\,j}^{k+1/2}\right) \\
&\quad + \frac{1}{\Delta x \Delta y}\left(h_{i+1\,j+\frac{1}{2}} v_{y,i+1\,j}^{k+1/2} - h_{i\,j+\frac{1}{2}} v_{y,ij}^{k+1/2} - h_{i+1\,j-\frac{1}{2}} v_{y,i+1\,j-1}^{k+1/2} + h_{i\,j-\frac{1}{2}} v_{y,i\,j-1}^{k+1/2}\right)\Bigg], \\
&\quad \left.- \frac{1}{\Delta x \Delta y}\left(h_{i+1\,j+\frac{1}{2}} v_{y,i+1\,j}^{k-1/2} - h_{i\,j+\frac{1}{2}} v_{y,ij}^{k-1/2} - h_{i+1,j-\frac{1}{2}} v_{y,i+1\,j-1}^{k-1/2} + h_{i\,j-\frac{1}{2}} v_{y,i\,j-1}^{k-1/2}\right)\right],
\end{aligned} \tag{6.105}$$

and

$$
\begin{aligned}
&\frac{v_{y,ij}^{k+1/2}-v_{y,ij}^{k-1/2}}{\Delta t}+g_0\frac{\eta_{i\,j+1}^{k}-\eta_{ij}^{k}}{\Delta y}\\
&=\frac{h_{i\,j+\frac{1}{2}}}{3\Delta t}\left[\frac{1}{\Delta x\Delta y}\left(h_{i+\frac{1}{2}\,j+1}v_{x,i\,j+1}^{k+1/2}-h_{i-\frac{1}{2}\,j+1}v_{x,i-1\,j+1}^{k+1/2}-h_{i+\frac{1}{2}\,j}v_{x,ij}^{k+1/2}+h_{i-\frac{1}{2}\,j}v_{x,i-1\,j}^{k+1/2}\right)\right.\\
&\quad-\frac{1}{\Delta x\Delta y}\left(h_{i+\frac{1}{2}\,j+1}v_{x,i\,j+1}^{k-1/2}-h_{i-\frac{1}{2}\,j+1}v_{x,i-1\,j+1}^{k-1/2}-h_{i+\frac{1}{2}\,j}v_{x,ij}^{k-1/2}+h_{i-\frac{1}{2}\,j}v_{x,i-1\,j}^{k-1/2}\right)\\
&\quad+\frac{1}{(\Delta y)^2}\left(h_{i\,j+\frac{3}{2}}v_{y,i\,j+1}^{k+1/2}-2h_{i\,j+\frac{1}{2}}v_{y,ij}^{k+1/2}+h_{i\,j-\frac{1}{2}}v_{y,i\,j-1}^{k+1/2}\right)\\
&\quad\left.-\frac{1}{(\Delta y)^2}\left(h_{i\,j+\frac{3}{2}}v_{y,i\,j+1}^{k-1/2}-2h_{i\,j+\frac{1}{2}}v_{y,ij}^{k-1/2}+h_{i\,j-\frac{1}{2}}v_{y,i\,j-1}^{k-1/2}\right)\right].
\end{aligned}
\tag{6.106}
$$

We rewrite these equations so that the quantities at the next time step (i.e., $k+1$) are on the left-hand side and the quantities at the previous time step (i.e., k) are on the right-hand side as follows:

$$
\begin{aligned}
\eta_{ij}^{k+1}=\eta_{ij}^{k}&-\frac{\Delta t}{\Delta x}\left(h_{i+\frac{1}{2},j}v_{x,i,j}^{k+1/2}-h_{i-\frac{1}{2},j}v_{x,i-1,j}^{k+1/2}\right)\\
&-\frac{\Delta t}{\Delta y}\left(h_{i,j+\frac{1}{2}}v_{y,i,j}^{k+1/2}-h_{i,j-\frac{1}{2}}v_{y,i,j-1}^{k+1/2}\right),
\end{aligned}
\tag{6.107}
$$

$$
\begin{aligned}
&-\frac{h_{i+1/2\,j}h_{i+3/2\,j}}{3(\Delta x)^2}v_{x,i+1\,j}^{k+1/2}+\left(1+\frac{\left(h_{i+1/2\,j}\right)^2}{3(\Delta x)^2}\right)v_{x,ij}^{k+1/2}-\frac{h_{i+1/2\,j}h_{i-1/2\,j}}{3(\Delta x)^2}v_{x,i-1\,j}^{k+1/2}\\
&\quad-\frac{h_{i+1/2\,j}h_{i+1\,j+1/2}}{3\Delta x\Delta y}v_{y,i+1\,j}^{k+1/2}+\frac{h_{i+1/2\,j}h_{i\,j+1/2}}{3\Delta x\Delta y}v_{y,ij}^{k+1/2}\\
&\quad+\frac{h_{i+1/2\,j}h_{i+1\,j-1/2}}{3\Delta x\Delta y}v_{y,i+1\,j-1}^{k+1/2}-\frac{h_{i+1/2\,j}h_{i\,j-1/2}}{3\Delta x\Delta y}v_{y,i\,j-1}^{k+1/2}\\
&=-\frac{h_{i+1/2\,j}h_{i+3/2\,j}}{3(\Delta x)^2}v_{x,i+1\,j}^{k-1/2}+\left(1+\frac{\left(h_{i+1/2\,j}\right)^2}{3(\Delta x)^2}\right)v_{x,ij}^{k-1/2}-\frac{h_{i+1/2\,j}h_{i-1/2\,j}}{3(\Delta x)^2}v_{x,i-1\,j}^{k-1/2}\\
&\quad-\frac{h_{i+1/2\,j}h_{i+1\,j+1/2}}{3\Delta x\Delta y}v_{y,i+1\,j}^{k-1/2}+\frac{h_{i+1/2\,j}h_{i\,j+1/2}}{3\Delta x\Delta y}v_{y,ij}^{k-1/2}\\
&\quad+\frac{h_{i+1/2\,j}h_{i+1\,j-1/2}}{3\Delta x\Delta y}v_{y,i+1\,j-1}^{k-1/2}-\frac{h_{i+1/2\,j}h_{i\,j-1/2}}{3\Delta x\Delta y}v_{y,i\,j-1}^{k-1/2}\\
&\quad+g_0\frac{\Delta t}{\Delta x}\left(\eta_{i+1\,j}^{k}-\eta_{ij}^{k}\right),
\end{aligned}
\tag{6.108}
$$

and

$$
\begin{aligned}
&-\frac{h_{i+1/2\,j}h_{i\;j+3/2}}{3(\Delta y)^2}v_{y,i\,j+1}^{k+1/2}+\left(1+\frac{\left(h_{i\;j+1/2}\right)^2}{3(\Delta y)^2}\right)v_{y,ij}^{k+1/2}-\frac{h_{i\;j+1/2}h_{i\;j-1/2}}{3(\Delta y)^2}v_{y,i\,j-1}^{k+1/2}\\
&-\frac{h_{i\;j+1/2}h_{i+1/2\,j+1}}{3\Delta x\Delta y}v_{x,i\,j+1}^{k+1/2}+\frac{h_{i\;j+1/2}h_{i-1/2\,j+1}}{3\Delta x\Delta y}v_{x,i-1\,j+1}^{k+1/2}\\
&+\frac{h_{i\;j+1/2}h_{i+1/2\,j}}{3\Delta x\Delta y}v_{x,i+1\,j-1}^{k+1/2}-\frac{h_{i\;j+1/2}h_{i-1/2\,j}}{3\Delta x\Delta y}v_{x,i-1\,j}^{k+1/2}\\
=&-\frac{h_{i+1/2\,j}h_{i\;j+3/2}}{3(\Delta y)^2}v_{y,i\,j+1}^{k-1/2}+\left(1+\frac{\left(h_{i\;j+1/2}\right)^2}{3(\Delta y)^2}\right)v_{y,i,j}^{k-1/2}-\frac{h_{i\;j+1/2}h_{i\;j-1/2}}{3(\Delta y)^2}v_{y,i\,j-1}^{k-1/2}\\
&-\frac{h_{i\;j+1/2}h_{i+1/2\,j+1}}{3\Delta x\Delta y}v_{x,i\,j+1}^{k-1/2}+\frac{h_{i\;j+1/2}h_{i-1/2\,j+1}}{3\Delta x\Delta y}v_{x,i-1\,j+1}^{k-1/2}\\
&+\frac{h_{i\;j+1/2}h_{i+1/2\,j}}{3\Delta x\Delta y}v_{x,i+1\,j-1}^{k-1/2}-\frac{h_{i\;j+1/2}h_{i-1/2\,j}}{3\Delta x\Delta y}v_{x,i-1\,j}^{k-1/2}\\
&+g_0\frac{\Delta t}{\Delta y}\left(\eta_{i\;j+1}^{k}-\eta_{ij}^{k}\right).
\end{aligned}
\tag{6.109}
$$

These finite difference representations have different forms from those of the linear long-wave equations or nonlinear long-wave equations. In the cases of linear long-wave equations and nonlinear long-wave equations, a (single) term at the next time step is represented using the values at the previous time step. In contrast, the left-hand side of Eq. (6.108) has seven terms represented by the seven values of the time step of k + 1/2, $v_{x,i+1\,j}^{k+1/2}$, $v_{x,ij}^{k+1/2}$, $v_{x,i-1\,j}^{k+1/2}$, $v_{y,i+1\,j}^{k+1/2}$, $v_{y,ij}^{k+1/2}$, $v_{y,i-1\,j}^{k+1/2}$, and $v_{y,i\,j-1}^{k+1/2}$. We cannot straightforwardly calculate these seven values just by substituting the values of the previous time steps. In order to calculate the quantities of the next time step, we need to solve the systems of the equations with respect to the unknown seven values at the next time step. To numerically solve the systems of the equations, we can use the Jacobi method or the Gauss-Seidel method. This scheme is referred to as an implicit scheme, while the schemes used in the linear long-wave equations and the nonlinear long-wave equations are referred to as explicit scheme.

6.3.3.1 Solving the Finite Difference Equations by the Jacobi Method

Here we explain the Jacobi method. Consider the simulation region represented by $n_x \times n_y$ cells in the (x, y) space. The cell and the grid points are located as shown in Fig. 6.11. We would like to obtain the unknown horizontal velocities of $v_{x,ij}^{k+1/2}$ and $v_{y,ij}^{k+1/2}$ ($i = 1, \ldots, n_x$ and $j = 1, \ldots, n_y$) that satisfy Eqs. (6.108) and (6.109). At the boundaries of the simulation region, the horizontal velocities are fixed in the simulation. For example, we set

$$\begin{aligned} v_{x,1\,j}^{k} &= 0\,\mathrm{for}\,j = 1,\ \ldots, n_y \\ v_{x,n_x\,j}^{k} &= 0\,\mathrm{for}\,j = 1,\ \ldots, n_y \\ v_{y,i\,1}^{k} &= 0\,\mathrm{for}\,i = 1,\ \ldots, n_x \\ v_{y,i\,n_y}^{k} &= 0\,\mathrm{for}\,i = 1,\ \ldots, n_x \end{aligned} \tag{6.110}$$

as the boundary conditions of the horizontal velocities. Thus, the number of variables to be calculated is $(n_x - 2) \times (n_y - 2)$ for $v_{x,ij}^{k+1/2}$ and $(n_x - 2) \times (n_y - 2)$ for $v_{y,ij}^{k+1/2}$ in Eqs. (6.108) and (6.109). The total number of unknown variables is thus $2 \times (n_x - 2) \times (n_y - 2)$. Since $i = 2, \ldots, n_x - 1$ and $j = 2, \ldots, n_y - 1$ for Eq. (6.108), the number of equations in Eq. (6.108) is $(n_x - 2) \times (n_y - 2)$. Similarly, the number of equations in Eq. (6.109) is $(n_x - 2) \times (n_y - 2)$. Thus, we have $2 \times (n_x - 2) \times (n_y - 2)$ equations for Eqs. (6.108) and (6.109) for solving $2 \times (n_x - 2) \times (n_y - 2)$ unknown variables of $v_{x,i,j}^{k+1/2}$ and $v_{y,i,j}^{k+1/2}$.

We here set the unknown variables of $v_{x,ij}^{k+1/2}$ and $v_{y,ij}^{k+1/2}$ at the time step $k + 1/2$ with x_i as

$$\begin{aligned} &x_1 = v_{x,2\,2}^{k+1/2}, x_2 = v_{x,3\,2}^{k+1/2}, \ldots, x_{n_x-2} = v_{x,n_x-1\,2}^{k+1/2} \cdots, \\ &x_{(n_x-2)(n_y-2)} = v_{x,n_x-1\,n_y-1}^{k+1/2}, x_{(n_x-2)(n_y-2)+1} = v_{y,2\,2}^{k+1/2}, \ldots, \\ &x_{2(n_x-2)(n_y-2)} = v_{y,n_x-1\,n_y-1}^{k+1/2} \end{aligned} \tag{6.111}$$

The column vector x_i has $2 \times (n_x - 2) \times (n_y - 2)$ elements. Then, setting $N = 2 \times (n_x - 2) \times (n_y - 2)$, we rewrite Eqs. (6.108) and (6.109) in the matrix form as

$$\begin{pmatrix} a_{11} & a_{12} & \cdots & a_{1N} \\ a_{21} & a_{22} & \cdots & a_{2N} \\ \vdots & \vdots & & \vdots \\ a_{N1} & a_{N2} & \cdots & a_{NN} \end{pmatrix} \begin{pmatrix} x_1 \\ x_2 \\ \vdots \\ x_N \end{pmatrix} = \begin{pmatrix} b_1 \\ b_2 \\ \vdots \\ b_N \end{pmatrix}, \tag{6.112}$$

or

$$a_{ij} x_j = b_i.$$

The coefficient of a_{ij} in Eq. (6.112) is given by the sea depth h_{ij} and the grid spacing Δx and Δy. These are independent of the time step k. The vector b_i is given by the dependent variables $v_{x,ij}^{k-1/2}$, $v_{y,ij}^{k-1/2}$, and η_{ij}^{k} at the previous time steps of k and $k - 1/2$. The vector b_i changes according to the time step k.

Equation (6.112) shows that the finite difference form (Eqs. (6.108) and (6.109)) has the form of the simultaneous equation with respect to the unknown values of x_j. The matrix a_{ij} is a square matrix of size $N = 2 \times (n_x - 2) \times (n_y - 2)$. Considering, for example, the simulation region represented by $n_x = 1000$ and $n_y = 1000$, the matrix a_{ij} is ~2×10^6 in size. This may be too large to numerically treat as an array in a simulation code.

The Jacobi method is a simple but effective method to numerically solve simultaneous equations with respect to the unknown variables, x_i, in particular, for a matrix of large size a_{ij} in Eq. (6.112). In the Jacobi method, we assume the variable x_i at the calculation of the lth iteration to be $x_i^{(l)}$. Using these values of $x_i^{(l)}$, we calculate the variable x_i at the calculation of $(l + 1)$th based on Eq. (6.112) as

$$\begin{aligned} x_1^{(l+1)} &= \left(b_1 - a_{12}x_2^{(l)} - a_{13}x_3^{(l)} \cdots - a_{1N}x_N^{(l)}\right)/a_{11} \\ x_2^{(l+1)} &= \left(b_2 - a_{21}x_1^{(l)} - a_{23}x_3^{(l)} \cdots - a_{2N}x_N^{(l)}\right)/a_{22} \\ x_N^{(l+1)} &= \left(b_N - a_{N1}x_1^{(l)} - a_{N2}x_2^{(l)} \cdots - a_{NN-1}x_{N-1}^{(l)}\right)/a_{NN} \end{aligned} \tag{6.113}$$

This calculation is iteratively done until the variables $x_i^{(l)}$ converge, as $x_i^{(l)} \rightarrow x_i$. This method is also suitable for parallel computing.

6.3.3.2 Flow of Solving Linear Dispersive Equations

We numerically solve the linear dispersive equations as shown in Fig. 6.16. The finite difference calculation starts with the tsunami height distribution at the time of $t = 0$ or the initial tsunami height distribution, $\eta_{ij}^0 = \eta(x_i, y_j, 0)$. Considering $k = 0$ in Eqs. (6.108) and (6.109), we then calculate the vertically averaged horizontal

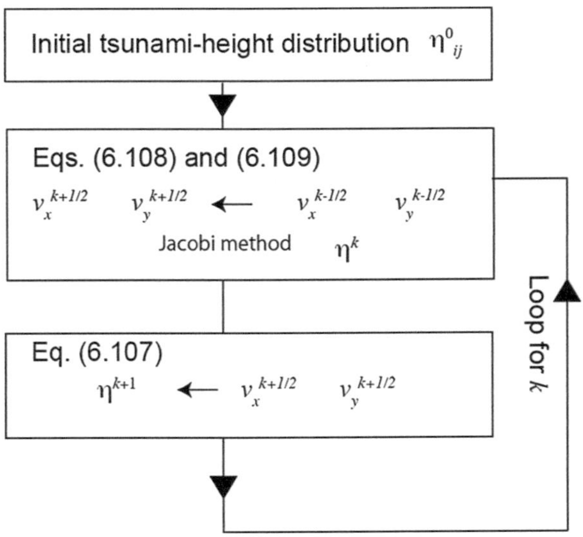

Fig. 6.16 Flowchart of the finite difference simulation using the linear dispersive equations

velocities $v_{x,ij}^{1/2}$ and$v_{y,ij}^{1/2}$ by the Jacobi method substituting $\eta_{ij}^{0} = \eta(x_i, y_j, 0)$ and $v_{x,i,j}^{-1/2} = v_{y,i,j}^{-1/2} = 0$ on the right-hand side of Eqs. (6.108) and (6.109). By using these calculated values of $v_{x,ij}^{1/2}$ and$v_{y,ij}^{1/2}$, we calculate the tsunami height at the time of $t = \Delta t$, $\eta_{ij}^{1} = \eta(x_i, y_j, \Delta t)$ considering $k = 0$ in Eq. (6.107). Similarly, we calculate the tsunami height at the time of $t = 2\Delta t$ from the tsunami height at the time of $t = \Delta t$. Taking this procedure iteratively, we obtain the tsunami height for the time of $t = 0, \Delta t, 2\Delta t, \ldots$.

In solving the finite difference form of the linear dispersive equations, we solve the simultaneous equations (Eqs. (6.108) and (6.109)) for each time step by the iterative method (Jacobi method). This generally takes more computational time than solving the explicit schemes of the linear and nonlinear long-wave equations.

6.3.4 Nonlinear Dispersive Equations

We have shown the numerical scheme for solving the nonlinear long-wave equations that are suitable for nearshore tsunamis (Sect. 6.3.2) and the scheme for solving the linear dispersive equations that are suitable for reproducing offshore short-wavelength tsunamis (Sect. 6.3.3). The following equations include both the nonlinear and dispersion effects. We referred to the following equations as nonlinear dispersive equations in this textbook:

$$\frac{\partial \eta}{\partial t} + \frac{\partial}{\partial x}\left[d v_x^{\mathrm{av}}\right] + \frac{\partial}{\partial \bar{y}}\left[d v_y^{\mathrm{av}}\right] = 0, \tag{6.114}$$

$$\begin{aligned}\frac{\partial v_x^{\mathrm{av}}}{\partial t} &+ v_x^{\mathrm{av}}\frac{\partial v_x^{\mathrm{av}}}{\partial x} + v_y^{\mathrm{av}}\frac{\partial v_x^{\mathrm{av}}}{\partial y} + g_0\frac{\partial \eta}{\partial x} \\ &= \frac{h}{3}\frac{\partial}{\partial t}\frac{\partial}{\partial x}\left(\frac{\partial}{\partial x}\left(h v_x^{\mathrm{av}}\right) + \frac{\partial}{\partial y}\left(h v_y^{\mathrm{av}}\right)\right) - C_f\frac{v_x^{\mathrm{av}}\sqrt{\left(v_x^{\mathrm{av}}\right)^2 + \left(v_y^{\mathrm{av}}\right)^2}}{d},\end{aligned} \tag{6.115}$$

$$\begin{aligned}\frac{\partial v_y^{\mathrm{av}}}{\partial t} &+ v_x^{\mathrm{av}}\frac{\partial v_y^{\mathrm{av}}}{\partial x} + v_y^{\mathrm{av}}\frac{\partial v_y^{\mathrm{av}}}{\partial y} + g_0\frac{\partial \eta}{\partial y} \\ &= \frac{h}{3}\frac{\partial}{\partial t}\frac{\partial}{\partial y}\left(\frac{\partial}{\partial x}\left(h v_x^{\mathrm{av}}\right) + \frac{\partial}{\partial y}\left(h v_y^{\mathrm{av}}\right)\right) - C_f\frac{v_y^{\mathrm{av}}\sqrt{\left(v_x^{\mathrm{av}}\right)^2 + \left(v_y^{\mathrm{av}}\right)^2}}{h + \eta},\end{aligned} \tag{6.116}$$

and

$$d(x, y, t) = h(x, y) + \eta(x, y, t). \tag{6.117}$$

We rewrite these equations in the finite difference form as follows.

$$\eta_{ij}^{k+1} = \eta_{ij}^{k} - \frac{\Delta t}{\Delta x}\left(d_{i+\frac{1}{2}j}^{k} v_{x,ij}^{k+\frac{1}{2}} - d_{i-\frac{1}{2}j}^{k} v_{x,i-1\,j}^{k+\frac{1}{2}}\right) - \frac{\Delta t}{\Delta y}\left(d_{i\;j+\frac{1}{2}}^{k} v_{y,ij}^{k+\frac{1}{2}} - d_{i,j-\frac{1}{2}}^{k} v_{y,i\,j-1}^{k+\frac{1}{2}}\right), \tag{6.118}$$

$$\frac{v_{x,ij}^{k+1/2} - v_{x,ij}^{k-1/2}}{\Delta t} = \left[-g_0 \frac{\partial \eta}{\partial x} - v_x^{\mathrm{av}} \frac{\partial v_x^{\mathrm{av}}}{\partial x} - v_y^{\mathrm{av}} \frac{\partial v_x^{\mathrm{av}}}{\partial y} - C_f \frac{v_x^{\mathrm{av}} \sqrt{(v_x^{\mathrm{av}})^2 + (v_y^{\mathrm{av}})^2}}{d}\right]_{ij} + \left[\frac{h}{3}\frac{\partial}{\partial t}\frac{\partial}{\partial x}\left(\frac{\partial}{\partial x}(h v_x^{\mathrm{av}}) + \frac{\partial}{\partial y}\left(h v_y^{\mathrm{av}}\right)\right)\right]_{i,j}, \tag{6.119}$$

and

$$\frac{v_{y,ij}^{k+1/2} - v_{y,ij}^{k-1/2}}{\Delta t} = \left[-g_0 \frac{\partial \eta}{\partial y} - v_x^{\mathrm{av}} \frac{\partial v_y^{\mathrm{av}}}{\partial x} - v_y^{\mathrm{av}} \frac{\partial v_y^{\mathrm{av}}}{\partial y} - C_f \frac{v_y^{\mathrm{av}} \sqrt{(v_x^{\mathrm{av}})^2 + (v_y^{\mathrm{av}})^2}}{d}\right]_{ij} + \left[\frac{h}{3}\frac{\partial}{\partial t}\frac{\partial}{\partial y}\left(\frac{\partial}{\partial x}(h v_x^{\mathrm{av}}) + \frac{\partial}{\partial y}\left(h v_y^{\mathrm{av}}\right)\right)\right]_{ij}, \tag{6.120}$$

where the bracket $[\cdots]_{ij}$ represents the finite difference form at the grid point of $v_{x,ij}^{k+1/2}$ in Eq. (6.115) and $v_{y,ij}^{k+1/2}$ in Eq. (6.116).

We numerically calculate these finite difference forms by a two-step method. By introducing a new parameter v_x^*, we divide Eq. (6.119) into two equations as

$$\frac{v_x^* - v_{x,ij}^{k-1/2}}{\Delta t} = \left[-g_0 \frac{\partial \eta}{\partial x} - v_x^{\mathrm{av}} \frac{\partial v_x^{\mathrm{av}}}{\partial x} - v_y^{\mathrm{av}} \frac{\partial v_x^{\mathrm{av}}}{\partial y} - C_f \frac{v_x^{\mathrm{av}} \sqrt{(v_x^{\mathrm{av}})^2 + (v_y^{\mathrm{av}})^2}}{d}\right]_{ij}, \tag{6.121}$$

and

$$\frac{v_{x,ij}^{k+1/2} - v_x^*}{\Delta t} = \left[\frac{h}{3}\frac{\partial}{\partial t}\frac{\partial}{\partial x}\left(\frac{\partial}{\partial x}(h v_x^{\mathrm{av}}) + \frac{\partial}{\partial y}\left(h v_y^{\mathrm{av}}\right)\right)\right]_{ij}. \tag{6.122}$$

If we add Eqs. (6.121) and (6.122), we obtain Eq. (6.119). Note that Eq. (6.121) has the same form as that of the nonlinear long-wave equations. Equation (6.121), hence, can be calculated using the explicit scheme used in Eq. (6.99). On the other hand, Eq. (6.122) has the same form as that of the linear dispersive equations. Equation (6.122) can be calculated using the implicit scheme used in Eq. (6.108). This also holds for Eq. (6.120) for the time evolution of $v_{y,ij}^{k+1/2}$.

6.3.4.1 Flow of Solving Nonlinear Dispersive Equations

By combining the explicit scheme used in the calculation of the nonlinear long-wave equations and the implicit scheme used in the calculation of the linear dispersive equations, we can solve the nonlinear dispersive equations as shown in Fig. 6.17.

Before starting the time step iteration with respect to k, the initial condition of d_{ij}^{0} is set as $d_{ij}^{0} = \eta_{ij}^{0} + h_{ij}$, where η_{ij}^{0} is the initial tsunami height distribution.

In the loop of the time step increment with respect to k, each cell is judged as to whether it is wet or dry according to the value of the water column height d_{ij}^{k}. We consider that when d_{ij}^{k} is larger than the very small positive value ϵ (e.g.,$= 0.01$ m), the cell is wet (sea). Otherwise, when $d_{ij}^{k} < \epsilon$, the cell is dry, and we set $d_{i,j}^{k} = 0$; in

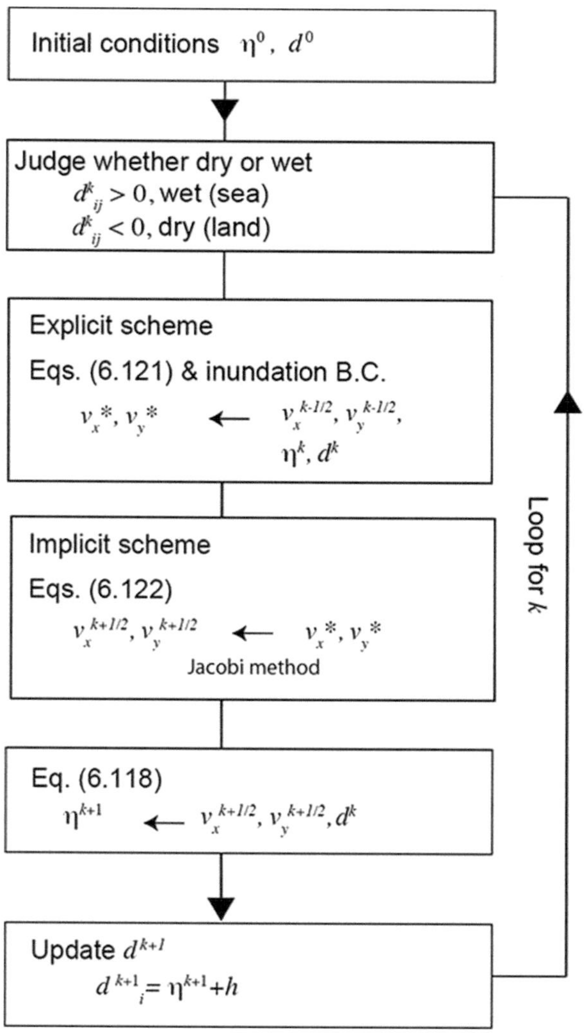

Fig. 6.17 Flowchart of the finite difference simulation using the nonlinear dispersive equations

other words, we set the tsunami height $\eta_{ij}^{k} = -h_{ij}$ in the dry cell to use Eq. (6.121) at the boundary between the land and sea. For the grids of wet cells, the horizontal velocities are calculated based on Eqs. (6.121) and (6.122).

Considering $k = 0$ in Eq. (6.121), we calculate the horizontal velocities v_x^*, substituting $\eta_{ij}^{0} = \eta(x_i, y_j, 0)$ and $v_{x,ij}^{-1/2} = v_{y,ij}^{-1/2} = 0$. Similarly, the value of v_y^* is also calculated. By using the values of v_x^* and v_y^* together with $\eta_{i,}^{0} = \eta(x_i, y_j, 0)$ and $v_{x,ij}^{-1/2} = v_{y,ij}^{-1/2} = 0$, we then calculate the horizontal velocities $v_{x,ij}^{1/2}$ and $v_{y,ij}^{1/2}$ by the Jacobi method based on Eq. (6.122) and the corresponding equation with respect to $v_{y,ij}^{1/2}$. By substituting $v_{x,ij}^{1/2}$ and $v_{y,ij}^{1/2}$ into Eq. (6.118), we calculate the tsunami height at the time of $t = \Delta t, \eta_{ij}^{1} = \eta(x_i, y_j, \Delta t)$. Using a similar procedure, we can calculate the tsunami height at the time of $t = 2\Delta t$ from the tsunami height at the time of $t = \Delta t$. Taking this procedure iteratively, we obtain the tsunami height for the time of $t = 0$, Δt, $2\Delta t$, ...

The nonlinear dispersive equations can describe both nonlinear effects near coasts and dispersive waves in deep oceans simultaneously. We need to use the implicit scheme (iterative calculation in each time step) in the simulation with higher-resolution sea-bottom bathymetry data. If we simulate tsunamis only for deep oceans, we do not need to use such high-resolution bathymetry data compared to simulations in shallow seas or near coasts, because the tsunami wavelength becomes longer in deep oceans. On the other hand, in order to correctly reproduce tsunamis near coasts, the bathymetry and topography data need to be highly resolved with small grid spacing. This requires a large number of grid points to represent the bathymetry and topography appropriately. The implicit scheme has much higher computational costs because of the increase in the grid points of the simulation. As a result, the nonlinear dispersive equation simulations usually have much higher computational costs than the nonlinear long-wave equations and dispersive equations for deep oceans.

References

Baba T et al (2016) Large-scale, high-speed tsunami prediction for the great Nankai trough earthquake on the K computer. Int J High Perform Comput Appl 30(1):71–84. https://doi.org/10.1177/1094342015584090

Cerjan C, Kosloff D, Kosloff R, Reshef M (1985) A nonreflecting boundary condition for discrete acoustic and elastic wave equations. Geophysics 50(4):705–708

Coastal Engineering Committee, Japan Society of Civil Engineers (1994) Kaigan Hado (in Japanese), Japan Society of Civil Engineers, pp 520

Fujii Y, Satake K, Sakai S, Shinohara M, Kanazawa T (2011) Tsunami source of the 2011 off the Pacific coast of Tohoku earthquake. Earth Planet Sp 63(7):55. https://doi.org/10.5047/eps.2011.06.010

Hwang LS, Butler HL, Divoky DJ (1972) Tsunami model: generation and open-sea characteristics. Bull Seismol Soc Am 62(6):1579–1596

Imamura F (1996) In: Yeh H, Liu P, Synolakis C (eds) Review of tsunami simulation with a finite difference method, in long-wave Runup models. World Scientific Publishing, Hackensack, pp 25–42

Inazu D, Saito T (2014) Two subevents across the Japan trench during the 7 December 2012 off Tohoku earthquake (Mw 7.3) inferred from offshore tsunami records. J Geophys Res Solid Earth 119(7):5800–5813. https://doi.org/10.1002/2013JB010892

Iwasaki R, Mano A (1979) Two-dimensional numerical simulation of tsunami runup in the Eulerian description (in Japanese). Proceedings of 26th conference on coastal engineering, JSCE, pp 70–74

Kirby JT, Shi F, Tehranirad B, Harris JC, Grilli ST (2013) Dispersive tsunami waves in the ocean: model equations and sensitivity to dispersion and Coriolis effects. Ocean Model 62:39–55. https://doi.org/10.1016/j.ocemod.2012.11.009

Koketsu K et al (2011) A unified source model for the 2011 Tohoku earthquake. Earth Planet Sci Lett 310(3):480–487. https://doi.org/10.1016/j.epsl.2011.09.009

Lotto GC, Nava G, Dunham EM (2017) Should tsunami simulations include a nonzero initial horizontal velocity? Earth Planet Sp 69:117. https://doi.org/10.1186/s40623-017-0701-8

Madsen PA, Sørensen OR (1992) A new form of the Boussinesq equations with improved linear dispersion characteristics. Part 2. A slowly-varying bathymetry. Coast Eng 18(3–4):183–204. https://doi.org/10.1016/0378-3839(92)90019-Q

Maeda T, Tsushima H, Furumura T (2016) An effective absorbing boundary condition for linear long-wave and linear dispersive-wave tsunami simulations. Earth Planet Sp 68(1):1–14. https://doi.org/10.1186/s40623-016-0436-y

Minoura K, Imamura F, Sugawara D, Kono Y, Iwashita T (2001) The 869 Jogan tsunami deposit and recurrence interval of large-scale tsunami on the Pacific coast of Northeast Japan. J Nat Dis Sci 23(2):83–88

Munk WH (1963) Some comments regarding diffusion and absorption of tsunamis. In Proceedings of tsunami meetings, tenth Pacific science congress, IUGG Monograph, no. 24, Paris, pp 53–72

Namegaya Y, Satake K (2014) Reexamination of the AD 869 Jogan earthquake size from tsunami deposit distribution, simulated flow depth, and velocity. Geophys Res Lett 41(7):2297–2303. https://doi.org/10.1002/2013GL058678

Oishi Y, Imamura F, Sugawara D (2015) Near-field tsunami inundation forecast using the parallel TUNAMI-N2 model: application to the 2011 Tohoku-Oki earthquake combined with source inversions. Geophys Res Lett 42(4):1083–1091. https://doi.org/10.1002/2014GL062577

Peregrine DH (1972) Equations for water waves and the approximations behind them. Waves on beaches and resulting sediment transport, pp 95–121

Saito T (2013) Dynamic tsunami generation due to sea-bottom deformation: analytical representation based on linear potential theory. Earth Planets Space 65(12):1411–1423

Saito T, Satake K, Furumura T (2010) Tsunami waveform inversion including dispersive waves: the 2004 earthquake off Kii peninsula, Japan. J Geophys Res Solid Earth 115:B06303. https://doi.org/10.1029/2009JB006884

Saito T, Ito Y, Inazu D, Hino R (2011) Tsunami source of the 2011 Tohoku-Oki earthquake, Japan: inversion analysis based on dispersive tsunami simulations. Geophys Res Lett 38(7). https://doi.org/10.1029/2011GL049089

Saito T, Inazu D, Tanaka S, Miyoshi T (2013) Tsunami coda across the Pacific Ocean following the 2011 Tohoku-Oki earthquake. Bull Seismol Soc Am 103(2B):1429–1443. https://doi.org/10.1785/0120120183

Saito T, Inazu D, Miyoshi T, Hino R (2014) Dispersion and nonlinear effects in the 2011 Tohoku-Oki earthquake tsunami. J Geophys Res Oceans 119:5160–5180. https://doi.org/10.1002/2014JC009971

Satake K (1995) Linear and nonlinear computations of the 1992 Nicaragua earthquake tsunami. Pure Appl Geophys 144:450–470. https://doi.org/10.1007/BF00874378

Satake K, Fujii Y, Harada T, Namegaya Y (2013) Time and space distribution of coseismic slip of the 2011 Tohoku earthquake as inferred from tsunami waveform data. Bull Seismol Soc Am 103(2B):1473–1492. https://doi.org/10.1785/0120120122

Sato H, Fehler MC, Maeda T (2012) Seismic wave propagation and scattering in the heterogeneous earth, vol 484. Springer, Berlin

Shi F, Kirby JT, Harris JC, Geiman JD, Grilli ST (2012) A high-order adaptive time-stepping TVD solver for Boussinesq modeling of breaking waves and coastal inundation. Ocean Model 43:36–51. https://doi.org/10.1016/j.ocemod.2011.12.004

Suzuki W, Aoi S, Sekiguchi H, Kunugi T (2011) Rupture process of the 2011 Tohoku-Oki mega-thrust earthquake (M9. 0) inverted from strong-motion data. Geophys Res Lett 38(7). https://doi.org/10.1029/2011GL049136

Van Dorn WG (1984) Some tsunami characteristics deducible from tide records. J Phys Oceanogr 13:353–363

Van Dorn WG (1987) Tide gage response to tsunamis. Part II: other oceans and smaller seas. J Phys Oceanogr 17:1507–1516

Chapter 7
Epilogue

Abstract This chapter summarizes tsunami generation and propagation. Also, we mention topics that we did not fully treat in this book but that are important for tsunami studies. The physics underlying tsunami has not yet been fully elucidated. Continuing this challenging research is important both for the advancement of natural science and for disaster prevention.

Keywords Tsunami propagation · Tsunami generation · Earth science · Disaster mitigation and prevention

7.1 Tsunami Propagation

Chapter 3 illustrated the mechanisms and features of tsunami propagation. By comparing the propagation of tsunami with that of seismic waves, we showed the features of tsunami propagation. With gravity as a restoring force, tsunami propagates by oscillating all or a major part of the sea layer. On the other hand, seismic waves propagate due to the elasticity of the earth medium. As a result, the propagation speeds and particle motions are considerably different between tsunami and seismic waves. Since seismic waves can contaminate tsunami signals at or near the tsunami source, it is also important to understand the behavior of seismic waves in order to correctly analyze the tsunami data inside an earthquake focal region.

Tsunami propagating across deep oceans is well described by linear theory. We usually suppose that the tsunami wavelength is much longer than the sea depth and that the tsunami velocity is independent of the wavelength. The propagation process is reproduced by linear long-wave theory. However, when the tsunami wavelength is not much greater than the sea depth, the tsunami becomes dispersive. We hence expect dispersive tsunami to be observed in association with small earthquakes whose source size may not be much larger than the sea depth. However, we should note that dispersive tsunamis have been observed not only in association with small earthquakes but also with large earthquakes (e.g., Fig. 6.6). This occurs when the earthquake fault reaches the sea bottom, which causes a steep sea-bottom

T. Saito, *Tsunami Generation and Propagation*, Springer Geophysics,
https://doi.org/10.1007/978-4-431-56850-6_7

deformation. The steep sea-bottom deformation then becomes a source of short-wavelength tsunamis. By analyzing such short-wavelength and dispersive tsunami carefully, we can estimate the high-resolution tsunami source. This would not be realized using only classical tide gauge records. Ocean-bottom pressure gauges that are now widely deployed in deep oceans have played a very important role in the development of dispersive tsunami analyses.

Offshore tsunami observations and reliable tsunami source estimation have enabled us to predict tsunami arrival times very accurately. As a result, we recognized the non-negligible contribution of the elasticity of the earth medium to tsunami propagation. We do not fully explain this point in this book. When a tsunami propagates across the Pacific Ocean, assuming fluid dynamics without considering the elasticity of the earth predicts the tsunami arrival slightly earlier than the observed arrival (e.g., Ward 1980; Inazu and Saito 2013; Tsai et al. 2013). By taking into account the contributions of seawater density stratification, the elasticity of the fluid and solid earth medium, and the gravitational potential change, we can reproduce the tsunami arrival time more accurately (e.g., Watada et al. 2014; Allgeyer and Cummins 2014; Baba et al. 2017). By using accurate tsunami velocity estimation, we can estimate the tsunami source location more precisely even if only far-field tsunami records are available (Yoshimoto et al. 2016).

Near coasts, tsunami shows a nonlinear nature. The important nonlinear features include energy dissipation and inundation. Chapter 6 describes simple methods for incorporating these features into numerical simulations. Nonlinear tsunami simulations can reproduce energy dissipation and inundation. In this book, the nondimensional bottom friction coefficient C_f was used to represent the energy dissipation phenomenologically. The physical mechanisms of energy dissipation were not treated in this book. Readers may learn the mechanisms in the textbooks of fluid dynamics (e.g., Landau and Lifshitz 1987). The energy dissipation is caused by the energy shift from large eddies to small eddies. The process is described by using the stochastic models of the velocity fluctuation in turbulent flow.

The inundation process is conspicuously complicated but is very significant. This book treated the inundation only by a simple numerical method in the 2-D tsunami simulation in Chap. 6. For the case when a tsunami inundates a coast with a constant slope, the analytical solution for the tsunami height was derived using simple geometry by Carrier et al. (2003). Synolakis (1991) theoretically investigated the run-up height using nonlinear and linear theories together with laboratory experiments. These solutions give us a fundamental basis for the inundation process. The numerical simulations have also been extensively developed and widely applied to field observations. Tsunami deposits are highly important for reconstructing past huge earthquakes (e.g., Namegaya and Satake 2014; Ioki and Tanioka 2016). It is critical to understand the mechanisms by which tsunami deposits are formed during tsunami inundation.

7.2 Tsunami Generation

This book considers the process in which the sea-bottom displacement causes the initial tsunami height distribution to be the generation process. Since the initial tsunami height distribution is excited even without gravity, the generation process is basically independent of gravity. Hence, the mechanisms of the generation and propagation are essentially different. This idea resulted directly from the analytical solution of the linear tsunami generation theory (Chap. 5). The solution (Eq. 5.34) indicates that the generation process and the propagation process are represented by different terms in the analytical solution.

If we focus only on the generation process, seismology and elastic dynamics play more important roles than fluid dynamics. Chapter 4 illustrates the earthquake source or fault characteristics. Kinematic earthquake fault motions are mainly treated in this book. Using the kinematic fault models in seismic wave simulations, we obtain a realistic tsunami source model that can be used in tsunami propagation simulations. This method enables us to simulate both the seismic waves and tsunami simultaneously (Fig. 4.22). This was developed for the records obtained inside the focal area where the seismic waves and tsunami coexist.

It should be noted that this book prescribes an earthquake fault at the seismic wave simulations in *Chap. 4: Earthquakes*. This is kinematic fault modeling. On the other hand, in earthquake physics, dynamic fault modeling has been developed in which the earthquake fault motion is simulated by using the equations of motion and the frictional constitutive law (e.g., Aki and Richards 2002; Fukuyama 2009). Geist and Dmowska (1999) showed kinematically prescribed uniform slip on a fault underestimates the tsunami height compared to nonuniform slip derived by mechanical fault modeling when the seismic moments of the two cases are the same. Although we did not fully illustrate these mechanical and dynamic approaches, they are important for understanding earthquake rupture process. Dynamic rupture simulations can provide deep understanding of the differences between the mechanisms of standard earthquakes and tsunami earthquakes (e.g., Ma and Hirakawa 2013; Kozdon and Dunham 2014; Lotto et al. 2017).

7.3 Earth Science and Disaster Prevention

A tsunami is a disastrous event. The 2011 Tohoku-Oki earthquake tsunami killed more than 15,000 people, and more than 2500 people are still missing, and also caused the Fukushima Daiichi Nuclear Power Station accident (e.g., Satake 2014). This devastating disaster was due to the overestimation of our earthquake disaster prevention capabilities such as anticipated earthquake scenarios, rapid earthquake magnitude estimation, and tsunami prediction. This was mainly a result of our inadequate knowledge regarding huge ($M_W \sim 9$) earthquakes and tsunamis.

After the Tohoku-Oki earthquake, we constructed a wide and dense seafloor observation network and huge coastal levees in order to prevent disasters due to future huge earthquakes. The hardware was promptly built in the 10 years following the earthquake. However, this is not sufficient for disaster prevention. We must make a persistent effort to better understand huge earthquakes and tsunamis. Our progress in understanding nature and further developing earth science might seem to occur much more slowly than the rapid growth of hardware in our society. Nevertheless, carefully observing and humbly learning about nature can help us live in harmony with nature and prevent us from overestimating our capability.

References

Aki K, Richards P (2002) Quantitative Seismology. University Science Books, Mill Valley

Allgeyer S, Cummins P (2014) Numerical tsunami simulation including elastic loading and seawater density stratification. Geophys Res Lett 41:2368–2375. https://doi.org/10.1002/2014GL059348

Baba T, Allgeyer S, Hossen J, Cummins PR, Tsushima H, Imai K, Yamashita K, Kato T (2017) Accurate numerical simulation of the far-field tsunami caused by the 2011 Tohoku earthquake, including the effects of Boussinesq dispersion, seawater density stratification, elastic loading, and gravitational potential change. Ocean Model 111:46–54

Carrier GF, Wu TT, Yeh H (2003) Tsunami run-up and draw-down on a plane beach. J Fluid Mech 475:79–99. https://doi.org/10.1017/S0022112002002653

Fukuyama E (ed) (2009) Fault-zone properties and earthquake rupture dynamics, vol 94. Academic, London

Geist EL, Dmowska R (1999) Local tsunamis and distributed slip at the source. Pure Appl Geophys 154:485–512. https://doi.org/10.1007/978-3-0348-8679-6_6

Inazu D, Saito T (2013) Simulation of distant tsunami propagation with a radial loading deformation effect. Earth Planets Space 65:835–842. https://doi.org/10.5047/eps.2013.03.010

Ioki K, Tanioka Y (2016) Re-estimated fault model of the 17th century great earthquake off Hokkaido using tsunami deposit data. Earth Planet Sci Lett 433:133–138. https://doi.org/10.1016/j.epsl.2015.10.009

Kozdon JE, Dunham EM (2014) Constraining shallow slip and tsunami excitation in megathrust ruptures using seismic and ocean acoustic waves recorded on ocean-bottom sensor networks. Earth Planet Sci Lett 396:56–65. https://doi.org/10.1016/j.epsl.2014.04.001

Landau LD, Lifshitz EM (1987) Fluid mechanics. In: Course of theoretical physics, vol 6. Butterworth-Heinemann, Oxford

Lotto GC, Dunham EM, Jeppson TN, Tobin HJ (2017) The effect of compliant prisms on subduction zone earthquakes and tsunamis. Earth Planet Sci Lett 458:213–222. https://doi.org/10.1016/j.epsl.2016.10.050

Ma S, Hirakawa ET (2013) Dynamic wedge failure reveals anomalous energy radiation of shallow subduction earthquakes. Earth Planet Sci Lett 375:113–122. https://doi.org/10.1016/j.epsl.2013.05.016

Namegaya Y, Satake K (2014) Reexamination of the AD 869 Jogan earthquake size from tsunami deposit distribution, simulated flow depth, and velocity. Geophys Res Lett 41(7):2297–2303. https://doi.org/10.1002/2013GL058678

Satake K (2014) Advances in earthquake and tsunami sciences and disaster risk reduction since the 2004 Indian Ocean tsunami. Geosci Lett 1(15):1–13. https://doi.org/10.1186/s40562-014-0015-7

Synolakis CE (1991) Tsunami runup on steep slopes: how good linear theory really is. In: Tsunami Hazard. Springer, Dordrecht, pp 221–234. https://doi.org/10.1007/978-94-011-3362-3_8

Tsai VC, Ampuero JP, Kanamori H, Stevenson DJ (2013) Estimating the effect of earth elasticity and variable water density on tsunami speeds. Geophys Res Lett 40:492–496. https://doi.org/10.1002/grl.50147

Watada S, Kusumoto S, Satake K (2014) Traveltime delay and initial phase reversal of distant tsunamis coupled with the self-gravitating elastic earth. J Geophys Res Solid Earth 119:4287–4310. https://doi.org/10.1002/2013JB010841

Ward SN (1980) Relationships of tsunami generation and an earthquake source. J Phys Earth 28(5):441–474. https://doi.org/10.4294/jpe1952.28.441

Yoshimoto M, Watada S, Fujii Y, Satake K (2016) Source estimate and tsunami forecast from far-field deep-ocean tsunami waveforms—the 27 February 2010 Mw 8.8 Maule earthquake. Geophys Res Lett 43:659–665. https://doi.org/10.1002/2015GL067181

Index

T. Saito, *Tsunami Generation and Propagation*, Springer Geophysics,
https://doi.org/10.1007/978-4-431-56850-6

MIX
Papier aus verantwortungsvollen Quellen
Paper from responsible sources
FSC® C105338

If you have any concerns about our products,
you can contact us on
ProductSafety@springernature.com

In case Publisher is established outside the EU,
the EU authorized representative is:
Springer Nature Customer Service Center GmbH
Europaplatz 3, 69115 Heidelberg, Germany

Printed by Libri Plureos GmbH
in Hamburg, Germany